Lecture Notes in Physics

377

W.J. Duschl S.J. Wagner
M. Camenzind (Eds.)

Variability of Active Galaxies

Proceedings of a Workshop of
the Sonderforschungsbereich 328
Held at Heidelberg, Germany
3–5 September 1990

Springer-Verlag
Berlin Heidelberg GmbH

Editors

Wolfgang J. Duschl
Institut für Theoretische Astrophysik, Universität Heidelberg
Im Neuenheimer Feld 561, W-6900 Heidelberg, Germany

Stefan J. Wagner
Max Camenzind
Landessternwarte, Königstuhl
W-6900 Heidelberg, Germany

This book was processed by the authors using the $\TeX$ macro package from Springer-Verlag

ISBN 978-3-662-13825-0 ISBN 978-3-540-46325-2 (eBook)
DOI 10.1007/978-3-540-46325-2

Originally published by Springer-Verlag Berlin Heidelberg New York in 1991
Softcover reprint of the hardcover 1st edition 1991

2153/3140-543210 – Printed on acid-free paper

Foreword

The fact that active nuclei of galaxies are the sites of the highest macroscopic energy densities is one of the major reasons for the attention attracted by the field of AGN research. This was noticed when it was found that the enormous amounts of energy radiated by the central engines vary on rather short time-scales, implying via the well-known light-travel argument that the radiating volumes are much smaller than the resolution limits of most ground-based telescopes.

Ever since this discovery, variability studies have not only offered, in effect, a magnifying glass for the unresolvable scales, but also improved our understanding of the processes that generate and reprocess the energy.

Although the basic phenomena of variability were already known in the early 1970s, progress in the field of variability studies was not as rapid as originally hoped for. Two of the few completely new aspects that were discovered during the first half of the 1980s were the extremely rapid flickering in the X-ray emission of AGNs and the temporal variability of the morphology of the very central cores observed with VLBI.

In the very recent past, however, interest in variability studies has again increased, triggered by empirical progress (both in the fields of observation and mathematical analysis) and theoretical advances in the stability of geometrical configurations (disks, jets, magnetic fields) and energy generation and reprocessing mechanisms.

This led to the idea that a conference on the specific topic of variability might be a timely one and ultimately resulted in the production of these proceedings.

All the various aspects of variability studies are intimately linked with each other, making logical order a desirable but unrealistic goal. The order of the contributions largely follows the order in which they were presented at the conference, broadly dividing between line- and continuum variability, with observational results, methodological approaches and theoretical models accompanying each other. The final part is devoted to the important aspect of propagation-induced variability.

We are grateful to the speakers and posterauthors for sending in the written versions of their contributions. Furthermore, we thank Drs. Klaus Meisenheimer, Thomas Schmutzler and Rainer Wehrse, who helped us to organize this conference. The financial assistance of the Sonderforschungsbereich 328 of the Deutsche Forschungsgemeinschaft and logistic help by its speaker, Prof. Appenzeller, are gratefully acknowledged. Last, but not least, thanks are due to Mr. K. Anton, Mrs. E. Bär, Mrs. M. Darr, Mr. J. Heidt and Mrs. B. Hoffmann for technical help during the conference and in the course of assembling these proceedings.

Heidelberg,
December 1990

Wolfgang J. Duschl
Stefan J. Wagner
Max Camenzind

Table of Contents

List of Participants

M. Abramowicz, SISSA, International School for Advance Studies, Strada Costiera 11, 34014 Trieste, Italy
E-Mail: abramowi at itssissa

K. Anton, Landessternwarte Königstuhl, 6900 Heidelberg 1, Germany

I. Appenzeller, Landessternwarte Königstuhl, 6900 Heidelberg 1, Germany
E-Mail: ct0 at dhdurz1

S. Appl, Landessternwarte Königstuhl, 6900 Heidelberg 1, Germany
E-Mail: bc7 at dhdurz1

K. Ballard, Royal Observatory, Blackford Hill, Edinburgh, EH9 3HJ, U.K.
E-Mail: krb at uk.ac.roe.star

P. Barr, EXOSAT Observatory, ESTEC, Postbus 299, 2200 AG Noordwijk, The Netherlands
E-Mail: exosat :: pb

B. Baschek, Institut für Theoretische Astrophysik, Im Neuenheimer Feld 561, 6900 Heidelberg 1, Germany
E-Mail: c07 at dhdurz1

P. Biermann, MPI für Radioastronomie, Auf dem Hügel 69, 5300 Bonn 1, Germany
E-Mail: p165bie at mpirbn.uucp

C.- I. Björnsson, Stockholms Observatorium, 13336 Saltsjöbaden, Sweden
E-Mail: bjornsso at astro.su.se

U. Borgeest, Hamburger Sternwarte, Gojenbergsweg 112, 2050 Hamburg 80, Germany

W. Brinkmann, MPI für Extraterrestrische Physik, 8046 Garching b. München, Germany
E-Mail: mpe :: wpb

H. Brunner, Astronomisches Institut, Waldhäuser Str. 64, 7400 Tübingen, Germany
E-Mail: ait :: brunner

A. G. de Bruyn, Radiosterrewacht, Postbus 2, 7990 AA Dwingeloo, The Netherlands
E-Mail: ger at nfra.nl

M. Camenzind, Landessternwarte Königstuhl, 6900 Heidelberg 1, Germany
E-Mail: ab4 at dhdurz1

J. Clavel, ESA-IUE Observatory, Apartado 50727, E - 28080 Madrid, Spain
E-Mail: jclavel at vilspa

T.J.-L. Courvoisier, Observatoire de Genève, 1290 Sauverny, Switzerland
E-Mail: courvois at cgeuge54

M. Däther, Institut für Theoretische Astrophysik, Im Neuenheimer Feld 561, 6900 Heidelberg 1, Germany
E-Mail: bf5 at dhdurz1

M. Dietrich, Universitäts-Sternwarte, Geismarlandstr. 11, 3400 Göttingen, Germany

E-Mail: mdietri at dgogwdgl

W.J. Duschl, Institut für Theoretische Astrophysik, Im Neuenheimer Feld 561, 6900 Heidelberg 1, Germany

E-Mail: cj0 at dhdurzl

R. Edelson, University of Colorado, Boulder, Colorado 80309-0391, USA

E-Mail: rick at jila or elroy :: rick

U. Erkens, Landessternwarte Königstuhl, 6900 Heidelberg 1, Germany

E-Mail: i61 at dhdurzl

H. Fink, MPI für Extraterrestrische Physik, 8046 Garching b. München, Germany

E-Mail: rosap hhf at dgaippls or mpe :: hhf

K.J. Fricke, Universitäts-Sternwarte, Geismarlandstr. 11, 3400 Göttingen, Germany

K.-D. Fritz, MPI für Kernphysik, Postfach 103980, 6900 Heidelberg 1, Germany

W. Glatzel, Universitätssternwarte Göttingen, Geismarlandstr. 11, 3400 Göttingen, Germany

E-Mail: wglatze at dgogwdgl

E. van Groningen, Uppsala Astronomical Observatory, Box 515, 751 20 Uppsala, Sweden

E-Mail: 21609 :: laban :: ernst

M. Haehnelt, Landessternwarte Königstuhl, 6900 Heidelberg, Germany

J. Heidt, Landessternwarte Königstuhl, 6900 Heidelberg 1, Germany

E-Mail: p20 at dhdurzl

K. Horne, STScI, Homewood Campus, San Martin Drive 3700, Baltimore, MD 21218, USA

E-Mail: horne at stsci

C.A. Hummel, MPI für Radioastronomie, Auf dem Hügel 69, 5300 Bonn 1, Germany

E-Mail: p479chu at mpirbn.uucp

N. Jackson, University of Manchester, Nuffield Radio Astronomy Laboratory, Jodrell Bank, Macclesfield, Cheshire, SK11 9DL, U.K

E-Mail: njj at starlink.jodrell-bank.manchester.ac.uk

J. Kirk, MPI für Kernphysik, Postfach 103980, 6900 Heidelberg 1, Germany

R. Khanna, Landessternwarte Königstuhl, 6900 Heidelberg 1, Germany

W. Kollatschny, Universitäts-Sternwarte, Geismarlandstr. 11, 3400 Göttingen, Germany

E-Mail: wkollat at dgogwdgl

T.Krichbaum, MPI für Radioastronomie, Auf dem Hügel 69, 5300 Bonn 1, Germany

E-Mail: p459kri at mpirbn.uucp

H. Lesch, Landessternwarte Königstuhl, 6900 Heidelberg, Germany

H. Lorenz, Zentralinstitut für Astrophysik, 1591 Potsdam, Germany

B. McBreen, Physics Dept., University College, Sillorgan Road, Dublin 4, Ireland
E-Mail: bmcbreen at irl.earn
K. Meisenheimer, MPI für Astronomie, Königstuhl, 6900 Heidelberg 1, Germany
E-Mail: meise at dhdmpisv
S. Molendi, MPI f. extraterr. Physik, Karl-Schwarzschild-Str.1, 8046 Garching b. München, Germany
E-Mail: mpe :: sil
H. Netzer, School of Physics, Tel Aviv University, Tel Aviv 69978, Israel
E-Mail: h31 at taunos
P. O'Brien, Dept. of Phys., University College London, Gower Street, London WC1E 6BT, U.K.
E-Mail: pto at starlink.ucl.ac.uk or plvad :: zouad :: pto
I. Papadakis, Department of Physics, Queen Mary and Westfield College, Mile End Road, London E1 4NS, U.K.
E-Mail: jhep at uk.ac.qnc.star
H. Paul, Zentralinstitut für Astrophysik, 1591 Potsdam, Germany
M. Penston, Royal Greenwich Observatory, Madingley Rd., Cambridge CB3 OEZ, U.K.
E-Mail: mvp at starlink. astronomy.cambridge.ac.uk
E. Pérez, Instituto de Astrofisica de Canarias, 38200 La Laguna, Tenerife, Spain
E-Mail: 28844 :: epj or 28846 :: epj
B. Peterson, Dept. of Astronomy, 174 West 18th Avenue, Columbus, OH 43210-1106, USA
E-Mail: peterson at ohstpy
A. Quirrenbach, MPI für Radioastronomie, Auf dem Hügel 69, 5300 Bonn, Germany
A. Robinson, Institute of Astronomy, Madingley Rd., Cambridge CB3 OEZ, U.K.
E-Mail: ar at starlink.astronomy.cambridge.ac.uk
H.-J. Röser, MPI für Astronomie, Königstuhl, 6900 Heidelberg 1, Germany
E-Mail: roeser at dhdmpisv
E. Rokaki, Institute d'Astrophysique, 98bis, Blvd. Arago, 75014 Paris, France
E-Mail: iapobs :: rokaki or rokaki at friap51
J. Roland, Institute d'Astrophysique, 98bis, Blvd. Arago, 75014 Paris, France
E-Mail: roland at friap51
R. Schaaf, Institut für Astrophysik, Auf dem Hügel 71, 5300 Bonn 1, Germany
E-Mail: k412rei at mpirbn.uucp
T. Schmutzler, MPI für Kernphysik, Postfach 103980, 6900 Heidelberg 1, Germany
T. Snijders, Astronomisches Institut, Waldhäuser Str. 64, 7400 Tübingen, Germany
E-Mail: aitmyx :: snijders

H. Sponholz, Institut für Theoretische Astrophysik, Im Neuenheimer Feld 561, 6900 Heidelberg 1, Germany
E-Mail: g82 at dhdurz1

M. Stickel, MPI für Astronomie, Königstuhl, 6900 Heidelberg 1, Germany
E-Mail: stickel at dhdmpisv

G. Stirpe, Osservatorio Astronomico di Bologna, Via Zamboni 33, 40126 Bologna, Italy
E-Mail: astbo3 :: giovanna or giovanna at astbo3.cineca.it

H. Störzer, Institut für Theoretische Astrophysik, Im Neuenheimer Feld 561, 6900 Heidelberg 1, Germany
E-Mail: y98 at dhdurz1

A. Treves, Dip. di Fisica, Universita di Milano, Via Celoria 16, 20133 Milano, Italy
E-Mail: vaxmi :: treves

M.-H. Ulrich, ESO, Karl-Schwarzschild-Str. 2, 8046 Garching b. München, Germany
E-Mail: marie at dgaeso51

E. Valtaoja, Turku University Observatory, SF - 21500 Piikkiö, Finland
E-Mail: valtaoja at kontu.utu.fi

L. Valtaoja, Turku University Observatory, SF - 21500 Piikkiö, Finland
E-Mail: valtaoja at kontu.utu.fi

S. Wagner, Landessternwarte Königstuhl, 6900 Heidelberg 1, Germany
E-Mail: w08 at dhdurz1

J. Wambsganss, MPI Astrophysik, Karl-Schwarzschild-Str. 1, 8046 Garching b. München, Germany
E-Mail: jkw at dgaipp1s

I. Wanders, Uppsala Astronomical Observatory, Box 515, 751 20 Uppsala, Sweden
E-Mail: plafys :: laban :: ignaz

M. J. Ward, Dept. of Astrophysics, University of Oxford, Keble Road, Oxford, U.K.

R. Wegner, MPI für Radioastronomie, Auf dem Hügel 69, 5300 Bonn 1, Germany

R. Wehrse, Institut für Theoretische Astrophysik, Im Neuenheimer Feld 561, 6900 Heidelberg 1, Germany
E-Mail: b28 at dhdurz1

A. Witzel, MPI für Radioastronomie, Auf dem Hügel 69, 5300 Bonn 1, Germany
E-Mail: p103wzl at mpirbn.uucp

Have All Questions Been Asked?

Peter L. Biermann

Max Planck Institut für Radioastronomie, Bonn, Germany

Abstract: My question will be specifically whether energetic neutrinos might play a role in the broad emission line regions of Active Galactic Nuclei. I derive a possible neutrino source from arguments about the heating of the circumnuclear dusty disk which has recently been deduced to exist in most AGN. In normal quasars, the heating of the circumnuclear dusty disk is an unsolved problem. I argue that a flux of relativistic particles, protons and heavy nuclei, with a hard spectrum and an intensity comparable to the flux in electromagnetic radiation is plausible. These particles can easily provide disk heating. Such a flux of energetic particles also invariably leads to an increased flux of neutrinos, with a hard spectrum and extending to very high energies. I note that stars on the main sequence are optically thick to energetic neutrinos, with optical depths of order unity or larger. Then I demonstrate, that a rather moderate flux of neutrinos, at a level of only a few percent to a few tens of percent of the electromagnetic flux changes the internal structure of low mass stars drastically. The stars expand, due to internal heating and begin to look like subgiants or giants. Such stars, exposed to the strong external flux of both electromagnetic radiation and relativistic particles then produce a wind, just as has been argued already for a stellar component in a former compact binary system, of which we now observe only a millisecond pulsar. These winds may contribute to the broad emission line regions in quasars. In fact, the effect of neutrinos is so strong, that it limits the lifetime of low mass stars due to mass loss; such lifetime arguments might be usable to actually put limits on the neutrino flux in quasars.

1. Introduction

The organizers of this meeting asked me to start with a talk that would provoke you, and so we settled on the title "Have all questions been asked?". I believe it is obvious that such a situation is unlikely ever to occur, since there will always be questions that nobody thought of asking, or that were asked but nobody listened to and took notice. Today I will try to provoke you by asking whether energetic neutrinos might play a role in broad emission line regions. The answer, as I will try to convince you, is that they might play an important role indeed, and to dismiss their possible influence at the very least will be an interesting exercise.

I start with some recent discoveries about normal quasars. Normal quasars were thought to be usually radioquiet, or as is now known, at least very weak in their radio emission (Kellermann et al. 1989); as a corollary it used to be accepted that a large part of their emission is nonthermal, especially in the farinfrared (e.g. Edelson and Malkan 1986). Since the discovery of the rather low level of mm-emission in quasars (Chini et al. 1989a, b), originally reported at the Santa Cruz Meeting on AGN in the summer of 1988, it has become commonly accepted, that not only Seyfert galaxies (Edelson et al. 1988), but also quasars are dominated throughout their infrared to ultraviolet emission by thermal processes (Sanders et al. 1989). The argument rests to a large degree on the difficulty of understanding the sharp turnoff from 100μm to 1.3mm wavelength together with the spectrum shortwards of 100μm; this spectral behaviour is readily fitted only with heated dust (Chini et al. 1989a), although fairly complicated models based on nonthermal emission have not totally been ruled out yet (de Kool & Begelman 1989, Schlickeiser et al. 1990).

Thus the only wavelength regions with any evidence of nonthermal processes are the radiorange and, by virtue of Compton models, the X-ray range. I note that the two well known examples of active galactic nuclei with strong hard X-ray emission, 3C273 (Bezler et al. 1984), and Cen A (von Ballmoos et al. 1987) actually show more luminosity in the hard X-ray range than at any other wavelength of the electromagnetic spectrum; this has been observed, however, only at one epoch of time. The conclusion that most the far infrared emission from quasars is actually thermal brings immediately the question to the foreground of what can possibly provide the heating, since at no other well observed wavelength range do we detect emission which is sufficient.

However, due to the strong interstellar absorption at ultraviolet wavelengths the UV/soft X-range is a possible candidate: It is commonly deduced from suitable upturns of the emission both at the extreme observed hard UV as well as the extreme soft X-ray range (Arnaud et al. 1985, Elvis et al. 1989, Reimers et al. 1989), that there is an emission bump connecting these two upturns, which would then dominate the entire emission by a substantial factor. This emission (Gondhalekar 1990) is the one that provides most of the heating and ionization in the normal photoionization models for the broad emission line region. This bump is commonly interpreted as emission from the innermost part of the accretion disk (Malkan 1983, Madau 1988, Band and Malkan 1989, Laor and Netzer 1989, Ulrich 1989, Laor et al. 1990), dominating all other easily observable wavelength ranges; this strong emission is energetically clearly capable of providing the heating for the circumnuclear dusty disk.

The problem with such an interpretation is that the extended narrow emission line regions observed in many active nuclei clearly outline a biconical radiation field (Pérez-Fournon and Wilson 1990, Acosta-Pulido et al. 1990, Haniff, Ward and Wilson 1991) which does not seem to be geometrically capable to strike the disk anywhere; this difficulty could be circumvented by introducing a very strong warp in the disk already at small distances from the nucleus (Sanders et al. 1989). On the other hand, it is not clear at all that such extreme warps at distances of out to a hundred parsecs from the nucleus are really realistic. Also, it is difficult to see how the radiation can be scattered out of its cone into the disk in such a manner over the various radii from less than a parsec to several hundred parsecs as to emit nearly equal luminosities in the range $100\mu m$ to about $12\mu m$. In conclusion it is not clear at all, from where the circumnuclear dusty disk obtains its energy which is radiated so profusely in the far infrared:

Here we note that relativistic particles can help. Many current models to explain the X-ray emission involve a population of highly relativistic electrons. Since it is normally much easier to accelerate protons and heavier nuclei than electrons, we assume in the following that nuclei are accelerated in the regions above and below, but close to the central part of the accretion disk. These energetic particles are easily scattered in the plasma of the circumnuclear region above and below the disk, so geometry becomes rather less relevant. A particle energy distribution with an energy density similar to the other forms of radiation is common in normal galaxies, be it our Galaxy or M82 (Kronberg et al. 1985, Schaaf et al. 1989), for which it is well known that the emission is totally dominated by dust emission from regions heated by hot stars, and yet the energy density of the relativistic particles is close to that of thermal matter. The one crucial difference between galaxies like M82 and the circumnuclear region of quasars is likely to be the spectrum of the energetic particles, not the energy density ratio between relativistic particles on the one hand and thermal radiation and thermal matter on the other.

We then introduce the hypothesis that indeed a spectrum of relativistic particles is produced in the innermost parts of quasars with a number spectrum near to E^{-2} in energy E. Biermann and Strittmatter (1987) demonstrated that based on first order Fermi acceleration at shocks proton energies can be expected of about $10^{10}\ (U/c)/\sqrt{B}$ GeV, where B is the magnetic field measured in Gauss and U/c is the shockvelocity relative to the speed of light. The particles are assumed to have an energy density similar to that of the UV radiation field to within a small number of powers of ten. An immediate consequence is obviously copious Gamma-ray emission which is likely to be just detectable in bright quasars with GRO. Here we want to pursue another argument: These relativistic particles produce a very strong neutrino flux through their interaction both with the ambient photon field as well as the thermal matter, again with a hard spectrum, and also of an energy density similar to that of the particle flux. The energies of these neutrinos extend to many powers of ten beyond the TeV range (Biermann 1989); from a single interaction of a proton with a photon of suitable laboratory energy neutrinos emerge with an energy of a few percent of the primary proton energy, while the proton can undergo many such interactions.

These neutrinos in turn may exert a strong influence on low mass stars which should be rather abundant in the innermost regions of quasars, producing the gravitational well in the distance range of parsecs to hundreds of parsecs, if not something else. It is known from compact X-ray binary and millisecond pulsar explorations that neutrinos can influence the structure of stars (Gaisser et al. 1986, 1987). Based on these ideas J. MacDonald, T. Stanev from Bartol and I explored the consequences of such a neutrino flux on low mass stars. In the following I will summarize the results of our investigations (MacDonald, Stanev and Biermann 1990) and related arguments. In section two I will report on the influence of the neutrinos on the internal structure of stars, and in section 3 I will describe what effect these changed stars may have in the core environment of quasars.

2. The influence of neutrinos on the structure of low mass stars

We note first that the interaction cross section of neutrinos with matter is given by

$$\sigma_v = 0.7 \cdot 10^{-38} E_\nu / \{1 + (E_\nu/\epsilon_W) ln(E_\nu/50GeV)\}$$

where the cross section is given in cm^2 and E_ν in GeV (Gaisser and Stanev 1985); ϵ_W is 3.5 TeV. At 3.5 TeV this cross section has its maximum with the value of $5 \cdot 10^{-36}$ cm^2, which corresponds to a column density of about $4 \cdot 10^{11}$ g/cm^2. H.C. Thomas (priv.comm. 1988) has kindly calculated the column density through various stars. Main sequence stars have the following column densities in the same units through the center and through an offcenter distance of 20% of their radius, respectively: $2.4 \cdot 10^{12}$ and $1.0 \cdot 10^{12}$ for a 1 solar mass star, $2.1 \cdot 10^{12}$ and $0.8 \cdot 10^{12}$ for a 3 solar mass star, $1.2 \cdot 10^{12}$ and $0.67 \cdot 10^{12}$ for a 10 solar mass star, and, finally, $0.85 \cdot 10^{12}$ and $0.54 \cdot 10^{12}$ for a 30 solar mass star. For giant stars I compare the column density through the center and at 10^{-3} of their radius: The column densities are then $4 \cdot 10^{14}$ and $1.5 \cdot 10^{11}$ for a 1 solar mass red giant, $3.3 \cdot 10^{15}$ and $1.2 \cdot 10^{7}$ for 5 solar mass red supergiant, and, correspondingly, $3.9 \cdot 10^{14}$ and $4 \cdot 10^{9}$ for a 25 solar mass red supergiant. It follows immediately that main sequence stars are optically thick to energetic neutrinos at energies beyond a few TeV through most of their mass, while red giants, albeit optically thick right through their center, have a small cross section. It ensues that neutrinos can influence the internal structure of a star by their energy deposition.

The question is then what neutrino fluxes to explore. We used the UV fluxes given by Gondhalekar (1990) of 10^9 to 10^{13} erg/cm^2/sec, which bracket most observed cases, as our reference, and explore the same range in fluxes for neutrinos. The calculations then span a range of stellar models of zero age of 0.25, 0.5, 0.8 and 1.0 solar mass, and produce generalized main sequences with the neutrino flux as a parameter. The results demonstrate that for increasing neutrino fluxes of 10^{10} to 10^{11} erg/cm^2/sec the stars expand, first become bluer and brighter, and then, for yet higher neutrino fluxes move to the red giant region. It is instructive to compare the internal structure of an evolved 0.8 solar mass star with that of a zero age 1 solar mass star exposed to a neutrino flux of about $3 \cdot 10^{12}$ erg/cm^2/sec. This comparison shows that the neutrino heating has taken over completely the task of providing the stars' luminosity and that the central regions of the star have expanded such as to lower the central density, for example, by eight powers of ten. The structure of the neutrino heated star resembles rather well the outer layers of the evolved star. Thus the binding energy of the surface layers is substantially reduced and prone to external influences. This bloating of stars due to neutrinos is drastically different from that done by electromagnetic radiation (Edwards 1980, Penston 1988, Voit and Shull 1988, Tout et al. 1989) because neutrinos act from the very central region of the star. Using then the prescription of van den Heuvel and van Paradijs (1988) of how an external hard radiation field can cause a wind to develop, we can calculate the mass loss due to heating of the outer layers by an effective flux of hard photons and energetic particles $F_{el,eff}$; the time scale to mass loss is then given by

$$\tau = (GM^2)/(\pi R^3 F_{el,eff})$$

This timescale, to take a 0.25 solar mass star as one example, is about $2 \cdot 10^7$ years for a neutrino flux of only 10^{10} and an effective particle and hard photon flux of $3 \cdot 10^{11}$ erg/cm^2/sec. This relation is derived by setting the energy loss due to a wind at surface escape velocity equal to the effective energy gain. This demonstrates that at only 3% of the effective electromagnetic flux the neutrino flux can have a drastic influence by expanding the star and so making it prone to mass loss.

3. The possible effect on the broad emission line region

What is then the relevance of stellar winds for the broad emission line region? Recently the idea, that the emission lines are due to stellar winds (e.g. Mathews, 1983) has again found interest (Scoville and Norman 1988, Kazanas 1989); the old argument against this, that compact binary stars do not show related effects (Verbunt et al. 1984) is superseded by the observation of millisecond pulsars and their possible history (Kluzniak et al. 1988, van den Heuvel and van Paradijs 1988, Grindlay and Bailyn 1988). If this idea should turn out to be the correct understanding of the broad emission line region, then it becomes of interest to study the origin of stellar winds. It is in this context that we have studied the possible origin of stellar winds above; with a sufficient flux of energetic neutrinos all low mass stars may have winds, and so there is a very large number of stellar candidates, since low mass stars are usually the most abundant stars. We summarize:

1) Given a sufficient flux of energetic neutrinos low mass stars turn off their nuclear reactions and no longer change their internal chemical structure.
2) The stars become red giants or subgiants to outward appearances, with loosely bound envelopes.
3) External hard radiation readily generates a wind in these envelopes.
4) Stellar evolution becomes dominated by stellar mass loss.
5) Low mass stars, being the most sensitive, move off the main sequence due to the neutrinos, and so the main sequence population is peeled off from the low mass end.
6) The stellar winds, thus caused, may be the origin of the observed broad emission lines under the influence of the ultraviolet radiation field.

7) Given this model it may become possible to constrain the neutrino flux in quasars, and thus, possibly, rule out the importance of energetic neutrinos altogether. This would also be an interesting limit.
8) We have not even tried to speculate on the evolution of a stellar cluster under the influence of steady mass loss from the stars, with emphasis on the low mass end of the stellar distribution.

Thus, in conclusion, I hope to have provoked you into thinking that, maybe, neutrinos are relevant after all for the physics of the broad emission line region.

Acknowledgements: I wish to thank Drs. J. MacDonald and T. Stanev for permission to report a summary of our common paper, which will be published with all relevant details; I also wish to thank Dr. H.C. Thomas for the calculations of stellar column densities, already done in late 1988, and his permission to quote them. This whole work was started when Dr. M.M. Shapiro invited both T. Stanev and myself to an Erice meeting early 1988 about gamma and neutrino astronomy; I owe him many thanks for his generous invitation and the inspiring environment at Erice.

References

Acosta-Pulido. J.A., Pérez-Fournon, I., Calvani, M., Wilson, A.S. 1990: Ap.J. (In press)
Arnaud, K.A., Branduardi-Raymont, G., Culhane, J.L., Fabian, A.C., Hazard, C., McGlynn, T.A., Shafer, R.A., Tennant, A.F., Ward, M.J. 1985: MNRAS 217, 105
von Ballmoos, P., Diehl, R., Schönfelder, V. 1987: Ap.J. 312, 134
Band, D.L., Malkan, M.A. 1989: Ap.J. 345, 122
Bezler, M., Kendziorra, E., Staubert, R., Hasinger, G., Pietsch, W., Reppin, C., Trümper, J., Voges, W. 1984: Astron.&Astroph. 136, 351
Biermann, P.L., Strittmatter, P.A. 1987: Ap.J. 322, 643
Biermann, P.L. 1989: Proc. Erice Meeting "Cosmic Gamma-Rays, Neutrinos and Related Astrophysics", Eds. M.M. Shapiro, J.P. Wefel, Kluwer, p.21
Chini,R., Kreysa, E., Biermann, P.L. 1989a: Astron.&Astroph. 219, 87
Chini, R., Biermann, P.L., Kreysa, E., Gemünd, H.-P. 1989b: Astron.&Astroph. Letters 221, L3
Edelson, R.A., Malkan, M.A. 1986: Ap.J. 308, 59
Edelson, R.A., Gear, W.K.P., Malkan, M.A., Robson, E.I. 1988: Nature 336, 749
Edwards, A.C. 1980: MNRAS 190, 757
Elvis, N., Wilkes, B.J., McDowell, J.C. 1989: Proc. Coll. EUV Astronomy, Berkeley, Eds. R.F. Malina, S. Bowyer, (Preprint)
Gaisser, T.K., Stanev,T. 1985: Phys.Rev.Letters 54, 2265
Gaisser, T.K., Stecker, F.W., Harding, A.K., Barnard, J.J. 1986: Ap.J. 309, 674
Gaisser, T.K., MacDonald, J., Stanev, T. 1987: Proc. 12. ICRC, Moscow, vol.2, 272
Gondhalekar, P.M.1990: MNRAS 243, 443
Grindlay, J.E., Bailyn, C.D. 1988: Nature 336, 48
Haniff, C.A., Ward, M.J., Wilson, A.S. 1991: Ap.J. (Feb 10)
van den Heuvel, E.P.J., van Paradijs, J. 1988: Nature 334, 227
Kazanas, D. 1989: Ap.J. 347, 74
Kellermann, K.I., Sramek, R., Schmidt, M., Shaffer, D.B., Green, R. 1989: A.J. 98, 1195

Kluzniak, W., Ruderman, M., Shaham, J., Tavani, M. 1988: Nature 334, 225
de Kool, M., Begelman, M.C. 1989: Nature 338, 484
Kronberg, P.P., Biermann, P.L., Schwab, F.R. 1985: Ap.J. 291, 693
Laor, A., Netzer, H. 1989: MNRAS 238, 897
Laor, A., Netzer, H., Piran, T. 1990: MNRAS 242, 560
Madau, P. 1988: Ap.J. 327, 116
MacDonald, J., Stanev, T., Biermann, P.L. 1990: Preprint
Malkan, M.A. 1983: Ap.J. 268, 582
Mathews, W.G. 1983: Ap.J. 272, 390
Penston, M.V. 1988: MNRAS 233, 601
Pérez-Fournon, I., Wilson, A.S. 1990: Ap.J. 356, 456
Reimers, D., Clavel, J., Groote, D., Engels, D., Hagen, H.J., Naylor, T., Wamsteker, W., Hopp, U. 1989: Astron.&Astroph. 218, 71
Sanders, D.B., Phinney, E.S., Neugebauer, G., Soifer, B.T., Matthews, K. 1989: Ap.J. 347, 29
Schaaf, R., Pietsch, W., Biermann, P.L., Kronberg, P.P., Schmutzler, T. 1989: Ap.J. 336, 722
Schlickeiser, R., Biermann, P.L., Crusius-Wätzel, A. 1990: Astron.&Astroph. (submitted)
Scoville, N., Norman, C. 1988: Ap.J. 332, 163
Tout, C.A., Eggleton, P.P., Fabian, A.C., Pringle, J.E. 1989: MNRAS 238, 427
Ulrich, M.-H. 1989: Proc. Coll. "Theory of Accretion Disks", Eds. F. Meyer, W.J. Duschl, J. Frank, E. Meyer-Hofmeister, Kluwer, p.3
Verbunt, F., Fabian, A.C., Rees, M.J. 1984: Nature 309, 331
Voit, G.M., Shull, J.M. 1988: Ap.J. 331, 197

Line and Continuum Variability in NGC 4151

M. A. J. Snijders

Astronomisches Institut Tübingen

Abstract: We present results on the variability of the absorption lines, the emission lines and the X-ray to optical continuum of NGC 4151. The ultraviolet and optical continua vary together and the emission lines follow these continuum changes with delays depending on the line studied. The strength of the absorption lines depends on the continuum strength too but the variability pattern is rather complicated. When the ultraviolet/optical continuum is bright there is no correlation with the X-rays. When this continuum is faint the X-ray flux and the optical/ultraviolet continuum are strictly correlated. Results from Principle Component Analysis of the IUE spectra are discussed.

1. Introduction

The Seyfert 1 galaxy NGC 4151 is the nearest and brightest Seyfert 1 galaxy known. It is therefor the most extensively studied Seyfert 1 galaxy in the optical, ultraviolet, X-ray and Gamma-ray regions of the spectrum. The results which I am going to present are largely based on the work of an international collaboration which has frequently observed the spectrum of NGC 4151 with the IUE satellite (Boggess et al., 1978) and in addition has a large amount of nearly simultaneous optical and X-ray data. Our main results have been published in Bromage et al. (1985), Clavel et al. (1987), Clavel et al. (1990), Penston et al. (1981), Perola et al.ts (1982), Perola et al. (1986), Ulrich et al. (1984), Ulrich et al. (1985) with related results published by Fiore, Perola and Romano (1990), Gill et al. (1984), Leech et al. (1987), Mittaz, Penston and Snijders (1990) and Penston et al. (1979). The spectrum of NGC 4151 varies continuously in both the lines and the continuum in the optical and in the ultraviolet regions and there is extensive material on the, often very rapid, X-ray variability (eg Perola et al. 1982, 1986). Qualitative models based on the assumptions that the optical and/or ultraviolet continua provide a good idea of the state of the flux in the Lyman continuum and that this ionising continuum drives the variability of both the Broad Emission Line Region (eg Ulrich et al. 1984, Clavel et al. 1987, Clavel et al. 1990) and the absorption lines (Bromage et al. 1985) give a good description of short term variability.

I was asked to present results on the variability of the spectral lines of NGC 4151 but as two of my collaborators will present results on the emission lines too (Clavel 1990, Ulrich 1990) I will start with a discussion of results on the continuum variability as the line variability appears driven by the changes in the continuum and insight in the continuum characteristics is essential for understanding the behaviour of the spectral lines.

NGC 4151 is a key object. Most of the variability patterns observed in the ultraviolet and optical continua and emission lines of NGC 4151 were later found to occur in the spectra of other, fainter Seyfert 1 galaxies too. However as NGC 4151 is so near, cz = 990 km/sec, it is much easier obtain high quality results for this object.

2. The optical, ultraviolet and X-ray continuum variability

While it has been clear almost from the beginning of the IUE observations that the optical and ultraviolet spectral variability is very closely related (Penston et al. 1981, Perola et al. 1982) the connection between the optical and ultraviolet spectra and the X-ray data is fairly complicated. When the source is bright in the optical and ultraviolet there is no correlation at all with the X-ray flux (Perola et al. 1982) but when the source is faint in the optical and ultraviolet the correlation between the optical and ultraviolet continua and the X-ray flux is excellent (Perola et al. 1986). When the source is faint the lagg (if any) between the UV/optical continuum and the X-ray flux must be very small, it is certainly substantially less then the 4 day spacing with which our simultaneous IUE and X-observations were obtained. There is a very close correlation between the variability of the optical and ultraviolet continua (Perola et al. 1982, Gill et al. 1984). The correlation between the emission line variability and the ultraviolet/optical continuum shows a phase lagg due to the light travel time between the continuum emitting nucleus and the material which emits the emission lines (Ulrich et al. 1984, Clavel et al. 1987, Clavel et al. 1990). Analysis of these phase laggs will, in principle, give the distance between the line emitting clouds and the continuum source through reveberation mapping (Blandford and McKee, 1982). The major problem is to obtain observations at a reasonable, if possible equally spaced, frequency. In view of the short variability time scale spectra should be obtained at least every 4 days and perhaps every 3 or even 2 days (Clavel et al., 1990). During our 1979 observing campaign significant variability in the CIV emission line and the ultraviolet continuum was observed over periods of only 2 days (19, 21, 23 and 25 May 1979). A second problem of a rather fundamental nature concerns the uncertainty in the derived correlations and the physical interpreation of the various measured laggs between lines and continua (Maoz and Netzer, 1989; Edelson and Krolik 1988).

In view of the very close relation between the IUE and the optical continua it is possible to use optical observations to fill in gaps in the IUE light curve of NGC 4151 without the need to use the highly oversubscribed IUE satellite. Unfortunately the Seyfert nucleus in NGC 4151 is at the low end of the luminosity distribution for Seyfert galaxies and consequently a substantial fraction of the detected flux, in the apertures used for optical photometry, is due to stars and not to the active nucleus (eg Penston et al. 1971). In addition the available optical data were obtained with a variety of instruments and apertures. This requires for the construction of a self consistent optical light curve the derivation and application of both aperture and color corrections to the various sets of observations (Gill et al., 1984). Our starting point are the two large sets of B-band photometry obtained since around 1970 by Lyutyi and various co-workers using photoelectric B-band photometry through a 27" aperture (eg Lyutyi and Oknyanskii 1988) and the photographic B-band photometry at Royal Greenwich Observatory (Gill et al. 1984, Snijders and Penston 1991). Both groups of photometry cover about 300 epochs and contain sufficient simultaneous observations to derive a magnitude dependent combined aperture and color correction. Other shorter stretches of data, covering parts of the last 22 years, for instance the data of Antonucci and Cohen (1983) for the period 1980/1981 and the FES optical magnitudes (Snijders and Penston 1991), were then tied into the combined data base. The resulting lightcurve (Snijders and Penston 1991) is shown in Fig. 1.

Fig. 1. together with the historical lightcurve of Lyutyi and Oknyanskii (1988) shows a number of features not generally realised. First the deep minimum of 1981 (Ulrich et al. 1984, Antonucci and Cohen 1983) is far from unique. A photometrically, reasonably well documented earlier minimum did occur in 1976 but none of the minima between 1970 and 1980 was as deep as the later minima. The historical light curve from Lyutyi and Oknyanskii shows that comparable, deep minima are very rare. From 1984 to November 1988 NGC 4151 has been almost continuously very faint. The only exception known is between November 1984 and March 1985 and even then the source was clearly fainter then in 1983 or the bright periods from 1978 to 1980. Since November 1988 the source seems to be slowly increasing in brightness. The only time, since IUE was launched, that the source was as bright as during the early nineteen seventies was in April 1978 and the strength and width of the high ionisation CIV emission doublet in April and May 1978 are impressive. Compared to the much longer light curve reconstructed by Lyutyi and Oknyanskii (1988) from old photographic plates it is clear that the brightest epochs from the period 1970-1980 are not partical bright at all and that the only period the source was as faint as during the 1984 minimum did occur around 1910! It should be realised that the magnitude scale used by Lyutyi and his colleagues is about half a magnitude brighter than ours as their larger aperture includes more starlight. We have no RGO magnitudes for really bright states of the nucleus, the 1945-1950 period for instance, and the necessary colour correction is consequently not known. It is however worthwhile to remember a fact our collaboration learned the hard way over the past 12 years of IUE observations: whenever the source was observed at a fainter state than observed previously new phenomena were discovered. It will be very interesting to obtain spectroscopy in a really bright state and to see what happens than.

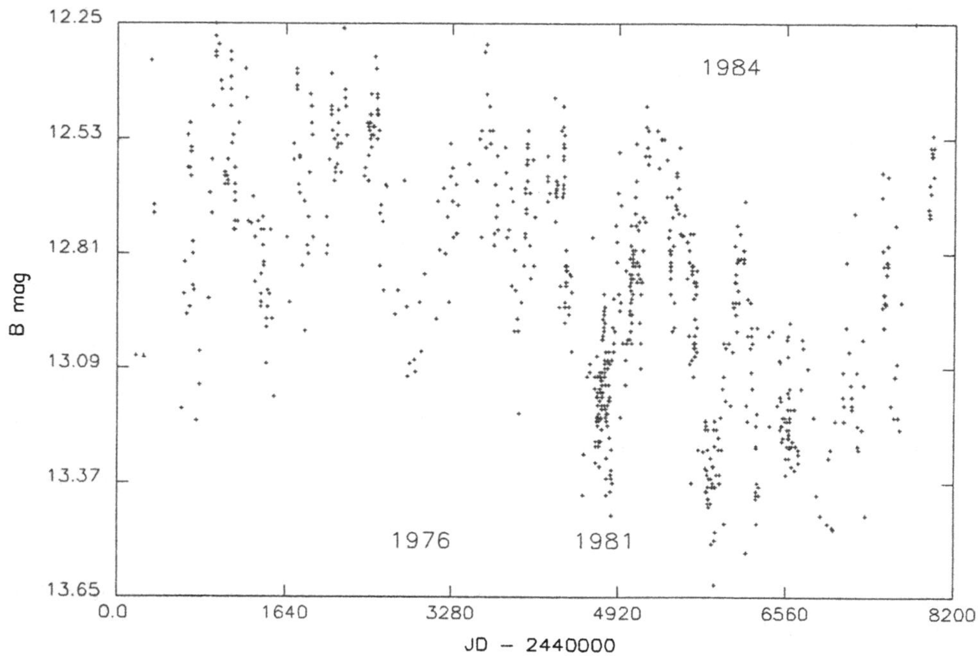

Fig. 1. The optical B-band lightcurve of NGC 4151 for the last 20 years. The position of the 1976, 1981 and 1984 minima is indicated.

3. Variability of the ultraviolet spectra

In Fig. 2 we show 2 representative IUE short wavelength spectra of the source in the bright (1 June 1979) and in the faint state (2 April 1984). Both spectra are averages over all IUE spectra obtained on these two dates. There is actually very little the 2 spectra have in common! The continuum between the SiIV +OIV] blend at 1400Å and the NIV] 1486Å line is virtually unaffected by lines over a 30Å stretch and a good measure for the continuum variability even when the source is at the brightest stage observed during the IUE lifetime. The 1455Å continuum is a factor 9 brighter in the first spectrum and the bright state continuum is much harder (eg Perola et al. 1982, Edelson, Krolik and Pike 1990).

On a first inspection of Fig. 2. the wealth of detail immediately draws attention. A comparison with the, by now, extensive literature on the IUE spectra of Seyfert 1 galaxies confirms this (eg. Edelson, Krolik and Pike 1990; Clavel and Joly 1984; Chapman, Geller and Huchra 1985). The large number of well defined high and low ionisation absorption lines (Bromage et al. 1985) is unique so far. The IUE spectra of NGC 3516 are of lower quality and show fewer absorption lines (eg. Walter et al. 1990). The excellent S/N of the low state spectrum makes many, in other IUE Seyferts not detectable faint emission lines visible. In particular the weak OI 1302Å and CII 1335Å emission lines, which are in the bright phases hidden under the corresponding absorption lines are clearly present. While the absorption lines appear to be absent in the faint state this is at least partly due to the low resolution of these spectra. High resolution IUE spectroscopy during low states (Leech et al. 1987, Leech et al. 1990) shows that absorption components are still present in the detectable permitted lines: Ly-α, and the CIV and MgII resonance doublets. The problem with the limited resolution of the IUE satellite in the low resolution mode affects lines in the bright phases too. Bromage et al. showed that their average bright phase HeII 1640Å profile possibly contained an absorption component and the excited fine structure absorption lines of SiII and FeII are clearly visible on their mean high resolution spectrum. This makes detailed analysis of the observed absorption lines rather difficult. For accurate numerical analysis of the absorption HST data with improved resolution and S/N are required. There is however one group of lines whose presence alone immediately gives a lower limit to the density of the absorbing material. We definitly detect in the IUE spectra absorption arising from excited metastable levels: CIII 1175Å, SiIII 1294-1303Å and HeI 2945Å (Bromage et al. 1985). The presence of the CIII 1175Å multiplet leads to a lower limit for the density in the absorbing region $log(Ne) >$ 8.5. Anderson (1974) derived from optical spectroscopy of HeI absorption lines $log(Ne) > 6.0$. A further problem in the analysis of the absorption lines are the weak emission lines present on all faint state spectra. If these emission lines arise partly in the Narrow Emission Line Region (NELR) they will be present in the bright states as well and affect our equivalent width measurements noticably. Both high and low resolution data show a blue shift of the absorption lines with respect to the emission lines of up to 1000 km/s (Penston et al. , 1979, Bromage et al. 1985, Leech et al. 1987).

The decline in the strength and the width of the emission lines when the ultraviolet continuum declines is large and shows a noticable lagg with respect to the continuum. The only lines not affected are the forbidden [OII] 2470Å and [NeIV] 2423Å lines which are emitted only by the narrow emission line region. It affects not only the permitted lines: Ly-α, CIV 1549Å, HeII 1640Å, MgIIÅ and in the optical the Balmer lines (Antonucci and Cohen, 1983) but also the intercombination lines. The effect is easiest to see from the NIV] 1486Å, line but OIV] + SiIV at 1400Å, OIII] 1663Å, NIII] 1750Å and CIII] 1908Å show the effect too.

The broad component of "CIII]" is of particular interest: its apparent blue shift and width are due to a blend with the SiIII] 1892Å intercombination line. As the SiIII] and CIII] broad components have nearly equal strength the gas emitting these lines must have a density intermediate between the critical densities of these two lines, well in excess of the classical value of log(Ne) 9.5. This is however in good agreement with recent work based on the size of the broad line region and the estimated ionization parameter which suggests higher densities too (eg. Clavel et al. 1990).

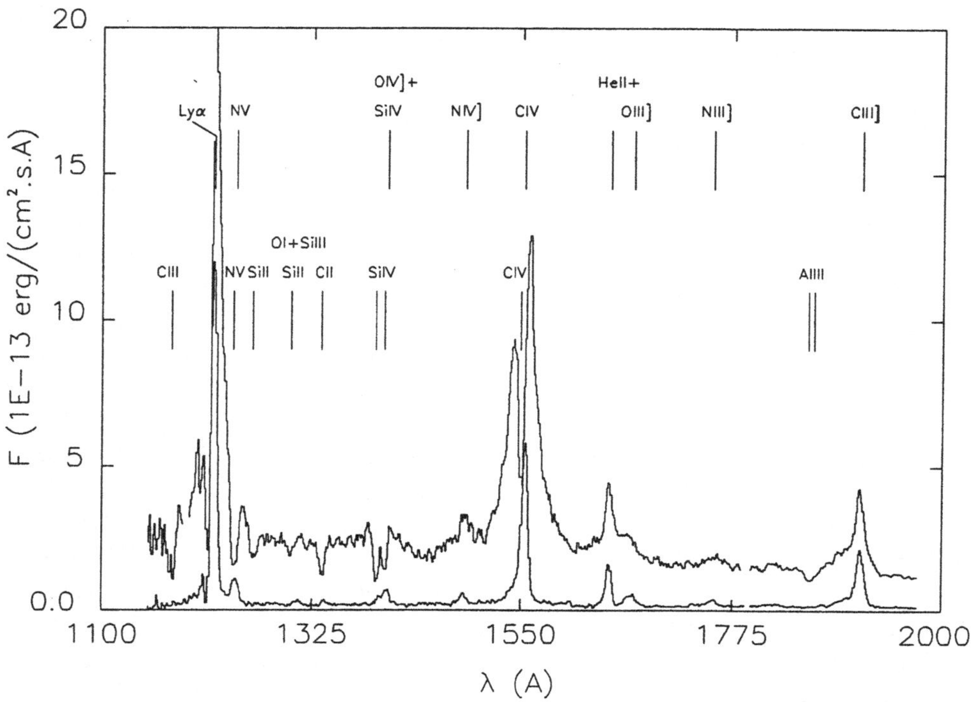

Fig. 2. IUE spectra for a representative high (1 June 1979) and a representative low state (2 April 1984). The line identifications in the top row are for the emission lines and those in the lower row are for the absorption lines.

4. Principle component analysis

The availability of a large number of spectra homogeneously reduced and on common wavelength and flux scales made a principle component analysis of the available IUE data possible (Mittaz, Penston and Snijders, 1990, hereafter MPS). So far astronomers have mainly been guided by intuition in their search for physical relations between the various variable components. A number of correlations, anticorrelations and temporal relations between the various emission and absorption lines and the continuum components have been proposed (eg Bromage et al., 1985, Clavel et al., 1987) but a systematic approach has been lacking. Principle component analysis has long been used to determine common underlying factors in data sets (eg. Murtagh and Heck, 1987). The aim is to obtain the minimum number of independent, orthogonal components necessary for recreating the observations. Each spectrum is considered as a vector with 1375 wavelength points and for every epoch there is one vector. We search the correlation matrix of the spectra for the eigenvectors and their eigenvalues after the mean of all the spectra has been subtracted. The eigenvector with the largest eigenvalue represents the major source of variability in the full data set and gives the first principle component. The eigenvector with the second largest eigenvalue specifies the next largest source of variability and the second principle component etc. Details concerning error estimates and the treatment of gaps in the data are given by MPS. MPS found that the first 9 to 11 principle components, depending on the type of statistical criterion used, appear statistically significant. While this shows that 9 to 11 independent components are sufficient to describe the data it does not give clear information on the nature of these components. Detailed analysis of the spectral eigenvectors has to clarify this. While strongly reinforcing a number of known ideas about UV variability derived from the normal way of studying the IUE spectra a number of surprises show up. The well known correlations between continuum brightness, spectral slope, the width and the strength of the strong CIV and MgII emission lines and strength of the well known UV absorbtion lines (eg NV, CIV, SiIV, MgII, FeII and SiII) are very clearly present in the first principle component. This first component alone accounts for 53% of the variability (Table 3 in MPS). The three first principle components together account for 76% of the total variability and the last 6 to 8 components for the remainder. The complete absence of the forbidden [NeIV] and [OII] lines in all principle components, while they are strongly present in the mean spectrum, is directly explained by the non-variability of these lines, which are formed in the extended Narrow Emission Line Region (NELR). They also serve as a powerfull check on the whole method and the accuracy of the data reduction process. The presence of these lines in the principle components would have been highly surprising and almost certainly have been due to errors in either the data reduction or the way the principle component analysis was executed.

Among the surprises are the clear presence of strong broad bands of FeII emission in the first principle component between 2250Å and 2650Å and a weaker band of FeII emission between 2900Å and 3000ÅThe broad MgII wings from 2700Å to 2900Å might hide additional FeII emission. The perfect removal of a number of strong narrow emission lines which do not vary makes FeII features for the first time easy to distinguish.

The CII] 2326Å and the CIII] 1908Å intercombination features are both strongly present in the mean spectra. The strong narrow component of the CII] 2326Å multiplet is completely absent from the first and higher principle components but a weak broad, variable CII] feature could be hidden in the local broad, variable FeII band. The strength of the variable CIII] 1908Å (and SiIII] 1892Å) feature in the first principle component is a surprise. The variability of this line in NGC 4151 has long been doubted (eg. Ulrich et al. 1984). The very strong narrow component from the NELR makes detection of the underlying weak broad variable component difficult but since the work of Clavel et al. (1987) its variability can no longer be questioned. Nevertheless its strength and width in the first principle component are unexpected.

Lastly there are the narrow, highly variable, features in the CIV line wings only observed when the nucleus is in its faint state (Ulrich et al., 1985). The source of these lines has been a long standing puzzle and the proposed mechanisms run from possible emission from jets to unidentified but otherwise normal emission lines. Their unusually fast variability generates problems for most models (Ulrich et al., 1985) and the principle component analysis adds a second problem. These lines are absent from all the lower principle components which describe the bulk of the variable behaviour. However principle components 5 to 8 show a variety of strong features at the wavelengths of these unidentified lines and very little else (MPS). Or whatever physical mechanism is behind these emission features is independent of the normal source of variability in the emission and absorption lines. Unidentified but otherwise normal emission lines from the broad line region are now rather unlikely.

References

Anderson, K.S., 1974, Astrophys.J., **189**, 195.
Antonucci, R.R.J. and Cohen, R.D., 1983, Astrophys.J., **271**, 564.
Blandford, R.D. and McKee, C.F., 1982, Astrophys.J., **255**, 419.
Boggess, A., et al., 1978, Nature, **275**, 372.
Bromage, G.E., et al., 1985, Monthly Notices Roy. Astr. Soc., **215**, 1.
Chapman, G.N.F., Geller, M.J. and Huchra, J.P.,1985, Astrophys.J., **297**, 151.
Clavel, J., et al., 1987, Astrophys.J., **321**, 251.
Clavel, J., et al., 1990, Monthly Notices Roy. Astr. Soc., **246** , 688.
Clavel, J., 1990, in "Heidelberg Conference on Variability in Active Galactic Nuclei" ed. S. Wagner, in press.
Edelson, R.A., Krolik, J.H., 1988, Astrophys.J., **333**, 646.
Edelson, R.A., Krolik, J.H. and Pike, G.F., 1990, Astrophys.J., **359**, 86.
Fiore, F., Perola, G.C. and Romano M., 1990, Monthly Notices Roy. Astr. Soc., **243**, 522.
Gill, T.R., Lloyd, C., Penston, M.V. and Snijders, M.A.J., 1984, Monthly Notices Roy. Astr. Soc., **211**, 31.
Leech, K., Penston, M.V., Snijders, M.A.J. and Gull, T., 1987, Monthly Notices Roy. Astr. Soc., **225**, 837.
Leech, K., Penston, M.V., Snijders, M.A.J., Ward, M. and Gull, T., 1990, in "Evolution in Astrophysics", ESA-SP 310, in press.
Lyutyi, V.M., Oknyanskii, V.L., 1988, Soviet Astron., **31**, 245.
Maoz, D. and Netzer, H., 1989, Monthly Notices Roy. Astr. Soc., **236**, 21.
Mittaz, J.P.D., Penston, M.V. and Snijders, M.A.J., 1990, Monthly Notices Roy. Astr. Soc., **242**, 370.
Murtagh, F. and Heck, A., 1987, in "Multivariate Data Analysis", Reidel Publishing Company, Dordrecht, p13.
Penston, M.V., et al., 1971, Monthly Notices Roy. Astr. Soc., **153**, 29.
Penston, M.V., et al., 1979, Monthly Notices Roy. Astr. Soc., **189**, 45P.

Penston, M.V., et al., 1981, Monthly Notices Roy. Astr. Soc., **196**, 857.
Perola, G.C., et al., 1982, Monthly Notices Roy. Astr. Soc., **200**, 293.
Perola, G.C., et al., 1986, Astrophys.J., **306**, 508.
Snijders, M.A.J. and Penston, M.V., 1991, in preparation.
Ulrich, M.H., Boisson. C., 1983, Astrophys.J., **267**, 515.
Ulrich, M.H., et al., 1984, Monthly Notices Roy. Astr. Soc., **206**, 221.
Ulrich, M.H., et al., 1985, Nature, **313**, 747.
Ulrich, M.H., 1990, in "Heidelberg Conference on Variability in Active Galactic Nuclei" ed. S. Wagner, in press.
Walter, R., Courvoiser, T.J.L., Ulrich, M.H., 1990, Astron.Astrophys., **233**, 53.

The Optical, Ultraviolet and X-ray Variations in NGC 4151

Marie-Hélène Ulrich

European Southern Observatory, Karl-Schwarzschild-Str. 2, 8046 Garching bei München, Germany

Abstract: We propose two processes to produce quasi simultaneous UV/optical variations in AGNs: instabilities in the inner part of the disk and irradiation of the disk surface by the variable X-ray source. We are led to this proposition by the quasi simultaneous variations observed in the ultraviolet and optical ranges in NGC 4151. Such variations are not consistent with the continuum in these wavelength ranges, being produced only by accretion flow process, and therefore other components must contribute significantly to the UV/optical flux.

1. Introduction

Since the launch of the IUE satellite in 1978, the ultraviolet spectrum of the Seyfert galaxy NGC 4151 has been observed on 125 days. The observations were grouped in campaigns lasting 15 to 60 days. Each campaign brought new results on the many diverse aspects of the complex spectrum of NGC 4151 (Ulrich *et al.* 1990 and references therein).

Considering the large amount of data obtained by our group, I thought, at the end of the latest campaign in April 1990, that we had reached a privileged position where we could compare the results from the different campaigns, follow the evolution of the nucleus on time scales of years rather than days, and identify its permanent properties. I re-measured the 480 low dispersion spectra in a consistent fashion and re-examined the results from the different campaigns. The results on the continuum emission and the emission lines which emerge from this analysis are presented and discussed in a forthcoming article by Ulrich *et al.* (1990) to which no further reference will be made here. The present paper is a summary of the results on the continuum.

2. Results

2.1. The light curves

The light curve of the continuum at 1455 Å is shown in Fig. 1 (top). In Fig. 1 (bottom) is plotted the light curve in the B band resulting from the photographic monitoring of NGC 4151 conducted independently of the IUE observations by the Royal Greenwich Observatory (Gill *et al.* 1984; Snijders 1990). The corresponding histograms are presented in Fig. 2. Note that the UV observations were not conducted at random since most were grouped in campaigns. The campaigns however were scheduled without a priori knowledge of the state of the nucleus. The peak at the left of Fig. 2 (UV continuum) corresponds to 2 extended campaigns (in 1981 and 1984) during which the nucleus was at a minimum.

2.2. Simultaneous optical, UV and X-ray observations

The Fine Error Sensor (FES) on board IUE provides measurements of the optical flux through an unfiltered S20 photocathode. The conversion of the FES (Imhoff and Wasatonic 1987) counts into erg cm^{-2} s^{-1} $Å^{-1}$ is uncertain by a constant factor which we cannot determine because we do not have at our disposal an absolutely calibrated optical spectrum of NGC 4151 taken during one of our IUE observations. The contribution of the Balmer continuum to the FES flux is negligible. That of the rapidly varying wings of the emission lines is estimated to be less than 20% (Boksenberg *et al.* 1975). Thus the observed variations of the FES flux (see below) are caused by a variable continuum.

The flux in the optical (FES) band and in the UV at 3040 Å and 1455 Å is shown in Fig. 3a,b for two recent campaigns. We note the good correlation between the flux in these 3 wavelength ranges. The variations are simultaneous with our time sampling of 4 to 5 days. The cross correlation between FES flux and the flux at 1455 Å for the campaign of 29 November 1988 – 30 January 1989 gives a delay less than 2 days in either direction (Clavel *et al.* 1990).

The non-variable part of the flux (at least not variable on a time scale of a year) has been evaluated from the data of 2–11 April 1984, the epoch of the lowest observed minimum of NGC 4151. The flux in April 1984 was (in units of 10^{-14} erg cm^{-2} s^{-1} $Å^{-1}$): 2 and 5.1 for the UV and FES flux respectively.

The modulation of the UV and optical flux defined by the ratio of the maximum flux to the minimum flux measured above the values of April 1984 is as follows:

Table 1. Modulation of the UV and optical flux: f_{min}/f_{max}

Campaign	Dec 84 – Jan 85	Nov 88 – Jan 89	Feb – April 90
UV (1455 Å)	2.7	8.5	2.3
optical (FES)	1.7	1.9	1.3

The modulation of the optical flux is always smaller than that of the UV flux even after subtraction of the "pedestal" of April 1984. This difference points to the presence in the FES window of a component which varies slowly with time (and which was ~ 0 in April 1984). Fig. 4 gives the FES flux in function of the flux at 1455 Å. There is a correlation between the two fluxes during each campaign, but this correlation is not the same for these 3 campaigns.

The modulation and time scale of the optical flux (FES) variations in Fig. 3a are consistent with those observed in a continuum window centered at 4735 Å by Maoz *et al.* during their ground based monitoring of NGC 4151 in the period January – July 1988.

Simultaneous campaigns of observations of NGC 4151 in the UV and the X-ray ranges have been conducted only during two periods: 7–19 November 1983 and 16 December 1984 – 2 January 1985. The results are plotted in Fig. 5 and 6 and show that there is a remarkable similarity between the variations in the UV and those in the 2–10 keV ranges (EXOSAT–ME observations: see Perola *et al.* 1986). This correlation, on only 9 dates, could of course be a fluke; the probability of this correlation being due to chance is 2.5×10^{-6} (Perola *et al.* 1986). We would like to stress that simultaneous UV/X-ray observations or UV/optical observations at other isolated epochs do not follow the relation defined by the data of 7–19 November 1983 and 16 December 1984 – 2 January 1985: The corresponding points fall below and to the right (see Penston 1986). For example, the point representing the simultaneous IUE/Einstein Observatory observations of 19–21 May 1979 is plotted in Fig. 6 (Perola *et al.* 1982, 1986).

2.3. The UV spectral energy distribution

Two spectral indices α_1 and α_2 defined by $f_\nu \propto \nu^{-\alpha}$ were calculated between 1455 and 1715 Å and between 1715 Å and 3042 Å respectively. For the calculations of α_1 and α_2 the fluxes were corrected for the galactic reddening ($E_{B-V} = 0.05$; Perola *et al.* 1986, extinction curve of Seaton 1979). The values of α_1 and α_2 are plotted against the flux at 1455 Å in Fig. 7. The dispersion in α_1 is large because of the smallness of the wavelength interval in which it is calculated. We find a trend for α_1 to increase when the continuum increases. This is however not statistically significant. Dividing the values of $f_\lambda(1455)$ into three intervals and calculating the mean and σ of α_1 in each interval gives the following results in the table below:

flux	mean α_1	standard deviation
$f_\lambda < 6 \times 10^{-14}$erg cm^{-2} s^{-1} Å^{-1}	–0.35	0.54
$6 \leq f_\lambda < 12$	–0.19	0.65
$12 \leq f_\lambda$	0.06	0.41

In contrast, α_2 (which is shown alone on a different scale in Fig. 7, middle panel) decreases when the continuum increases, a result which was already found by

Perola *et al.* (1982). The different values and variability character of α_1 and α_2 with f_λ indicate that two different continuum components contribute to the flux at 1455–1715 Å and 1715–3042 Å (Perola *et al.* 1982). In the first interval we are probably seeing the thermal emission from an accretion disk. In the second interval, the continuum flux may be due mostly to Fe II blends and Balmer continuum (e.g. Wills, Netzer and Wills 1985) plus some contribution from the stellar population near 3000 Å.

Figure 7 (lower panel) shows α_1 and α_2 during the large amplitude decrease and increase of f_λ of 1988 December – 1989 January. We do not see any systematic change of α_1 with f_λ.

The values of α_2 before and after October 1983 are shown with different symbols in Fig. 7 (middle panel). The values after October 1983 fall systematically below those of earlier epochs. This may be due to a genuine evolution of the spectrum. However no conclusion is reached here for the following reason: the data after 1983 October come from the LWP camera for which we have applied no correction for sensitivity changes. If the sensitivity of the LWP camera were lower by 20% than the value assumed here, the spectral index α_2 after 1983 October would fall in the same range as before this date.

3. Discussion

It has been shown that the long viscous time scale of accretion disks rules out that optical/UV variability in AGNs is caused by variable fuelling (Clarke 1987, 1988). Accretion disk instabilities in the inner region are however compatible with the observed UV variability in NGC 4151 (Siemiginowska and Czerny 1989). We suggest that the small high-temperature region which corresponds to such an instability and which emits mostly in the extreme UV also contributes to the optical flux through the low energy tail of its spectrum. In this case, one expects perfect simultaneity between the UV and optical variations. The amplitudes of the variations in the optical and UV have the potential to give strong constraints to this model.

Alternatively, the UV/optical variations could be due to the irradiation of the disk by the variable X-ray source. Such a model has been proposed by Czerny, Czerny and Grindlay (1986) for low mass binaries. We propose that, in some cases, the X-rays can alter the distribution of temperature vs disk radius with respect to that produced by a pure accretion flow process so that the disk surface can produce the UV and optical continuum.

In this case, one expects the UV/optical variations to be correlated with the X-rays variations. The X-ray data of 7–19 November 1983 and 16 December 1984 – 2 January 1985 (Fig. 6) are consistent with irradiation. [The effect of the X-ray source on the disk surface has been considered by Collin-Souffrin and Dumont (1990; see also Dumont and Collin-Souffrin 1990) but only as a source of heating of the dense gas emitting the Fe II lines and the Balmer lines].

The correlation of Fig. 6 taken together with the "discrepant" data such as those of May 1979 suggests to us that several processes contribute to the UV emission. Periods when there is a good UV/X-rays correlation indicate that, at this time, irradiation makes a major contribution to the UV flux. The absence of such a correlation (with *larger UV flux levels* than expected from the correlation) would indicate that instabilities are the main cause of the simultaneous optical/UV variations and that their effect masks any variations due to the X-ray source.

Clearly the relative importance of the two processes will define the optical/UV/X-ray flux ratio as well as the regime of variabilities in these 3 energy ranges. Furthermore other processes such as comptonization are likely to affect the energy distribution and the variability characteristics.

References

Boksenberg, A., *et al.* (1975): Monthly Notices Roy. Astron. Soc. **173**, 381
Clarke, C.J. (1987): Bull. American Astr. Soc. **19**, 732
Clarke, C.J. (1988): Monthly Notices Roy. Astron. Soc. **235**, 881
Clavel, J., *et al.* (1990): Monthly Notices Roy. Astron. Soc. in press
Collin-Souffrin, S., Dumont, A.M. (1990): Astron. Astrophys. **229**, 292
Czerny, B., Czerny, M., Grindlay, J.E. (1986): Astrophys. J. **311**, 241
Dumont, A.M., Collin-Souffrin, S. (1990): Astron. Astrophys. **229**, 302
Edelson, R.A., Krolik, J.H., Pike, G.F. (1990): Astrophys. J. in press
Gill, T.R., *et al.* (1984): Monthly Notices Roy. Astron. Soc. **211**, 31
Imhoff, C., Wasatonic, R. (1987): IUE-NASA Newsletter **32**, 45
Maoz, D., *et al.* (1990): preprint
Penston, M.V. (1986): in "Physics of Accretion onto Compact Objects", ed. K.-O. Mason, Springer, Berlin 1986
Perola, G.C., *et al.* (1982): Monthly Notices Roy. Astron. Soc. **200**, 293
Perola, G.C., *et al.* (1986): Astrophys. J. **306**, 508
Seaton, M.J. (1979): Monthly Notices Roy. Astron. Soc. **187**, 73P
Siemiginowska, A., Czerny, B. (1989): Monthly Notices Roy. Astron. Soc. **239**, 289
Snijders, M.A.J. (1990): private communication
Ulrich, M.-H., *et al.* (1990): Astrophys. J. submitted
Wills, B.J., Netzer, H., Wills, D. (1985): Astrophys. J. **288**, 94

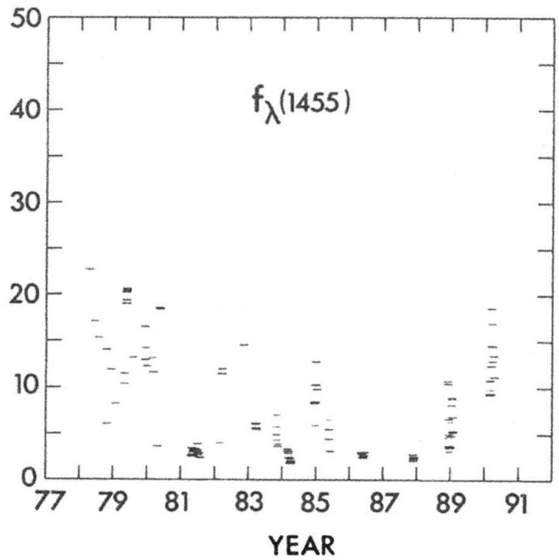

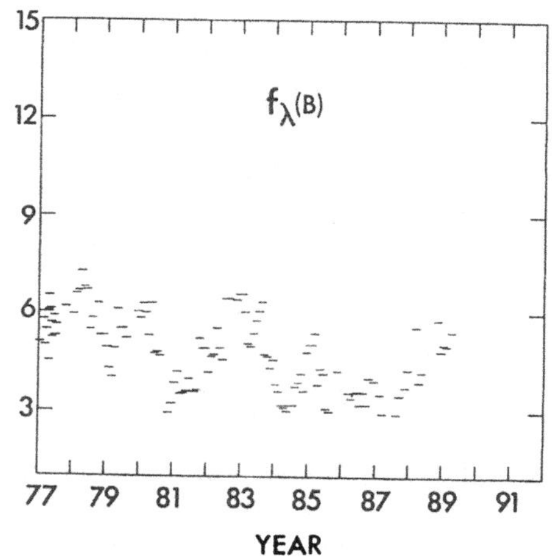

Fig. 1. *Continuum light curves* of NGC 4151. (Top): f_λ at 1455 Å; (bottom): the B magnitude defined by Gill *et al.* (1988) converted into f_λ at 4200 Å. Ordinates in 10^{-14} erg cm^{-2} s^{-1} $Å^{-1}$.

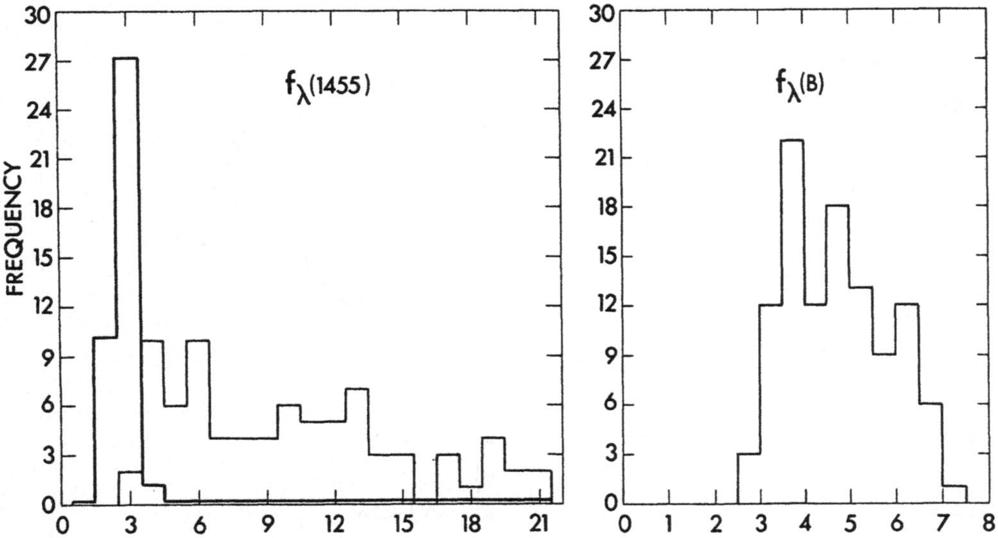

Fig. 2. Histogram of f_λ in 10^{-14} erg cm^{-2} s^{-1} Å^{-1} at 1455 Å, and in the B wavelength range of Gill *et al.* (1988).

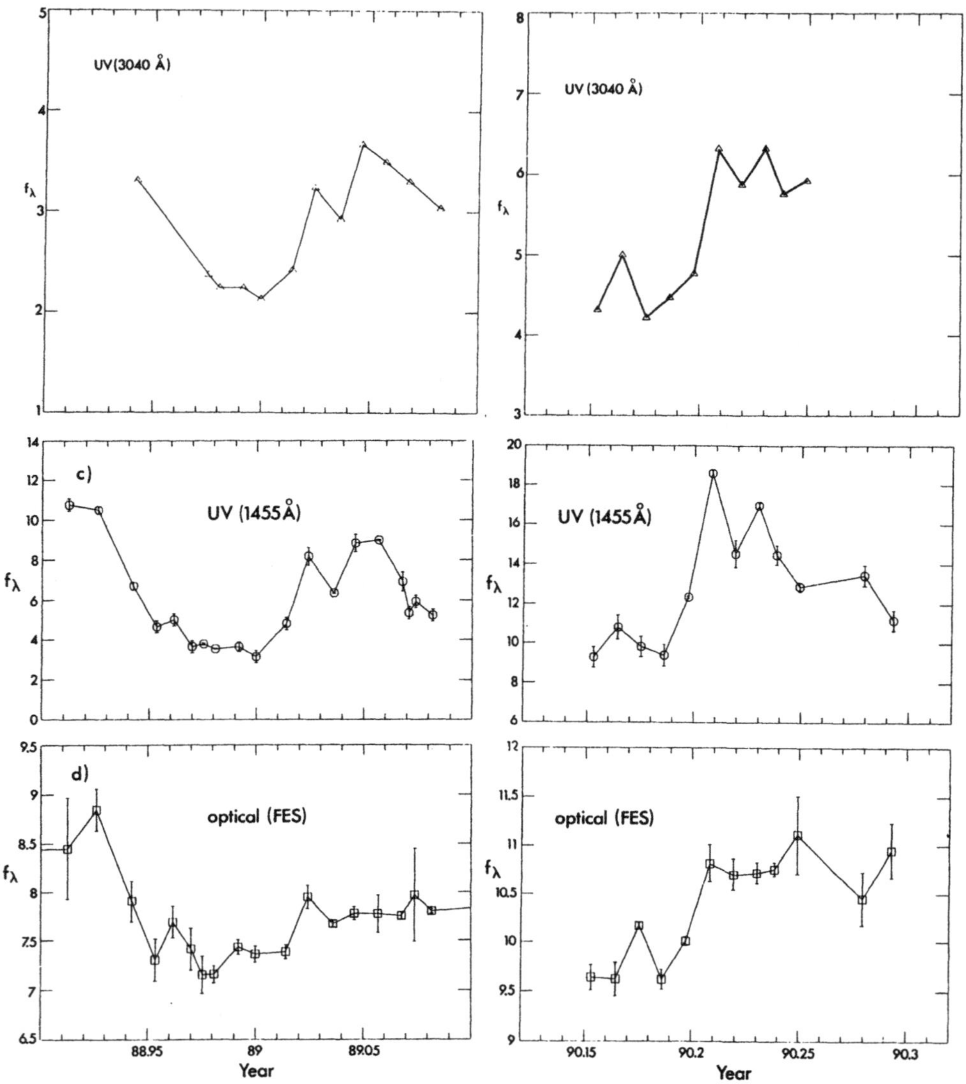

Fig. 3. The flux at 3040 Å and at 1455 Å and in the FES for 2 episodes (see Ulrich *et al.* 1990) (flux in 10^{-14} erg cm^{-2} s^{-1} Å^{-1}).

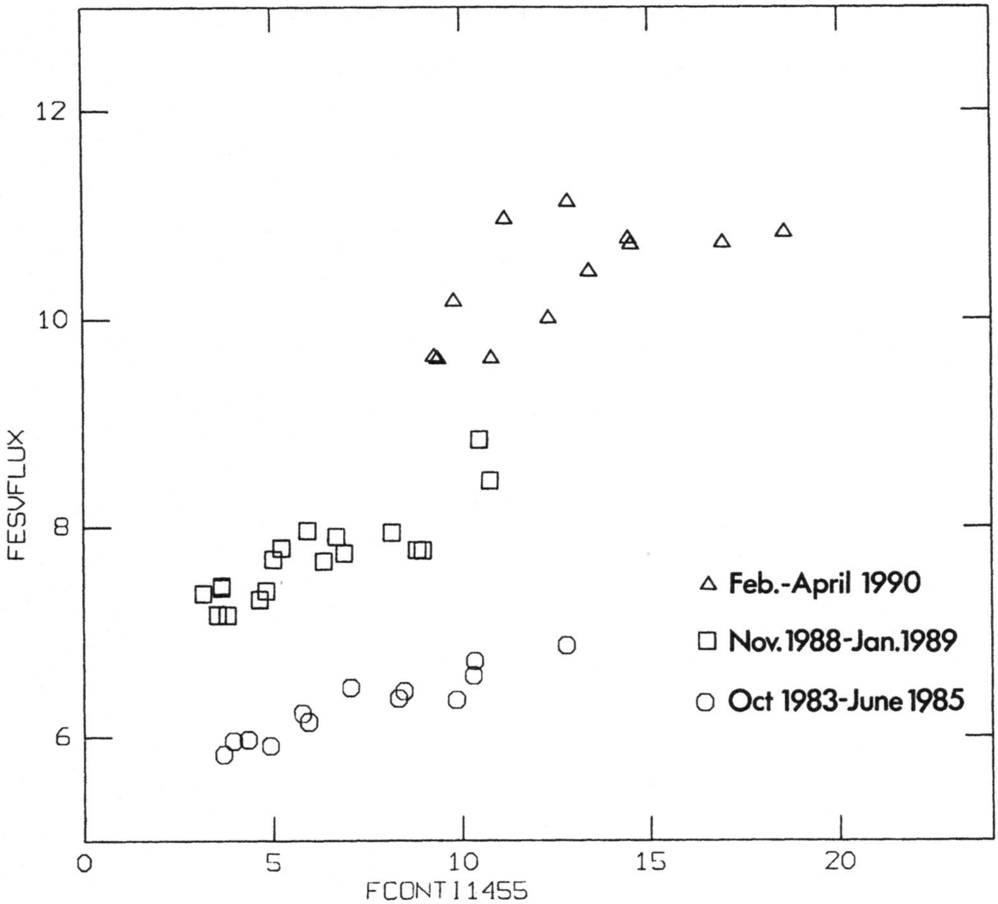

Fig. 4. The FES flux vs the flux at 1455 Å for three campaigns (in 10^{-14} erg cm^{-2} s^{-1} Å^{-1}).

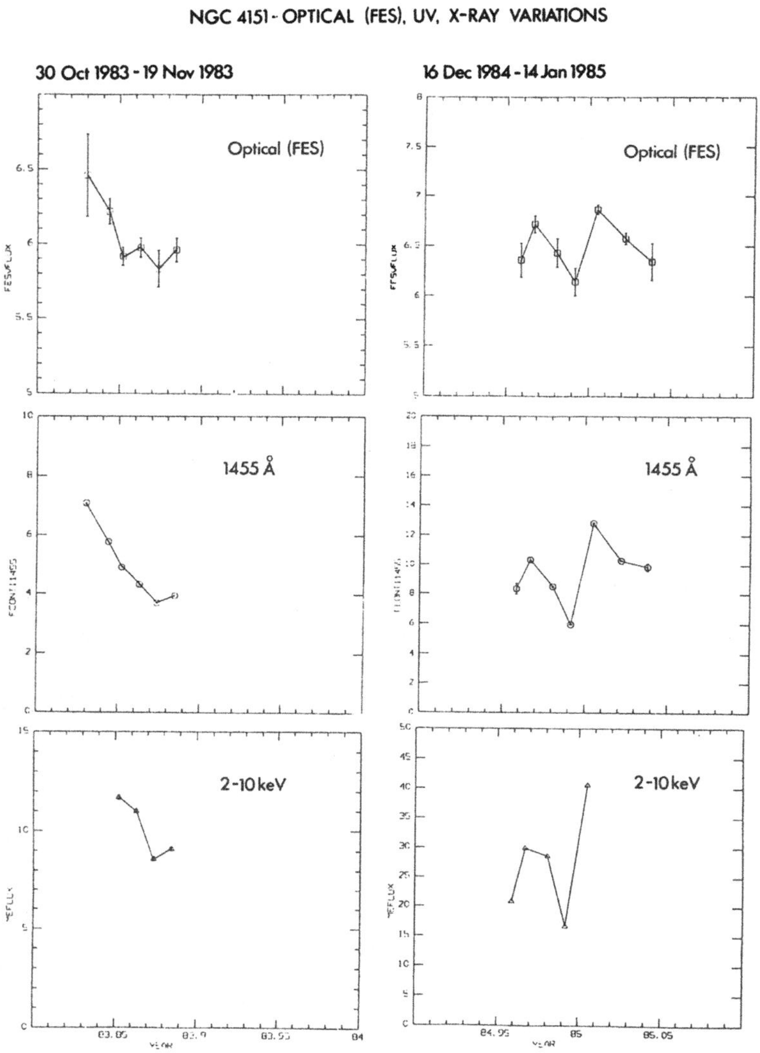

Fig. 5. The FES flux, UV flux at 1455 Å and 2–10 keV flux during two campaigns. Simultaneous observations with IUE and EXOSAT took place on 7, 11, 15, 19 November 1983 and 16, 20, 24, 28 December 1984 and 2 January 1985 (Perola *et al.* 1986).

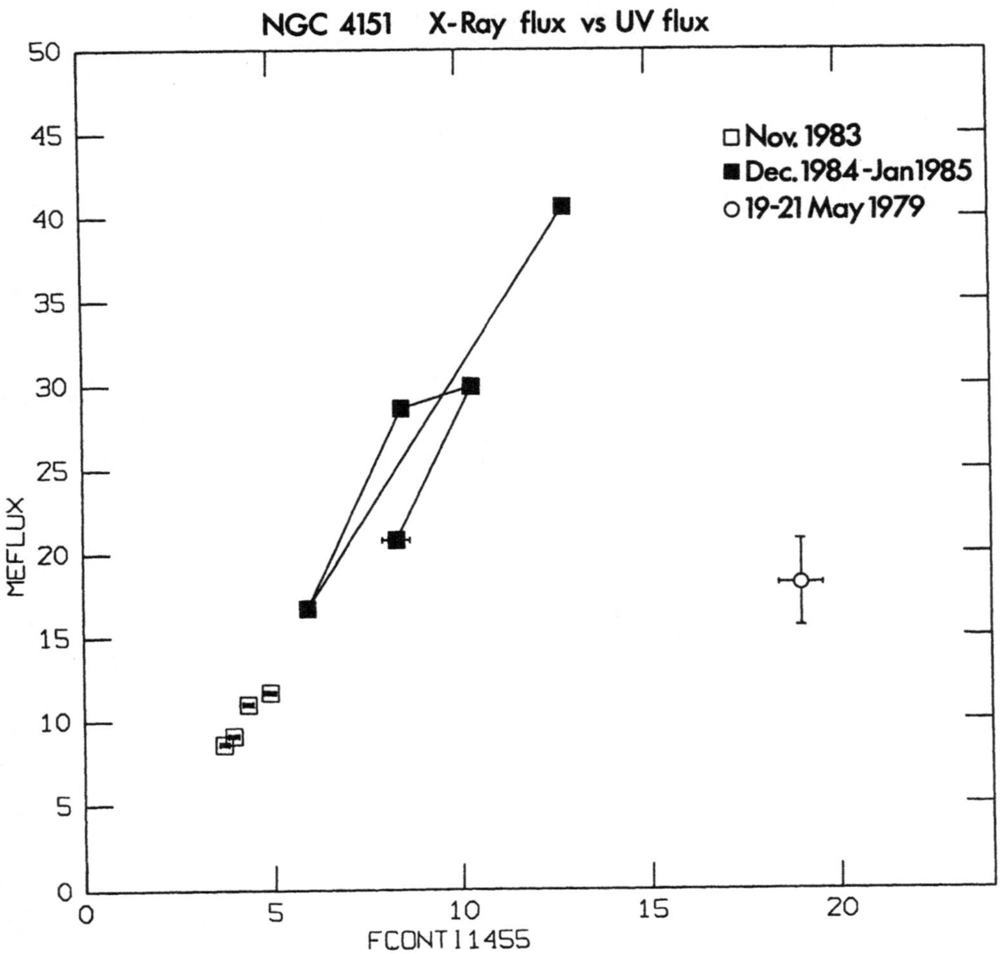

Fig. 6. Same UV/X-ray data as in Fig. 5 plus the point representing the UV flux and X-ray flux on 19–21 May 1979 (Perola *et al.* 1982, 1986).

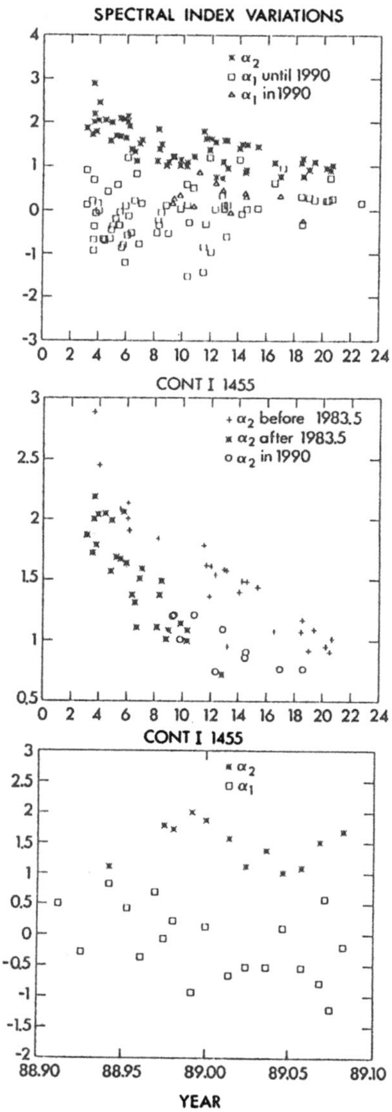

Fig. 7. Spectral indices (defined by $f_\nu \propto \nu^{-\alpha}$ in function of f_λ at 1455 Å. Index α_1 is the index between 1455 and 1715 Å, α_2 between 1715 and 3042 Å. Top panel: α_1 (squares) and α_2 (stars) at all epochs. Middle panel: α_2 only. Bottom panel: α_1 and α_2 for the campaign 1988 November – 1989 January. Abscissae in 10^{-14} erg cm^{-2} s^{-1} Å^{-1}.

Mapping the BLR in NGC 5548 & NGC 4151

Jean Clavel

ESA IUE Observatory, Apartado 50727, 28080 Madrid, Spain

Abstract: We present new results from intensive monitoring campaigns of the Seyfert I galaxies NGC 5548 and NGC 4151 with the IUE satellite which allow a reliable determination of the BLR dimension in these sources. Most of the gas is concentrated at about 12 lt-d from the continuum source in NGC 5548, and 4 lt-d in NGC 4151. Furthermore, there is a gradient in the degree of ionization of the gas in NGC 5548, the latter increasing inward. The picture is more complex in NGC 4151 because the density of the gas also increases toward small radial distances. In NGC 4151, the line wings respond more rapidly and with a larger amplitude than the "cores" ($V \leq 3000$ km·s^{-1}) to the continuum variations, therefore implying that the velocity dispersion of the gas also increases inward. A similar though shallower velocity gradient is also present in NGC 5548. The red and blue part of the line profiles vary simultaneously in both galaxies ruling-out the possibility that radial motions play a significant role in the velocity field of the BLR. The continuum variations are simultaneous from $\simeq$ 1300 Å to 5000 Å to within ±1 day. This pattern is difficult to reconcile with models where the "big-bump" represents thermal emission from an optically thick and geometrically thin accretion disk.

1. Introduction

For more than 20 years, it has been known that the fluxes of the continuum and broad emission lines of active galactic nuclei (AGN's) are variable. It was also realized early that the broad emission line and continuum variations are closely related, as would be expected if the gas responsible for these lines is photoionized and heated by the continuum source. If photoionization is indeed the correct explanation for the emission lines' source of energy, the local response to changes in the continuum flux takes < 1hr, but the line-emitting region is predicted to be much larger: light-days to many light-years across, scaling crudely as $L^{1/2}$, where L is the ionizing luminosity of the nucleus (Davidson and Netzer 1979). Therefore, one would expect emission line fluctuations to follow continuum fluctuations, but spread out and delayed by an amount comparable to the light travel time across the emission line region.

In principle, it is possible to deconvolve the emission line and continuum light curves to infer the detailed structure of the emission line region (Blandford and McKee 1982) It might be possible to obtain by this method the geometrical configuration and the run of physical conditions of the BLR gas as a function of distance from the central source. Such a deconvolution would amount to a direct test of the size prediction made by photoionization models. Not surprisingly therefore, much observational effort has been recently devoted to the spectrophotometric monitoring of active galaxies. However, most of these studies have suffered from at least one of the following problems: inadequate temporal resolution, insufficient span of observations, too many gaps in the time series, or low ratio of fluctuation amplitude to measurement uncertainty.

A large consortium (see Clavel *et al.* 1991a for the list of 56 collaborators) therefore initiated a program whose object was to circumvent all these problems. The choice of the target (NGC 5548) was mainly driven by practical considerations. A large apparent brightness, a long IUE observability window (eight months) and a documented history of strong line and continuum variability were the main selection criteria. For more than 10 years, a smaller group (see Clavel *et al* 1991b for the list of co-authors) had been monitoring with IUE another bright Seyfert I galaxy, NGC 4151. Important results had been achieved (Penston *et al* 1981; Perola *et al* 1982; Ulrich *et al* 1984 and 1985; Bromage *et al* 1985; Clavel *et al* 1987), but until recently the luck had hardly been our side and the galaxy entered a semi-quiescent state more or less each time we started monitoring it. In late 1988 however, we caught NGC 4151 during an episode of intense variability and were able to follow it with an adequate temporal resolution through 2 large maxima and 2 deep minima.

We report briefly here the results of these two monitoring campaigns. We concentrate on the essential results while the reader is referred to Clavel *et al.* (1991a and 1991b) for details of the observing procedure and data analysis techniques.

2. The data

The nucleus of NGC5548 was observed at 60 different epochs in the low resolution mode (1000 km.s^{-1}) and through the large apertures (10"× 20") of the IUE (Faelker, Gordon and Sandford 1987) spectrographs (1200–3200 Å) The continuum was measured at 1350 Å, 1840 Å and 2670 Å. We caution here that these continuum measurements unavoidably contain some emission line flux, in particular the 2670 Å flux 30% to 40% of which may in fact be due to FeII + BaC emission (Wamsteker *et al.* 1990). We have measured the flux of the strong H Ly$\alpha\lambda$1216, CIVλ1549 emission lines as well as that of the weaker MgIIλ2798, CIII]λ1909, HeIIλ1640, SiIV+OIV]λ1402 and NVλ1240 lines. No evidence was found for variability on a time scale of a few hours. Hence, the 25 pairs of SWP spectra obtained on the same day have been used to estimate the errors on the line and continuum flux measurements.

To investigate the behaviour of the line emissivity as a function of radial velocity, we have divided the CIVλ1549 emission line profile into three bins and measured the line flux in each bin independently. Hereafter, these bins will be referred to as the "core", the "blue" wing and the "red" wing. The core contains all the emission in the velocity range -3000 km.s^{-1} to +3000 km.s^{-1} from the line peak, while the blue and the red wing includes respectively all the flux at velocities smaller than -3000 km.s^{-1} and larger than +3000 km.s^{-1}.

The low luminosity Seyfert I galaxy NGC 4151 was observed at 19 different epochs with the IUE satellite during the period November 29, 1988 to January 30, 1989. We measured the continuum at 1455 and 1715 Å as well as the intensity of the CIVλ1549, CIII]λ1909 and MgIIλ2798 line. To quantify spectral variations, we computed the continuum "hardness ratio" H/R defined as the ratio of F_{1455} to F_{1715}. Again, no evidence was found for variability (at the 3% level) on time scales of a few hours. Hence, as for NGC 5548, the 14 pairs of SWP spectra and 8 pairs of LWP spectra obtained on the same day have been used to infer the error on the line and continuum flux at each epoch. We have also divided the CIVλ1549 and MgIIλ2798 line profiles into a "core", a "blue" and a "red" wing and measured their intensity at each epoch. Immediately before each exposure, the $\sim$ 5000 Å flux of NGC 4151 and NGC 5548 were also recorded with the FES star tracker on board IUE through its $10'' \times 20''$ aperture.

3. Results

3.1. NGC 5548

The light-curves of the continuum in the different ultraviolet and optical bands together with those of the strongest emission lines are shown in fig. 1. It is obvious that significant changes were observed in each of the lines and continua we measured. Table 1 summarizes the main characteristics of the variability for each quantity: its mean value over the entire period of observations (in units of 10^{-14} erg cm^{-2} s^{-1} and 10^{-14} erg cm^{-2} s^{-1} Å^{-1}, for the lines and continua, respectively), the fractional variation F_{var} defined as the ratio of the rms fluctuation to the mean flux, and the ratio of maximum to minimum flux, R_{max}.

The longer the wavelength, the smaller the amplitude of the continuum variations. F_{var} decreases gradually from $\simeq$ 0.30 at 1350 Å, to 0.26 at 1840 Å, 0.16 at 2670 Å and 0.08 near 5000 Å (FES). Since the variations are simultaneous in all four bands, this implies that the spectrum becomes systematically bluer when it gets brighter. The large dilution of the nuclear continuum by stellar emission accounts for most of the reduced amplitude of the variations in the optical. Much of the diminution in variability at the longer ultraviolet wavelengths is due to dilution by relatively steady FeII and Balmer continuum (BaC) emission. In the SWP, the full range of spectral index variations is $\Delta\alpha = 1.1 \pm 0.2$, with r.m.s. fluctuations $\Delta\alpha = 0.23$. For comparison, the average reddening corrected spectral index for a power-law fit to the SWP spectrum ($F_\nu \propto \nu^{-\alpha}$) is $\langle\alpha\rangle \simeq 0.95$. Blended FeII lines – provided their intensity does not vary more than that of MgIIλ2798 –

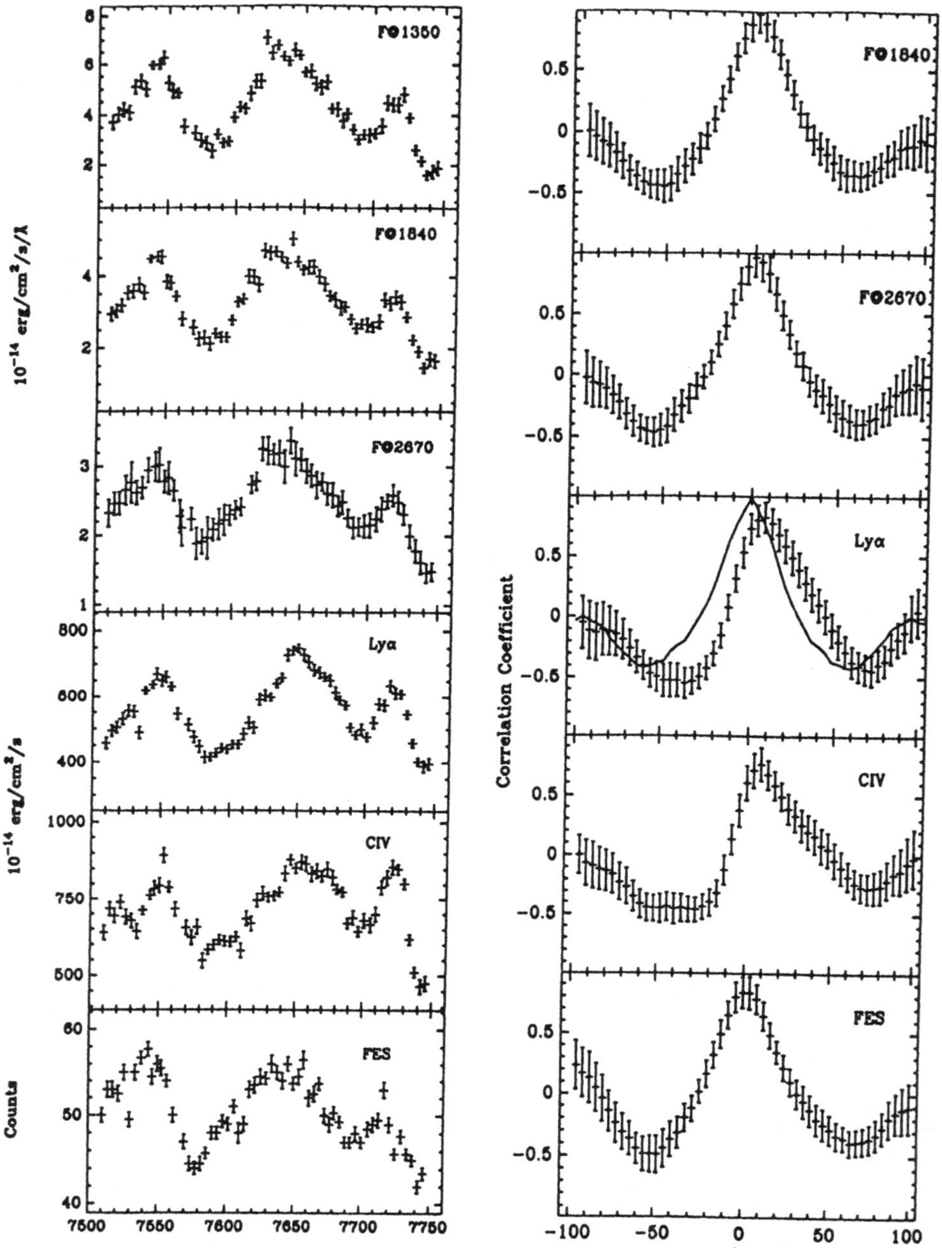

Fig. 1. Left panel: the light-curves of the continuum and strongest emission lines in NGC 5548: from top to bottom, the continuum at 1350 Å, 1840 Å, and 2670 Å, the Ly$\alpha\lambda$1216 and CIVλ1549 emission lines, the FES ($\simeq$ 5000 Å) count rate. Units are 10^{-14} erg cm^{-2} s^{-1} Å^{-1} for the continuum, 10^{-14} erg cm^{-2} s^{-1} for the lines, and counts for the FES. Right panel: the cross-correlations (CC) of the emission lines and continua with the continuum flux at 1350 Å. From top to bottom: the CC of the continuum at 1840 Å, at 2670 Å, of the Ly$\alpha\lambda$1216 and CIVλ1549 emission lines and of the FES.

Table 1. NGC 5548: variability parameters

Feature	Mean	F_{var}	R_{max}	Δt_{peak} (days)	Δt_{center} (days)	r_{max}	FWHM (days)
F_{1350}	4.36	0.30	4.46±0.53	...	...	...	...
F_{1840}	3.31	0.26	3.45±0.46	0	0	0.98	38
F_{2670}	2.49	0.16	2.30±0.25	0	0	0.97	40
FES	50.6	0.08	1.37±0.03	2	4	0.88	42
NVλ1240	89.7	0.49	56.0±243	4	4	0.79	34
HeII + OIII]	64.0	0.23	3.96±1.70	4	10	0.67	40
CIVλ1549	718.	0.14	1.92±0.11	8	16	0.66	38
Lyα + NV	556.	0.17	1.93±0.09	12	12	0.84	34
SiIV+OIV]	49.4	0.15	2.98±1.30	12	34	0.61	36
CIII]λ1909	118.	0.09	1.86±0.28	26	32	0.56	40
MgIIλ2798	139.	0.05	1.31±0.08	≃34 to 72		0.46	72

account for variations of ~ 0.3 in the spectral index if their contribution amounts to 10 % of the 1840 Å continuum flux when it is minimum. The remaining change in α could be due to genuine spectral variations of the underlying continuum (see Wamsteker *et al.* 1990).

All the emission line fluxes also varied significantly. However, the amplitude of these variations differs greatly from one line to the next. The high ionization lines NVλ1240 and HeIIλ1640 had variations whose amplitude was as great as or greater than that of the short wavelength continuum, with $F_{var} \simeq 0.5$ and 0.23 respectively. Lines of intermediate ionization had lower amplitudes, with F_{var} ranging from 0.09 to 0.17 for CIVλ1549, CIII]λ1909, Lyαλ1216, and SiIV+OIV]λ1402. The lowest ionization line in our data set, MgIIλ2798, exhibited the lowest amplitude fluctuations with $F_{var} \simeq 0.05$.

The overall resemblance of the continuum band and emission line light-curves is striking. All three major events are clearly recognizable in most light-curves. There are some noticeable differences however. For instance, during the second event, the 1350 Å continuum flux rose more rapidly than the Lyαλ1216 or CIVλ1549 emission line flux and reached its maximum 10 to 15 days earlier. This suggests that the variations of the lines are delayed with respect to those of the ultraviolet continuum, as predicted by the photoionization models. To quantify this delay, we computed the cross-correlation between the emission lines and the 1350 Å (and 1337 Å) continuum,. The results are also listed in Table 1 and the CC's of the continuum bands and of the Lyα1216 and CIVλ1549 lines are shown in fig. 1.

The uncertainty on the location of the maximum of these cross-correlations arises from two sources: measurement error and incomplete sampling. The uncertainty resulting from the measurement errors has been estimated by Monte-Carlo simulation. It turns out to be small: $\epsilon(\Delta t) = \pm 0.13$, and ± 0.17 days for the strong Lyαλ1216 and CIVλ1549 lines, respectively, while the weaker and noisier HeIIλ1640 feature yields $\epsilon(\Delta t) = \pm 1.8$ days. Unfortunately, there is no universally accepted prescription for estimating the larger uncertainty due to incomplete sampling. An additional complication arises because the cross-correlations are broad and smooth. To illustrate this difficulty, we have listed in Table 1 the position of the peak of the cross-correlations as well as the position of its "center". By "center" we mean the average of the two lags at which the cross-correlation passes through a level 0.3 below the peak. Taken together, these two give an order of magnitude estimate of the characteristic delay timescale. Any difference between them indicates the degree of skewness in the cross-correlation. The full-width at half-maximum (FWHM) of the cross-correlation function can, in principle, provide another estimate of the response timescale. If the line emission can be described as a linear convolution of the continuum light curve with a response function, the cross-correlation function is the convolution of the continuum autocorrelation function with the response function. Any difference between the FWHM of the line-continuum cross-correlation and the continuum autocorrelation would then indicate the width of the response function.

The peak amplitude of the cross-correlation is substantial for all continuum bands and emission lines. Only for MgIIλ2798 does the correlation amplitude at the peak fall below 0.5. It is clear from Table 1 that, within the uncertainties, the continuum variations are simultaneous in all 4 bands, from 1350 Å to $\simeq$ 5000Å. It is also obvious that the characteristic delays are quite short for most emission lines. Furthermore, there is a clear trend for the highest ionization lines to have the shortest response time. For both HeIIλ1640 and NVλ1240, the bulk of the cross-correlation is found in the neighborhood of 4 – 10 days, while for CIVλ1549 and L$\alpha\lambda$1216 the range is 8 – 16 days. SiIVλ1400 and CIII]λ1909 have cross-correlations concentrated between 12 and 34 days, while MgIIλ2798 is found from 34 to 72 days. For comparison, analysis of the optical data by Peterson *et al.* (1990) shows that the maximum of the cross-correlation for the H_β line occurs at a lag of $\simeq$ 24 days. Apart from MgIIλ2798, the cross-correlations presented here (as measured by their FWHM) are not much wider than the continuum autocorrelation. This implies that the line transfer functions are necessarily narrow.

Table 2. NGC 5548: CIVλ1549 line profile variability

Component	Mean	F_{var}	R_{max}	Δt_{Peak} (days)	Δt_{Center} (days)	r_{max}	FWHM (days)
Blue wing	198	0.14	2.20±0.27	...	...	...	...
Red wing	104	0.21	3.43±0.76	0	1	0.753	19
Sum wings	302	0.16	2.34±0.20	...	...	...	...
Line Core	415	0.14	1.77±0.06	2	3	0.843	20

Table 2 summarizes the results concerning the CIVλ1549 line profile in NGC 5548 and its variations. Here, we list the results of the cross-correlation of the red wing versus the blue wing and of the line core versus the sum of the 2 wings. The latter is also displayed in fig. 2. The CIV line is asymmetrical, the blue wing being twice as strong as the red wing on the average. More important is the fact that, within the uncertainties of the CC techniques, ***the blue wing and the red wing vary simultaneously***. On the average, the core contains ~ 30% more flux than the wings while the amplitude of variability is slightly larger for the wings than for the core. The variations of the core are delayed by 2–3 days with respect to those of the wings.

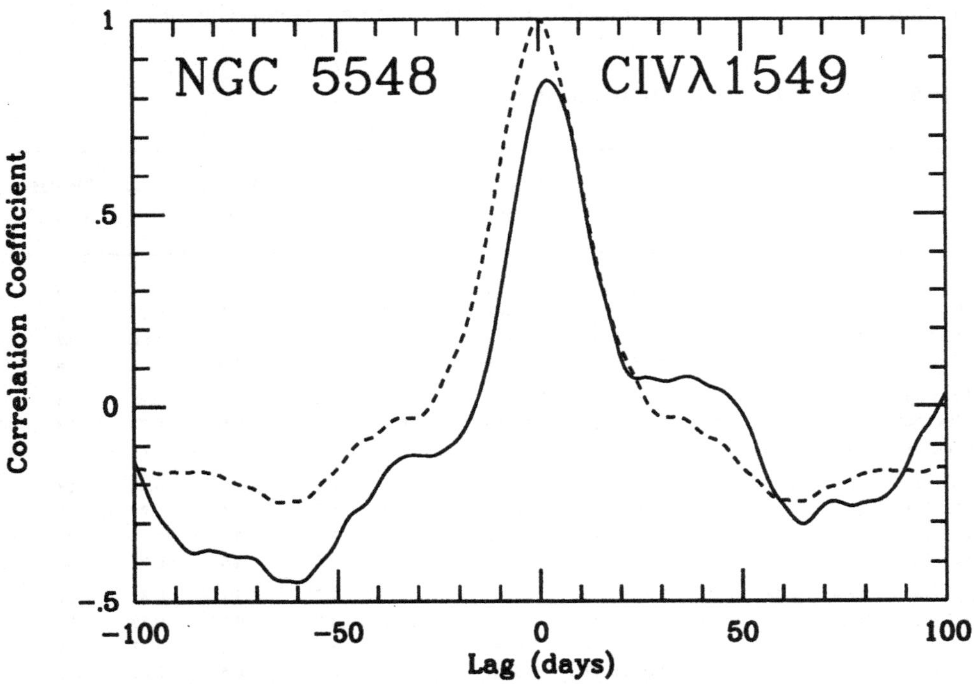

Fig. 2. NGC 5548: The cross-correlation of the CIVλ1549 emission line core ($-3000 < \Delta V < +3000$ km.s^{-1}) with respect to the wings ($\Delta V > 3000$ km.s^{-1}). The auto-correlation of the line core is also shown for reference (dashed line).

3.2. NGC 4151

The light-curves of the continuum in the different ultraviolet and optical bands together with those of the strongest emission lines are shown in fig. 3. Table 3 summarizes the main parameters which characterize the variability of the continuum and of the emission lines in NGC 4151 during our observing run: the unweighted mean flux (in units of 10^{-13} erg.cm^{-2}.s^{-1} for the lines and 10^{-14} erg.cm^{-2}.s^{-1}.Å^{-1} for the continuum), the reduced chi-square χ^2_ν (calculated under the assumption that the flux did not vary and remained equal to its average value), the fractional variation F_{var} as defined previously, the ratio of maximum to minimum flux, R_{max}, the two-folding time (Penston *et al.* 1981) i.e. the actual time it took for the flux to increase ($T_{\times 2\uparrow}$) or decrease ($T_{\times 2\downarrow}$) by a factor 2. When the amplitude of the *observed* variation did not reach a factor 2, the figure for the two-folding time has been put in parenthesis to indicate that it has a lower significance. There are no entries for features with R_{max} less than 1.5. The last 4 columns of Table 3 characterize the Cross-Correlations of the different continuum and emission lines with F_{1455}. The CC's of the CIVλ1549 and MgIIλ2798 emission lines are shown in fig. 3. As for NGC 5548, we list both the location of the peak of the cross-correlation, Δt_{peak}, and the position of its centre, Δt_{centre}. Table 3 also lists the correlation coefficient at the peak, r_{max}, and the FWHM of the cross-correlations.

Table 3. NGC 4151: Variability parameters

Feature	Mean	χ^2_ν	F_{var}	R_{max}	$T_{\times 2}\uparrow$ (days)	$T_{\times 2}\downarrow$ (days)	ΔT_{peak} (days)	ΔT_{center} (days)	r_{max}	FWHM (days)
F_{1455}	5.32	42.3	0.367	3.08±0.30	6.8±1.1	12.4±2.2	...	...	...	23
F_{1715}	4.08	110.	0.386	3.69±0.24	6.8±0.8	11.5±1.2	0	1.5	0.97	22
H/R	1.31	2.14	0.093	1.36±0.31	...	...	-11.5	-15.5	0.821	22
FES	160.6	40.3	0.057	1.24±0.01	...	...	1.5	3.0	0.79	25
CIVλ1549	165.7	98.0	0.218	2.30±0.08	21.3±1.7	22.2±1.4	2.5	5.0	0.89	26
MgIIλ2798	26.5	6.4	0.122	1.46±0.13	(39±12)	(64±17)	3.0	5.0	0.86	23
CIII]λ1909	23.1	0.86	0.040	1.22±0.08	...	...	...	...	...	...
CIV core	82.8	23.5	0.119	1.56±0.07	...	(50±5)	6.0	7.0	0.89	28
CIV blue	46.2	99.1	0.292	3.10±0.17	9.4±1.1	15.3±1.2	2.0	4.5	0.88	26
CIV red	37.0	161.	0.377	5.22±0.46	3.6±0.4	6.5±0.6	2.0	4.5	0.93	26
CIV wing	83.1	123.	0.326	3.73±0.24	6.1±0.6	10.1±0.8	2.0	4.5	0.91	26
MgII core	17.8	4.34	0.079	1.27±0.08	...	...	4.0	5.7	0.87	24
MgII blue	5.26	8.81	0.270	2.37±0.47	9.1±3.0	23.2±6.1	2.0	4.5	0.89	24
MgII red	3.38	1.09	0.157	1.78±0.54	(22±16)	(27±18)	3.0	5.0	0.77	23
MgII wing	8.64	3.64	0.220	2.06±0.49	(16±6)	23±10	3.0	4.5	0.87	23

The ultraviolet continuum underwent large variations during our observing run, the flux changing by a factor ≥ 3. To within the error, the amplitude R_{max} and the fractional variation F_{var} are the same at 1455 Å and 1715 Å. The two-folding times at 1455 Å and 1715 Å – $\simeq 7$ days for a rise and 12 days for a decline – are also equal. The dominant time scale of the variations – as measured by the FWHM of the auto-correlation (AC) of F_{1455} – is however larger, $\simeq 23$ days.

The variations at 1455 Å and 1715 Å are perfectly correlated ($r_{max} = 0.97$) and simultaneous. The average continuum spectral index for the episode – as defined by the mean "hardness ratio" H/R of Table 3 – is $\alpha \doteq -0.36$ ($F_\nu \propto \nu^\alpha$), interestingly close to the index of the *polarized* component of the optical continuum, $\alpha = -0.33$ (Schmidt and Miller 1980). It also agrees well with the mean ultraviolet spectral index from 1978 to 1983 ($\alpha = -0.19 \pm 0.15$; Clavel *et al.* 1987), showing that the continuum shape remains remarkably stable, on the average, over time scales of ~ 10 years. On the other hand, the hardness ratio underwent small but significant

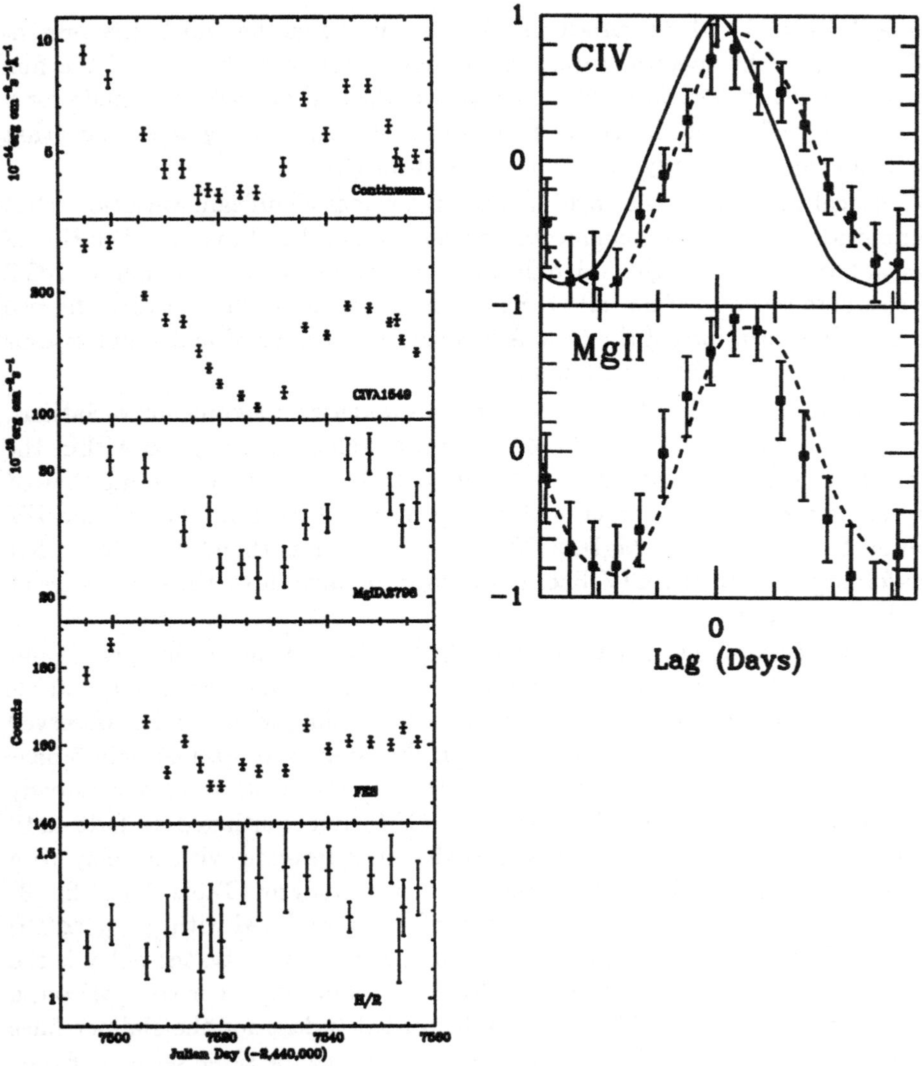

Fig. 3. Right panel: the light curves of the different continuum bands and emission lines in NGC 4151: from top to bottom: the 1455 Å continuum, the CIVλ1549 and the MgIIλ2798 emission lines, the FES count rate, and the hardness ratio, $H/R = F_{1455}/F_{1715}$. Units are 10^{-14} erg cm^{-2} s^{-1} Å^{-1} for the continuum fluxes, 10^{-13} erg cm^{-2} s^{-1} for the lines, and counts for the FES. The fluxes are represented with their associated errors. Right panel: the cross-correlations of the CIVλ1549 (top) and MgIIλ2798 (bottom) emission line with the 1455 Å continuum. The auto-correlation function (heavy line) of the continuum is also shown for comparison in the upper panel.

($P(\chi^2_\nu) \geq 2.14 = 3 \cdot 10^{-3}$) variations, during the 2 months which the present campaign lasted. The spectral variations do not correlate with those of the flux at zero lag, but they correlate strongly ($r_{max} = 0.82$) if allowance is made for a systematic delay of about -13 days. The fact that the delay is *negative* means that the spectral changes actually *precede* the flux variations.

The optical continuum near 5000 Å also underwent significant variations, but their amplitude is an order of magnitude smaller than in the ultraviolet. Dilution of the optical flux by stellar light – significantly stronger in NGC 4151 than in NGC 5548 – accounts for most if not all of the difference in amplitude between the two wavebands. The variations of the 5000 Å flux are well correlated and simultaneous with those of the ultraviolet continuum.

The CIVλ1549 and MgIIλ2798 emission lines underwent significant variations. By contrast, the flux of the CIII]λ1909 feature remained constant to within the accuracy of our measurements ($P(\chi^2_\nu) \geq 0.86 = 0.63$). The CIV$\lambda$1549 line showed the largest variation with a fractional change nearly two-third that of the UV continuum. By contrast, the MgIIλ2798 flux varied by less than 50 %. Note that the two-folding time of the CIV line, $\simeq$ 22 days, is only three times as large as that of the continuum.

The light-curves of the CIVλ1549 and MgIIλ2798 lines are strikingly similar to that of the continuum. If one looks carefully however, there are some notable differences. For instance, the continuum was already fading when we first observed it on November 29, whereas the CIV and MgII flux started to decline only 5 days later. Similarly, whereas by December 20 (JD7516) the continuum had already reached a minimum and levelled-off, the CIVλ1549 line continued to fade until December 31 (JD7527). This suggests that the lines respond with a delay to a change in the continuum flux. The cross-correlation analysis (Table 3 and fig. 3) confirms that the variations of the lines are perfectly correlated with ($r_{max} \simeq 0.9$) but lag behind those of the continuum. To within the uncertainty, the delay is the same for the CIVλ1549 and for the MgIIλ2798 line, 4$\pm$3 days. For comparison, a recent study by Maoz *et al.* (1991) yield a lag of 9 ± 2 days for the Balmer lines in NGC 4151. Given that both lines vary together and in the same sense as F_{1455}, but that the variations of CIV have a larger amplitude than those of MgII, the CIVλ1549/MgIIλ2798 ratio is also variable ($\chi^2_\nu = 6.4, F_{var} = 0.13$) and positively correlated with the continuum flux.

As outlined in our previous work, (Ulrich *et al.* 1984; Clavel *et al.* 1987), there is a hierarchy in the widths of the three main UV emission lines in NGC 4151. The CIII]λ1909 feature is the narrowest, with a FWHM of 1600 km.s^{-1} and a FWZI $\simeq$ 3500 km.s^{-1}. The MgIIλ2798 is broader than CIII], its FWHM varying around an average value of $\simeq$ 5000 km.s^{-1}. The CIVλ1549 profile has about the same FWHM as MgIIλ2798, but it has much more prominent and extended wings which are particularly strong during an outburst. For instance, during the maximum of January 17 (JD7544), the Full-width at zero-intensity (FWZI) of CIV is $\simeq$ 30000 km.s^{-1} whereas that of MgII is only 20000 km.s^{-1}

As can be judged from Table 3, the pattern of variability is quite similar for the blue and for the red wings, both in the CIVλ1549 and in the MgIIλ2798 line. Most important is the fact that the variations of the two wings appear to be equally delayed with respect to the continuum variations. A direct cross-correlation confirms that the variations of the red and the blue wings are perfectly correlated ($r_{max} = 0.97$ and 0.90 for CIV and MgII, respectively) and simultaneous ($\Delta T = 0.3 \pm 3$ days for both CIV and MgII). From Table 3, it is obvious that the bulk of the CIV variations is carried by the wings: the fractional variation of the wing flux is as large as that of the continuum, whereas the flux in the core only changes by $\simeq 50\%$. Photionization model calculations show that the Narrow Line Region contributes only for 3 % and 10 % to the total flux in the CIVλ1549 and MgIIλ2798 lines respectively (Ferland and Mushotzky 1982). This is consistent with the upper limit on the narrow CIVλ1549 flux which can be derived from the few low signal-to-noise ratio high resolution SWP spectra of NGC 4151 (Penston *et al.* 1979). Therefore, the difference in the amplitude of variability of the wings and of the core cannot be accounted for by dilution of the low velocity flux by constant NLR emission. Moreover, the core of CIV responds to the variations of F_{1455} with a delay of 6.5 ± 3 days, significantly larger than the response time of the wings, 3.2 ± 3 days. Identical results obtain when the cross-correlation is performed against F_{1715} instead of F_{1455}. Averaging the 2 estimates yields $\delta(\Delta T) = 3.2 \pm 3$ days for the difference in the delays of the core and of the wings of the CIVλ1549 line with respect to the ultraviolet continuum. A direct cross-correlation ($r_{max} = 0.98$) confirms that the core lags behind the wings by 2.5 ± 3 days. This is clearly illustrated in the wing versus core intensity diagram of fig. 4 where data-points corresponding to contiguous epochs have been connected. The points describe clockwise "hysteresis" loops, indicating that the variations in the core systematically lag behind those in the wings. A similar pattern is present in MgIIλ2798 line, but it is less pronounced than for CIVλ1549.

The average values of the CIV/CIII] and CIV/MgII flux ratios were 7.2 and 6.3 respectively during our campaign. However, as already mentioned, these lines ratios are highly variable and correlate positively with the continuum flux. Hence, during an outburst they reach even higher values, e.g. CIV/CIII] $= 11.0 \pm 0.5$ and CIV/MgII $= 8.0 \pm 0.4$ on December 4, 1989. If we restrict our analysis to the wings, the above line ratios become even much higher. The mean CIV/MgII flux ratio for the wings is 9.6 but occasionally reaches values as large as ~ 14 during an outburst. A conservative lower limit of 50 can be set on the CIV/CIII] ratio in the wings from the absence of detectable CIII]λ1909 emission at velocities larger than ~ 3500 km.s^{-1}. For comparison, the mean CIV/CIII] ratio for Seyfert I galaxies is ~ 5 (Wu, Boggess, Gull, 1983) and 2.5 for quasars (Wilkes, 1986) while for CIV/MgII, the corresponding figures are 3.7 and 2.8 respectively.

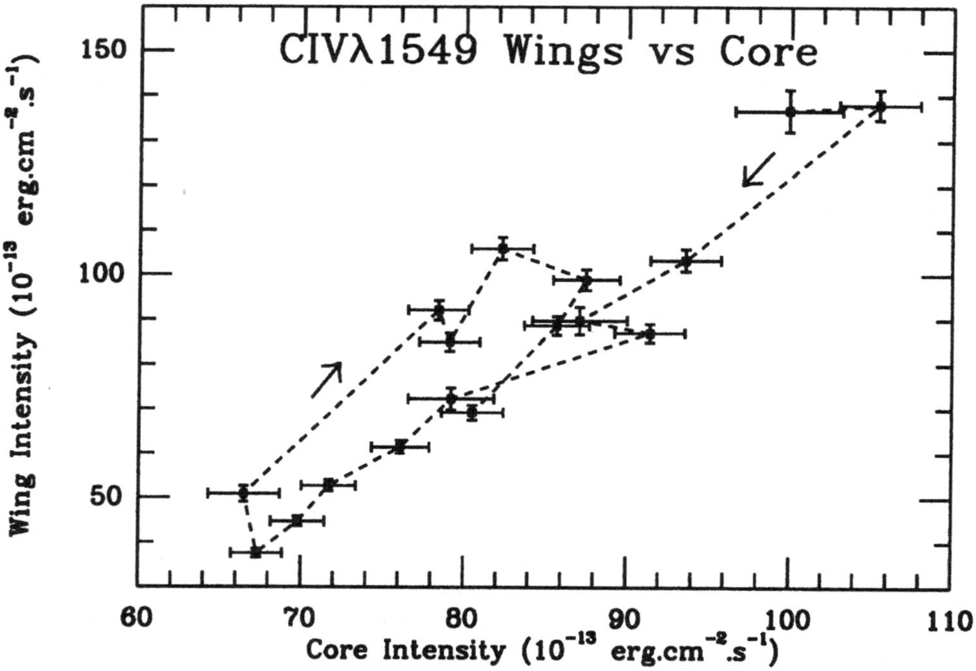

Fig. 4. NGC 4151: the CIVλ1549 wing versus core intensity diagram. Data-points which correspond to contiguous epochs have been connected and the sense of time is indicated by arrows. The points describe clockwise "hysteresis" loops indicating that the variations in the wings precede those in the core.

4. Comparison between NGC 5548 & NGC 4151

4.1. Continuum

In NGC 5548, the typical amplitude of the continuum variability decreases with increasing wavelength from 1350 Å to 5000 Å, making the continuum appear systematically bluer when it is brighter. While the strong FeII emission lines which contaminate the long wavelength continuum accounts for much of the spectral variation in the near UV, the changes in the spectral index below 1900 Å may, at least in part, represent a genuine property of the underlying continuum. This contrasts with the case of NGC 4151 where the typical amplitude of the variability is the same at 1715 Å and 1455 Å. It may be worth recalling here that, unlike NGC 5548, NGC 4151 emits virtually no optical FeII lines flux at all (Boksenberg *et al.* 1975). However, there appears to be genuine spectral variations of the UV continuum in NGC 4151 as well, but the spectral changes seem to *precede* the flux variations by about 13 days in this galaxy. In both galaxies, the reduced amplitude of the variability at 5000 Å is mostly the consequence of dilution by stellar emission.

In NGC 4151, the ultraviolet continuum can double its intensity in ~ 7 days. The two-folding time appears to be somewhat longer in NGC 5548, ~ 20 days, though on one occasion, the 1335 Å flux dropped by a factor 3 in 16 days. In both NGC 5548 and in NGC 4151, the continuum variations are simultaneous in all ultraviolet and optical bands, from $\simeq$ 1350 to 5000 Å. This sets strong constraints on theories which aim at explaining the origin of the "big bump" in AGN's, in particular on the geometrically thin accretion disk models. In a thin disk, the radius of the optical emitting region is much larger than that which emits in the ultraviolet, and the two zones have no mean of exchanging information faster than the sound speed. The corresponding time scale is orders of magnitude larger than the upper-limits on the lag between UV and optical variations reported here.

4.2. Emission lines

In NGC 5548, the amplitude of variation of a given emission line is directly related to the degree of ionization of its ion, the highest ionization lines NVλ1240 and HeIIλ1640 showing variations as large as those of the UV continuum, while the lowest ionization line MgIIλ2798 varies by 30% only. This contrasts sharply with the case of NGC 4151 where there is no clear pattern connecting the amplitude of variability and the ionization stage. However, in this galaxy as well the amplitude is different for different lines, the CIVλ1549 line having the highest fractional variation (0.22) followed by MgIIλ2798 (0.12) and the CIII]λ1909 line which does not vary on time scales from 3 days to 2 months.

In both galaxies, the flux of most emission lines correlate extremely well with those of the continuum if allowance is made for a systematic short delay, lending qualitative support to the view that photoionization by the nuclear continuum is responsible for driving the emission lines. In NGC 5548, the delay of a given line is a function of the degree of ionization of its ion, increasing from 4 days for the HeIIλ1640 and NVλ1240 lines to more than 34 days for MgIIλ2798. Again this is in sharp contrast with the case of NGC 4151, since in this object, the delay is the same for the CIVλ1549 and for the MgIIλ2798 emission line - $\Delta T = 4 \pm 3$ days - while any lag for the CIII]λ1909 line (which does not vary) has to be at least 23 days.

The fact that the CC's of the CIVλ1549 and MgIIλ2798 emission lines in NGC 4151 and of all the high ionization lines in NGC 5548 are not significantly broader than the continuum AC further implies that the transfer functions of these lines are narrow. This, together with the large fractional variations, suggests that the bulk of the BLR gas is concentrated at very small radial distances from the continuum source. From the delays in Table 1 and 3, we infer typical length scales l of about 12 lt-d for the high ionization gas in NGC 5548 and 4 lt-d in NGC 4151. Note that the two-folding time of the CIVλ1549 line flux in NGC 4151 independently sets an upper limit of 11 lt-d on l.

4.3. Line profiles

To within the uncertainty, the variations are simultaneous in the red and in the blue wings of the lines in both galaxies This rules-out the possibility that radial motions dominate the velocity field of the BLR. Maoz *et al.* (1991) reach the same conclusion from a study of the Balmer line profile variability in NGC 4151.

In NGC 4151, there is a close correspondence between the FWZI of a given line on the one hand, and the amplitude of its variability or the length of its delay on the other: the broader the line, the more highly variable it is and the shorter its response time. This is related to the fact that most of the variability is carried by the wings ($|V| \geq 3000$ km.s^{-1}) rather than the core. Moreover, in that galaxy the response time of the core ($|V| \leq 3000$ km.s^{-1}) of the CIVλ1549 line to a change in the flux of the UV continuum is larger than that of its wings by 3.2 ± 3 days. The same pattern, though less pronounced, is present in NGC 5548. The wings vary only slightly more than the core and the delay in the response of the core is only 2.5 days longer than that of the wings. This pattern strongly suggests that *the velocity dispersion of the BLR gas increases with decreasing radial distance*, the velocity gradient being steeper in NGC 4151 than in NGC 5548.

4.4. Implications for models

In both galaxies, there is ample evidence for a radial gradient in the physical conditions of the gas: different lines have different response time, different amplitude of variations, and in the case of NGC 4151, different profiles. In NGC 5548, the BLR stratification appears to be mainly a gradient in the degree of ionization of the gas since the delay in the response of a line increases systematically for decreasing ionization stages of the emitting ion at the same time as the fractional variation of the line decreases. This is consistent with the photionization scheme where the gas which lies further away from the continuum source is less ionized, varies less and responds with a larger delay than the gas which lies closer to the continuum source. In NGC 4151 the situation is not as simple: on the one hand, there are the CIV and MgII lines whose ions have widely different ionization potentials but which nevertheless respond with the same very short lag; on the other hand, there is the CIII]λ1909 line which does not vary and presumably originates from a region whose radius is at least 23 lt-d, though the ionization potential of C^{2+} is intermediate between that of C^{3+} and Mg^{+}. Since the CIII]λ1909 intercombination line is collisionally de-excited for densities in excess of 3×10^9 cm^{-3}, the most plausible explanation is that the gas density in the inner BLR of NGC 4151 is far above its canonical value. As a matter of fact, the unusually large CIV/CIII] intensity ratio in this object cannot be accounted for solely by invoking large values of the ionization parameter (Ferland and Persson 1990). Large densities are also needed. However, since the CIV/MgII and the CIV/CIII] line ratios are larger in the wings than in the core of the lines, and since the wings response time is shorter than that of the core, we conclude that *in NGC 4151, both the density and the ionization parameter of the gas increase toward small radii, while in NGC 5548, it is mainly the ionization degree which increases inward.*

The CIV/CIII] and CIV/MgII line ratios in NGC 4151 are much higher than in a typical AGN or quasar. Since average line ratios have been used to constrain the standard photoionization model of the BLR (e.g. Kwan and Krolik 1981), the results from such model calculations do not apply to NGC 4151. In particular, the distance at which the gas lies from the continuum source may be much smaller in that galaxy than allowed by the standard model. Hence, there is no conflict as yet between the very small BLR size inferred here and theoretical predictions. Krolik (1990) concludes that the size of the BLR in NGC 5548 as inferred from variability agrees reasonably well with that derived from a detailed photionization modelling of the line spectrum.

To conclude, we briefly compare the length scale l of the inner BLR in NGC4151, NGC 5548 and F9, the only other Seyfert galaxies for which this has been reasonably well determined, (Clavel, Wamsteker & Glass, 1989). For consistency, we use the *same line*, CIVλ1549, to define l in those 3 objects. The length-scales are in the ratio 1:3:35 for the sequence NGC4151:NGC5548:F9. The corresponding (outburst) luminosity ratios are 1:20:500. There is clearly a dependence of l on the luminosity L. Given the uncertainties involved, this dependence is not too different from the $L^{1/2}$ scaling law which one would naively expect on the basis of the overall similarity of the emission line spectra of Seyfert I galaxies and quasars as a class.

References

Blandford, R. D. and McKee, C. F. 1982, *Ap. J.*, **255**, 419.
Boksenberg, A. *et al.* 1975, *M. N. R. A. S.*, **173**, 381.
Clavel, J. *et al.* 1987, *Ap. J.*, **321**, 251.
Clavel, J.,Wamsteker, W. and Glass, I. S. 1989, *Ap. J.*, **337**, 236.
Clavel, J. *et al.* 1991a, *Ap. J.*, **366** (in press).
Clavel, J. *et al.* 1991b, *M. N. R. A. S.*, (in press).
Davidson, K. and Netzer, H. 1979, *Rev.Modern.Phys.*, **51**, 715.
Faelker, J.,Gordon, F. and Sandford, M. C. W. 1987, in *"Exploring the Universe with the IUE satellite"*, Y.Kondo *et al.* eds., Reidel, p.21.
Ferland, G. J. and Mushotzky, R. F., 1982. *Ap. J.*, **262**, 564
Ferland, G. J. & Persson, S. E. 1990. *Ap. J.* **347**, 656.
Krolik, J. H., 1990. Proceedings of the Atlanta Conference *"Variability of Active Galactic Nuclei"*, May 2-4, 1990, H.R. Miller & P.J. Wiita Eds., Cambridge University press.
Kwan, J. & Krolik, J. H., 1981. *Ap. J.* **250**, 478.
Maoz, D. *et al.*, 1991, *Ap. J.*, in press.
Penston, M. V. *et al.*, 1979. *M. N. R. A. S.*, **189**, 45
Penston, M. V. *et al.*, 1981. *M. N. R. A. S.*, **196**, 85
Perola, G. C. *et al.*, 1982. *M. N. R. A. S.*, **200**, 293
Peterson, B. M. *et al.*, 1991. *Ap. J.*, in press.
Schmidt, G. D. & Miller, J. S., 1980. *Ap. J.*, **240**, 759.
Ulrich, M. H. *et al.* 1984, *M. N. R. A. S.*, **206**, 221.
Ulrich, M. H. *et al.*, 1985. *Nature*, **313**, 745.
Wamsteker, W. *et al.* 1990, *Ap. J.*, **354**, 446.

Wilkes, B. J., 1986. *M. N. R. A. S.*, **218**, 331
Wills, B. J.,Netzer, H. and Wills, D. 1985, *Ap. J.*, **248**, 472.
Wu, C. C., Boggess, A. & Gull, T. R., 1983. *Ap. J*, **266**, 28

Ground-Based Studies of Emission-Line Variability: Recent Results for NGC 5548 and Future Plans

Bradley M. Peterson

Department of Astronomy, The Ohio State University

1. Introduction

Ever since emission-line variability was first detected in AGNs, it has been clearly understood that light-travel time effects in the broad-line region (BLR) afford a unique tool for studying the structure of these spatially unresolved regions (see Peterson 1988 for a review). However, only within the last few years has the observational problem become well defined in terms of the sampling rates and quality of data necessary to address the problem correctly. The amount of data required to extract structural information about the BLR from the continuum and emission-line light curves is so considerable that many of us were convinced that the most promising approach would be to combine our observational efforts. Our goal was to work together to produce a large, high-quality database which would then be released to individual investigators for more complete analysis. The cornerstone of this group effort, described in these proceedings by Clavel, was a joint NASA/ESA/SERC program to monitor the spectrum of NGC 5548 every four days with *IUE* from 1988 December through 1989 August (Clavel *et al.* 1990). A concurrent ground-based program was organized in an effort to enhance the scientific return on the project by extending the wavelength coverage and providing higher spectral resolution and signal-to-noise ratios than would be possible with *IUE*. It also turned out that the temporal resolution of the ground-based program was somewhat better than the temporal resolution of the *IUE* program, and the temporal baseline is longer and continues to grow. The database and initial results are presented by Peterson *et al.* (1991). In this contribution, I will summarize the important results of the ground-based program, describe some additional applications of the existing data, and make a few comments on what we can do in the future to expand on this work.

2. The NGC 5548 optical database

The ground-based study of the temporal behavior of NGC 5548 differed from the concurrent *IUE* project in that it was not a joint program undertaken with a single instrument. The basis of the ground-based program was an informal agreement that participants would obtain optical spectra, photometry, and CCD images of NGC 5548 whenever possible, although there were a few notable cases where observers devoted much of their observational effort for the year to this single project. The ground-based data are not evenly spaced, as AGN observers tend to be at the telescope during dark or gray time rather than bright time, and the data are not homogeneous, as they were obtained with many different instruments in a variety of configurations.

As a first step in the analysis of these data, we have examined the spectral variations in the vicinity of the Hβ emission line. The principal reason for this choice is the proximity of the [O III] $\lambda\lambda$4959, 5007 narrow emission lines, which do not vary over the time scales of interest and thus can be used for absolute flux calibration if the [O III] flux is determined accurately from observations under photometric conditions. Evidence for aperture effects is seen by comparing the data obtained with different instruments — larger spectrograph entrance apertures admit relatively more [O III] flux from the partially resolved narrow-line region as well as more starlight from the host galaxy. In practice, we find that we can apply empirically determined small nominal corrections to the data obtained with different instruments to put all of the measurements on a common flux scale.

The database for the study of continuum and Hβ variations in NGC 5548 consists of 177 spectra covering 129 independent dates between 1988 December 14 and 1989 October 10. Except for the final month of the campaign, when only two observations were obtained, the temporal coverage was extraordinarily good, with the average interval between observations being 3.1 days, and a median interval of only 1 day. There are no gaps in the coverage larger than 12 days, and gaps of a week or longer are rare. The typical uncertainty in the continuum measurements (at 4870Å, underneath the Hβ line) is $\sim$4.5%, and the mean uncertainty in the Hβ measurements is $\sim$3.7%. The continuum and emission-line light curves are shown in Fig. 1, which show that three significant continuum "events" were detected during this campaign.

3. Phase analysis: cross-correlation results

The primary scientific goal of this project was to determine the "lag" or phase shift between the continuum and emission-line light curves, and also to determine whether or not there is a lag between the ultraviolet and optical continuum variations. The lag between two light curves is determined by cross correlation, either by using the interpolation method (Gaskell and Sparke 1986) or the discrete correlation method (Edelson and Krolik 1988) — in this particular case, the light curves are so well sampled that it makes no difference which method is used.

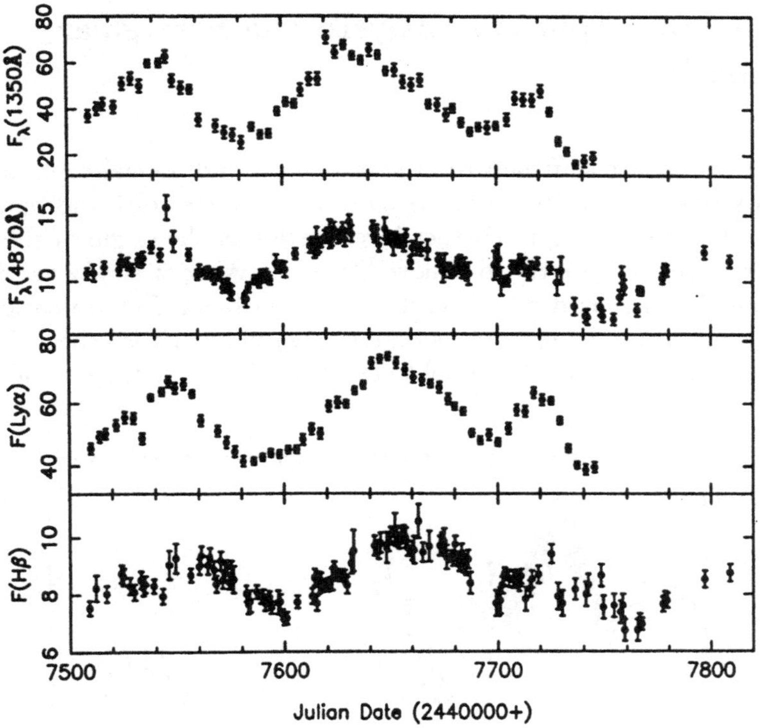

Fig. 1. The continuum light curves at 1350Å and 4870Å are shown, as well as the emission-line light curves for Lyα and Hβ. Continuum fluxes are in units of 10^{-15} ergs s^{-1} cm^{-2} $Å^{-1}$ and line fluxes are in units of 10^{-13} ergs s^{-1} cm^{-2} (from Peterson *et al.* 1991).

Cross correlation of the ultraviolet and optical continuum light curves shows that the optical continuum variations follow those in the ultraviolet by ~2 days; various methods of assessing the uncertainty in the cross-correlation lag all seem to indicate that the formal uncertainties in the results discussed here are 2 – 3 days, which indicates that the ultraviolet – optical continuum lag is not significantly different from zero, and thus should not be taken too seriously.

Cross correlation of the continuum and Hβ light curves gives a lag of ~20 days. This result is highly significant (the correlation coefficient at the peak of the cross-correlation function is $r \approx 0.86$), and is significantly different from the ~10-day lag found for Lyα from the *IUE* data; direct cross-correlation of the Lyα and Hβ light curves shows that Hβ lags Lyα by ~8 days. This appears to be consistent with the *IUE* results for other emission lines, which suggest that the high ionization lines respond more rapidly than low ionization lines to changes in the continuum flux. This has been interpreted as evidence for ionization stratification of the BLR.

4. Continuum amplitude analysis: the host galaxy contribution

The absence of a phase shift between the ultraviolet and optical continuum light curves shows that at least the ***variable*** parts of the ultraviolet and optical continuum have a common origin. Further investigation of the origin of the variable continuum requires that we try to remove the non-varying or slowly varying components from the measured flux in each continuum band. Determination of the stellar contribution to the optical continuum measurements is especially important in determining the relative amplitude of variability in the optical and the ultraviolet.

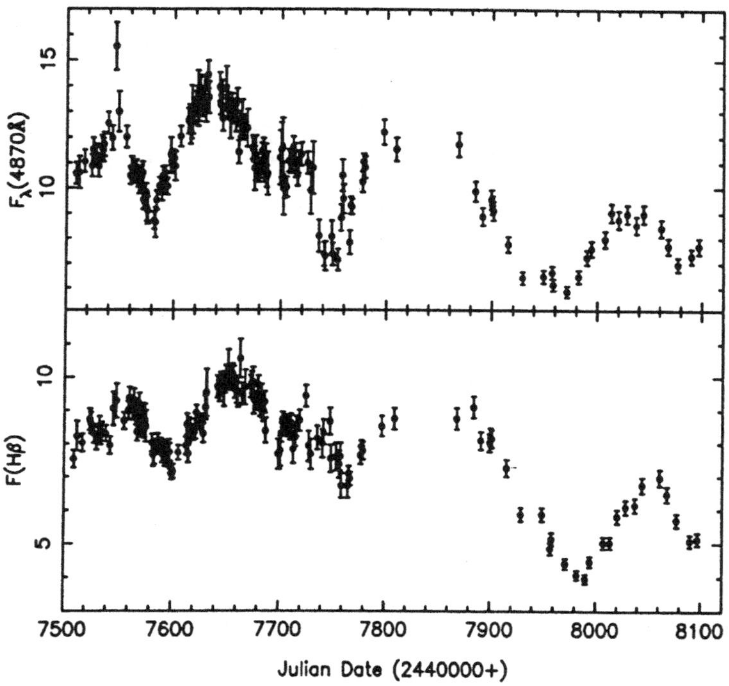

Fig. 2. The optical continuum and Hβ light curves for the two most recent observing seasons. Units as in Fig. 1.

Particularly useful in this regard are observations made during very faint states, as these provide firm upper limits to the flux in other components, and indeed the contributions due to underlying starlight and narrow emission lines, for example, become relatively prominent and easier to evaluate when the nonstellar continuum and broad emission lines are weak. We were therefore very fortunate to observe NGC 5548 in its faintest recorded state during the most recent observing season. We have continued to monitor NGC 5548, and a preliminary optical light curve, so far based on Ohio State data only, is shown in Fig. 2. This light curve shows that in 1990 March – April, NGC 5548 was remarkably faint. An *IUE* spectrum obtained close to minimum light (Reichert, Webb, and Crenshaw 1990) shows that the nucleus was at that time only half as bright as in its previously recorded faintest state.

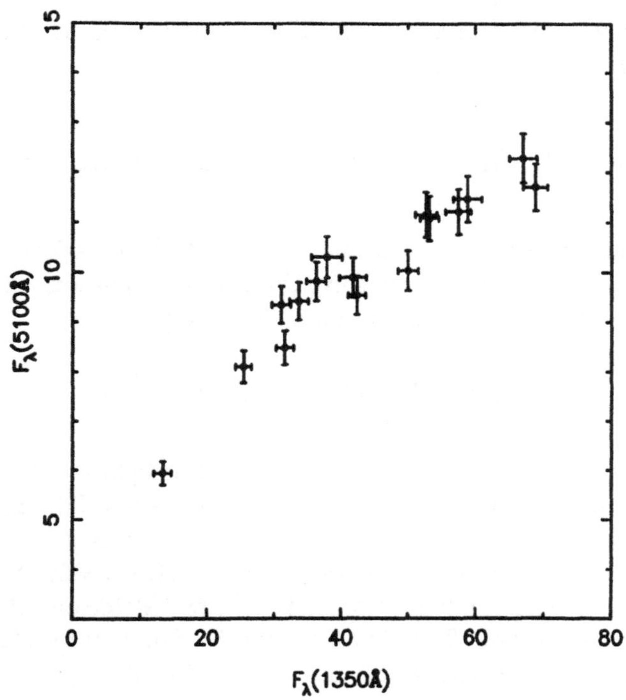

Fig. 3. The relationship between the total continuum flux at 5100Å, as measured through a 5×7.6 arcsec^2 aperture, and the total continuum flux at 1350Å. Fluxes are in units of 10^{-15} ergs s^{-1} cm^{-2} Å^{-1}.

There are a number of ways that the host-galaxy contribution to the optical continuum can be measured, and here we describe a very simple method. As shown in Fig. 3, we plot the optical continuum flux at 5100Å versus the simultaneously measured ultraviolet flux at 1350Å (note that we have restricted this analysis to the homogeneous Ohio State data, obtained through an entrance aperture of 5 arcsec × 7.6 arcsec, and we have used a more "line-free" wavelength than in the earlier analysis). If it is assumed that no other spectral components other than the variable continuum and the starlight (measurable only in the optical) contribute to these continuum bands, then the y-intercept of this relationship gives the starlight contribution at 5100Å, and any non-linearity indicates a change in spectral index with continuum level.

From Fig. 3, one can see that the best-fit function to these data will have some curvature, in the sense that the ultraviolet flux increases by a larger factor than does the optical flux. In other words, the spectrum gets harder as it gets brighter. By fitting a quadratic to these data, we find that the y-intercept of this function is $(3.37 \pm 0.54) \times 10^{-15}$ ergs s^{-1} cm^{-2} Å^{-1}. If we adopt this as the host galaxy flux[1] at 5100Å as observed through a 5 arcsec × 7.6 arcsec aperture and subtract this constant contribution from the optical flux measurements, we can determine a relationship between the spectral index between 1350Å and 5100Å and the continuum flux. Typically this index is about $\alpha \approx 0.5$ ($F \propto \nu^{-\alpha}$), although it increases to ~ 0.9 in the faintest states. This result is in good agreement with the findings of Wamsteker *et al.* (1990).

5. The Hβ transfer function and emissivity distribution

Because the sampling grid is very tight and the amplitude of variation is large relative to the noise level, it is possible to solve the integral equation which describes the response of the emission line to continuum variations, i.e.,

$$L(t) = \int_{-\infty}^{\infty} \Psi(t-\tau)C(\tau)d\tau,$$

where $C(t)$ and $L(t)$ are the continuum and emission-line light curves, respectively, and $\Psi(t)$ is the transfer function. The transfer function is essentially the emissivity-weighted geometrical distribution of the line-emitting gas as a function of time delay as seen by an external observer (Blandford and McKee 1982). A first attempt at solving for $\Psi(t)$ has been made by Horne, Welsh, and Peterson (1990) by using the maximum entropy method. The resulting transfer function for Hβ is shown in Fig. 4. The remarkable thing about this function is that it approaches zero near zero lag; by contrast, the transfer function for a thin spherical shell of radius R is a rectangular function which is non-zero in the range $0 \leq t \leq 2R/c$. The lack of immediate response of the Hβ emission line to continuum variations implies that the emissivity along our line of sight to the continuum source is very low. This points to a flattened geometry which we are observing more or less close to pole-on.

6. Work in progress

At this point, we have only scratched the surface of the information available in the NGC 5548 database. Some important additional studies which are currently underway include:

- Determination of the spectral energy distribution over a larger wavelength range. This includes attempting to refine the model for the underlying starlight distribution.
- Determination of the velocity field and kinematics of the BLR. As described in these proceedings, Kollatschny and others are examining the high-S/N,

[1] A stringent upper limit on this quantity is provided by the *total* flux at minimum light, $F_\lambda = 5.46 \times 10^{-15}$ ergs s^{-1} cm^{-2} Å^{-1}.

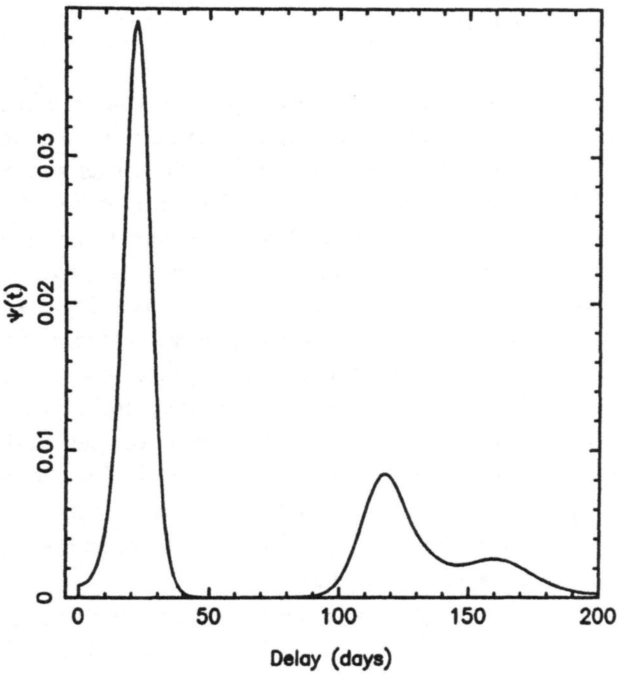

Fig. 4. The transfer function for the Hβ emission line (from Horne, Welsh, and Peterson 1990).

high-resolution optical spectra to look for line-profile changes. Crenshaw and Blackwell (1990) have examined the behavior of the C IV λ1549 line during the period at the end of the *IUE* campaign when the continuum and the lines decreased rapidly in flux. On the basis of their analysis, they conclude that the red side of the line profile follows the continuum more rapidly than the blue side, which argues for net infall of the BLR gas. Unfortunately, we do not see any such gross effect when we analyze all of the C IV λ1549 data for the campaign, so it is not yet clear what this result means. Efforts are also being made to recover the velocity-dependent transfer function from the optical data.

- Measurement of the variability of the blue bump. The mean ultraviolet-optical NGC 5548 spectrum of Wamsteker *et al.* (1990) indicates that ~ 1/3 of the emission from the BLR comes out in the "small bump", and obviously this important source of cooling needs to be better understood before detailed photoionization models can be attempted.
- Measurement of other optical emission lines. Additional important broad lines which can be examined in the optical spectra include other Balmer series lines, He II λ4686, He I λ5876, and the optical Fe II blends. These will require more effort to analyze as the flux calibration away from [O III] λλ4959, 5007 must be checked very carefully, and some of these lines are very low contrast features which are often badly blended with other lines.

7. Impact on future studies

It is clear from the success of the NGC 5548 project that similar projects can and should be pursued in the future. Therefore, I think it is instructive to reflect briefly on some of the reasons that this particular effort was successful (see also Alloin 1990). First of all, we were lucky — NGC 5548 underwent three significant outbursts during the first year of the study, and continues to undergo pronounced variations. In contrast, in the Wise Observatory project of 1988, only 3 of about a dozen sources varied significantly during an entire observing season (see Maoz 1988). A second bit of good fortune was that the power spectrum of the continuum variations turned out to be steep, something like $1/f^2$. The continuum outbursts were smooth and well resolved, which led to accurate cross-correlation results. What this all comes down to was that the right target was selected for the study — NGC 5548 is apparently bright and well placed in the sky for a long observability window for *IUE*. It had been observed frequently as part of earlier projects (e.g., Wamsteker *et al.* 1990 and Peterson *et al.* 1990) which indicated variability time scales well suited to the capabilities of *IUE*.

Both the satellite-based and ground-based programs were remarkable in terms of the resources required. Overall, over 100 individuals were involved directly and are co-authors on the Clavel *et al.* (1990) and Peterson *et al.* (1991) papers. The *IUE* project required scheduling $\sim$ 60 half-shifts at four-day intervals, and this was possible only because of the cooperative spirit between Vilspa and Goddard. The ground-based program involved about 20 telescopes from all over the northern hemisphere. The great support this project enjoyed was, I believe, because it was the right project at the right time — the idea for this effort came about at the Segovia meeting in 1987, and gained momentum at the Georgia State meeting a few weeks later. Organization of the concurrent ground-based effort began in earnest at IAU Symposium 134 the following summer. Each of these meetings featured lively debates on interpretation of spectral variations. Even though the results of the earlier studies were often inconclusive, it was obvious that there was great hope for learning about the inner structure of AGNs given enough closely spaced observations during periods of significant activity.

There were also three other circumstances which contributed to the success of this effort. The first of these was that *IUE* was a mature project. It has been an enormously successful workhorse which has more than fulfilled its original goals. At this point in its mission, it was ripe to undertake something rather innovative. Moreover, the climate for this project was right as the Europeans had been using *IUE* for similar albeit somewhat less intensive studies for years (which may account for what I would characterize as the early enthusiasm of ESA/SERC for the project in contrast to the early caution of NASA). The second circumstance was the relatively recent proliferation of CCD spectrographs and cameras on moderate-aperture and smaller telescopes. The ground-based results would have been far less spectacular if the typical uncertainties had turned out to be $\sim$10% or worse rather than better than $\sim$5%. The third important development was in our ability to communicate rapidly. The electronic mail networks, which had come into widespread use only shortly before planning for this project began, were critical in organization and execution of the effort.

In light of the results of this project, what are the most interesting problems for future studies in emission-line variability? Certainly one of the most important questions is the radius – luminosity relationship, which is predicted to be $r \propto L^{1/2}$ in the simplest models. This relationship is fundamentally important in refining photoionization models, extending our understanding to high-luminosity quasars, and explaining correlations such as the Baldwin effect.

It is also important to continue studying NGC 5548 to determine if the transfer function is stationary, i.e., does either the geometry or the emissivity distribution of the BLR change with time? Netzer and Maoz (1990) have already opened this question by noting that the three individual outbursts shown in Fig. 1 have different lags.

The one issue that I find the most confusing is the relationship between the observable continuum at energies less than 1 Ryd and the high-energy continuum. X-ray variability data show different characteristics than we see at lower energies — the variations in the X-ray are apparently faster and their power spectrum is flatter than what we see at lower energies. It is extremely important that we understand the variability in these different wavebands, since the X-rays contribute substantially to the energy input to the BLR.

If future massive monitoring efforts are to be undertaken, I think that the most important lessons to be learned from the recent campaigns are the following:

1. The different response time for the various emission lines show that it is important that *all* of the lines are measured simultaneously. One cannot infer the temporal behavior of one line from the behavior of another.
2. For reasons we do not yet understand, some AGNs (e.g., NGC 5548) are "reliably" variable, i.e., the probability of observing continuum and emission-line variability over the course of one observing season is quite high. Other sources seem to vary only episodically (e.g., Akn 120). Undoubtedly, observing only the most variable AGNs will introduce some kind of selection effect, perhaps related to the inclination of the BLR or the purported accretion disk; however, at this early stage in reverberation mapping it is most important to get some return on the investment of so much observational effort. It is important to choose a source that one is quite confident will vary during the campaign.
3. We can also see from Fig. 2 that the amplitude of variation is often not very large, even though the total range in brightness can be a magnitude or more. For the light-curve data to be useful, it is essential that the amplitude of variability greatly exceed the measurement errors. It is important that the mean measurement uncertainties are smaller than $\sim 5\%$. Unless large telescopes are used in future campaigns, this probably restricts massive monitoring programs to apparently brighter AGNs.

Acknowledgements: I wish to thank the National Science Foundation and NASA for support of this work under grants AST-8915258 and NAG5-1366, respectively, and the organizers of this conference for their hospitality. I gratefully acknowledge travel support from the Graduate School of The Ohio State University. I also thank Gail Reichert, Jim Webb, and Mike Crenshaw for allowing me to use their recent low-state ultraviolet continuum measurement in Fig. 3 and Hagai Netzer for comments on this manuscript.

References

Alloin, D. 1990, in *Evolution in Astrophysics: IUE Astronomy in the Era of New Space Missions*, ESA Publication ESA-SP-310, in press.

Blandford, R.D., and McKee, C.F. 1982, *Ap. J.*, **255**, 419.

Clavel, J., *et al.* 1990, *Ap. J.*, in press.

Crenshaw, D.M., and Blackwell, J.H., Jr. 1990, *Ap. J. (Letters)*, **358**, L37.

Edelson, R.A., and Krolik, J.H. 1988, *Ap. J.*, **333**, 646.

Gaskell, C.M., and Sparke, L.S. 1986, *Ap. J.*, **305**, 175.

Horne, K., Welsh, W.F., and Peterson, B.M., 1990, preprint.

Maoz, D. 1988, in *Active Galactic Nuclei*, ed. D.E. Osterbrock and J.S. Miller, (Dordrecht: Kluwer), p. 100.

Netzer, H., and Maoz, D. 1990, preprint.

Peterson, B.M. 1988, *Pub. A.S.P.*, **100**, 18.

Peterson, B.M., Reichert, G.A., Korista, K.T., and Wagner, R.M. 1990, *Ap. J.*, **352**, 68.

Peterson, B.M., *et al.* 1991, *Ap. J.*, in press.

Reichert, G.A., Webb, J., and Crenshaw, D.M. 1990, private communication.

Wamsteker, W., Rodríguez-Pascual, P., Wills, B., Netzer, H., Wills, D., Gilmozzi, R., Barylak, M., Talavera, A., Maoz, D., Barr, P., and Heck, A. 1990, *Ap. J.*, **354**, 446.

Variability of Line Profiles in NGC 5548

Wolfram Kollatschny and Matthias Dietrich

Universitäts-Sternwarte Göttingen, Geismarlandstraße 11, 3400 Göttingen, Germany

Abstract: High resolution profiles of the Hα and Hβ emission lines of NGC 5548 have been investigated as part of the international UV/optical monitoring campaign. The Hα and Hβ difference profiles show strong variations. The broad line difference profiles consist of three components. The main component is centered at the same wavelength as the narrow component. The intensity of this component is correlated with the UV continuum. It is symmetric and has constant width at zero intensity during the variations. Furthermore a blue and red component – best visible during low active states – varies independently.

1. Introduction

Variability studies of the broad emission lines in active galactic nuclei give some information on the structure and the velocity field of the region where they originate.

For twenty years it has been known that the broad emission line profiles of NGC 5548 are complex (Andersson, 1971). This Seyfert galaxy was one of the first targets for the study of the variability of the UV-emission lines with the IUE satellite. During the last 8 years many optical monitoring programs have been performed showing the variability of the optical Balmer lines. NGC 5548 was monitored in the UV with the IUE satellite every four days from Dec. 88 until Aug. 89 (Clavel et al.,1990; and these proceedings). Parallel to the UV observations B. Peterson (Peterson et al.,1991; and these proceedings) organized an international ground based monitoring program for variability studies of the optical continuum and optical line intensities. As part of this international campaign we tried to get as many high resolution profiles of the optical emission lines – especially of the Balmer lines – as possible at Calar Alto Observatory. Here we present preliminary results of the variability of Balmer line profiles in NGC 5548.

2. Observations

For the study of the optical emission line profiles of NGC 5548, spectra with high signal to noise ratio and high spectral resolution, that is 1 - 4 Å resolution, are needed. We made our observations at Calar Alto Observatory with the 3.5m and 2.2m telescopes. Typical exposure time per spectrum was 30 to 60 minutes. Further high resolution spectra were obtained as part of the international optical campaign from A. Filippenko, B. Peterson and P. Smith. Altogether we got high resolution spectra of Hβ for 35 epochs from Dec. 88 until Aug. 89. Twenty-one of these spectra were obtained at Calar Alto Observatory and 14 spectra at other observatories. For Hα we got high resolution spectra at 21 epochs; 17 of these were obtained at Calar Alto.

3. Results

Emission line profile variations of Hβ and Hα are shown in Fig. 1a and 2a for two epochs. The continuum has been subtracted from the original spectra. The Hβ line flux has been normalized with respect to the [OIII] 4959,5007 lines which are known to be constant at least for periods of years. The [OIII] lines originate in a compact region with spatial extent of less than two arcsec. The flux of the Hβ broad line component varied by 47 % during a time interval of 90 days. The calibration of the Hα line is more difficult because there are no strong forbidden emission lines in this wavelength range as there are for Hβ. Whenever possible the broad Hα line has been normalized to all forbidden lines in this range, namely the [SII], the [NII], narrow Hα and the [OI] lines. The flux of the broad Hα line component varied by 45 % during an interval of 100 days.

Figure 3 shows the variation of the Hβ profile after subtraction of the continuum from Dec. 88 until Aug. 89. The intensity of the profiles is given by the different colours. The observed spectra are plotted from left to right, from 4800 Å to 5200 Å. The narrow Hβ component is visible on top of the broad component. Furthermore, the blue and the red wings can be seen as well as the [OIII] 4959 and 5007 lines, which are constant in flux. All the observed high resolution Hβ spectra from Dec. 88 until Aug. 89 are plotted from the bottom to the the top with intervals of one day. For those time intervals where no observations exist the Hβ profiles, which are closest in time, have been extrapolated across the interval.

Finally the image was smoothed with a ten by ten pixel gaussian filter. One can recognize the three maxima of the Hβ line intensity – two stronger ones and a weaker maximum (e.g. Peterson et al.,1991 and Fig. 4). Further, one can see that the Hβ profile is not symmetric. There is an additional blue component which seems to be more pronounced when the Hβ line is strong and which is relatively weaker when the total flux is lower.

In the following we will discuss the differences of line profiles relative to an epoch when the emission line was in a low state. In all cases the Hβ profiles are refered to the spectrum taken during the second minimum.

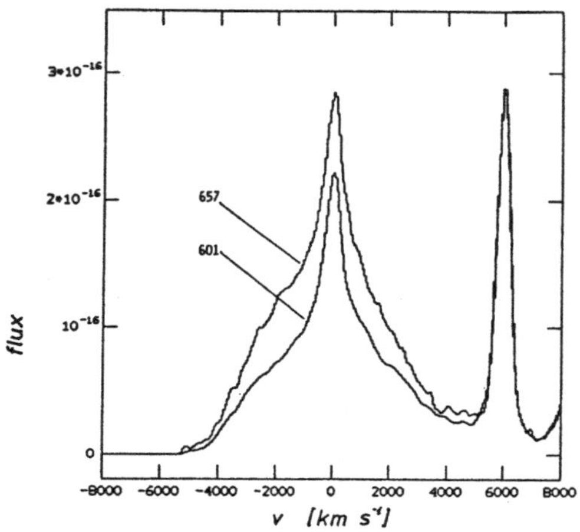

Fig. 1a. Normalized Hβ line profiles after subtraction of the continuum for the epochs 244 7657 (max) and 7601 (min)

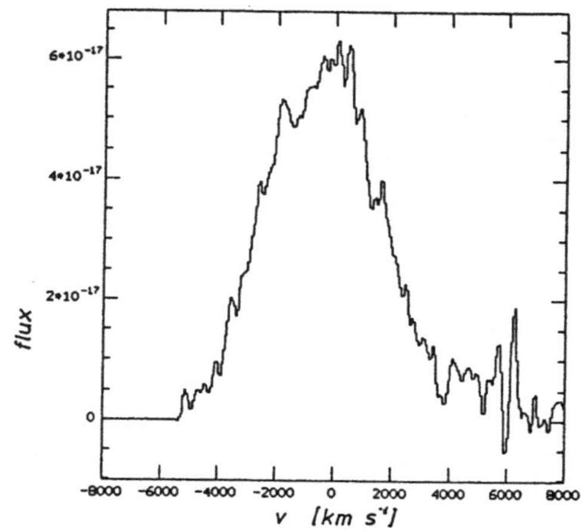

Fig. 1b. Hβ difference line profile for the epochs of Fig. 1a

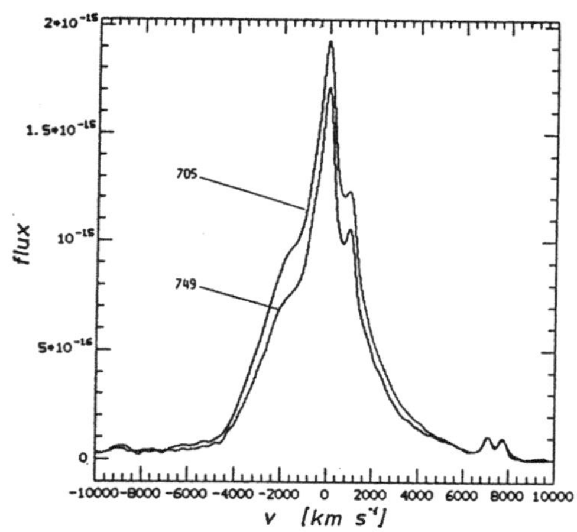

Fig. 2a. Normalized Hα line profiles after subtraction of the continuum for the epochs 244 7705 (max) and 7749 (min)

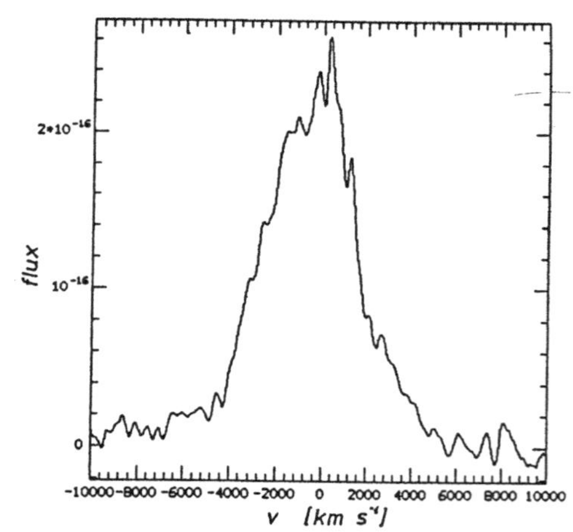

Fig. 2b. Hα difference line profile for the epochs of Fig. 2a

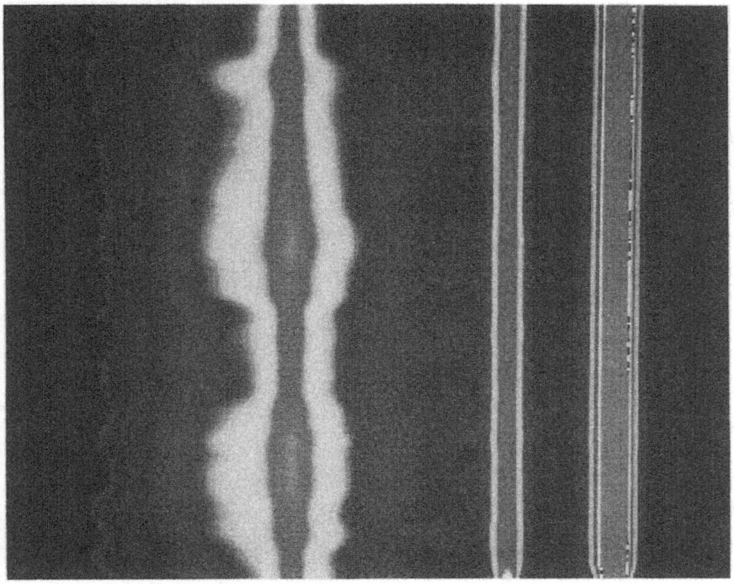

Fig. 3. Variations of the normalized Hβ line profile from Dec. 88 (bottom) until Aug. 89 (top)

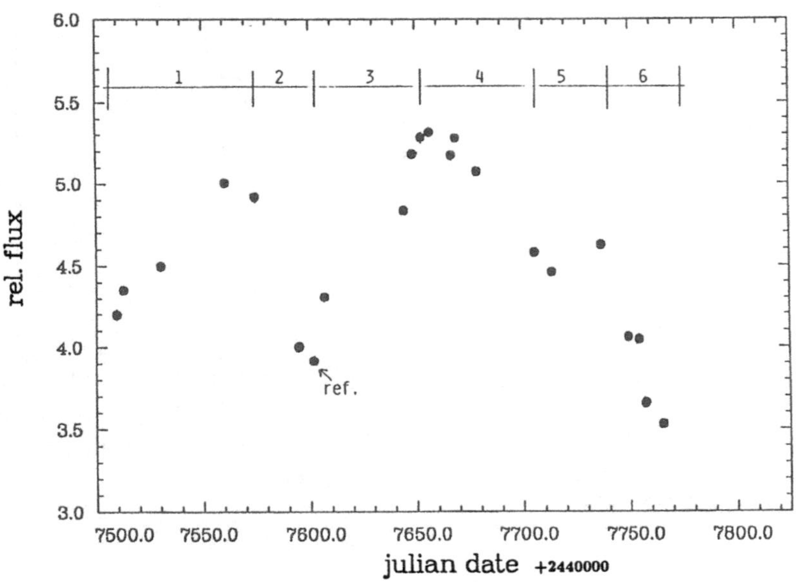

Fig. 4. Variations of the Hβ line flux for epochs with high quality spectra. The Hβ difference spectra in Fig. 5 and 6 refer to JD 2447601

Figure 1a shows the Hβ line profile during the second minimum and the line profile during the second maximum; Fig. 1b shows the difference profile. The line profiles are shown as functions of velocity; zero velocity refers to the narrow line component system. We are not concerned here with the underlying profile during the minimum state. We only want to consider the changes of the line profiles during the 8 months period relative to the existing emission line profile during the minimum at Julian date 2440000 + 7606. Taking into account only the difference profiles has the advantage that the influence of the varying continuum flux can be studied very carefully with respect to the broad line changes.

In the following Figures (5,6) the changes of the Hβ difference profiles for some high quality spectra during the first two increasing and decreasing epochs that is period 1 to 4 (see Fig. 4), have been plotted. During the first minimum the Hβ difference spectra show only a blue component (JD 2447509, 2447512) (Fig. 5a). A few days later a more central component with respect to the narrow Hβ component (JD 2447530, 2447549) is increasing. During the first maximum this central component is most pronounced, while the blue component did not increase significantly. After the first maximum the central component is decreasing as the continuum decreases, and during the second minimum (JD 2447594) only a weak blue component is visible in the difference profile (Fig. 5b). During the second increasing and decreasing period the components of the Hβ profile show the same behavior as during the first outburst.

Sometimes an additional red component is present (Fig. 5c). This red component decreases soon after the minima relative to the blue component. Most of the time the central Hβ component is dominant, but during the minimum epochs the blue component can be the strongest. This blue component seems to be constant for some weeks, but it is at least variable on short time scales (a few days) during the minimum. Immediately after the minimum a red component having nearly the same intensity as the blue component is present for a short time.

Figure 6 shows the variation of the central Hβ difference profile after subtraction of the blue component of JD 2447509 during the second increasing and decreasing period. These difference profiles are centered exactly at the velocity of the narrow component and are nearly symmetric. Furthermore, they have constant width at zero intensity throughout the observed epochs independent of the variable difference profile intensities.

Our preliminary results of the Hα line show many similarities to the Hβ line. Figure 2b shows the Hα difference profile of the minimum (JD 2447749) and maximum (JD 2447705) flux. One can immediately detect – even more clearly than in the Hβ difference profile – a blue and a red component. Furthermore, during the decrease in intensity after the second outburst the profiles also have constant width at zero intensity.

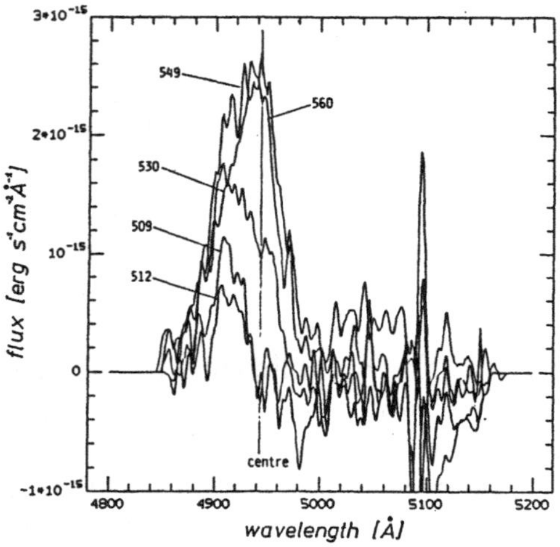

Fig. 5a. Hβ difference profiles for the first increasing period from JD 2447509 – 2447560

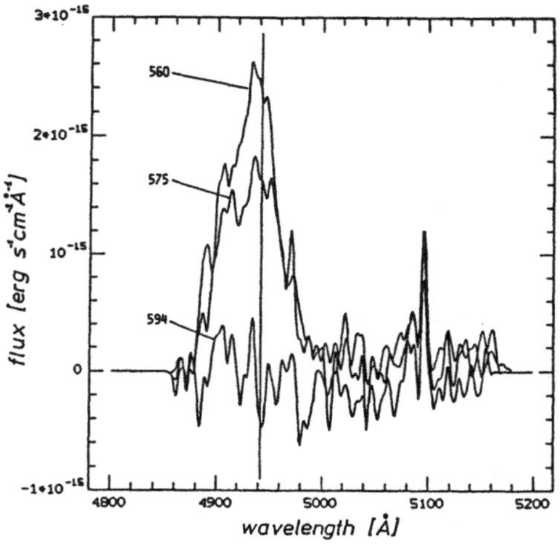

Fig. 5b. Hβ difference profiles for the first decreasing period from JD 2447560 – 2447594

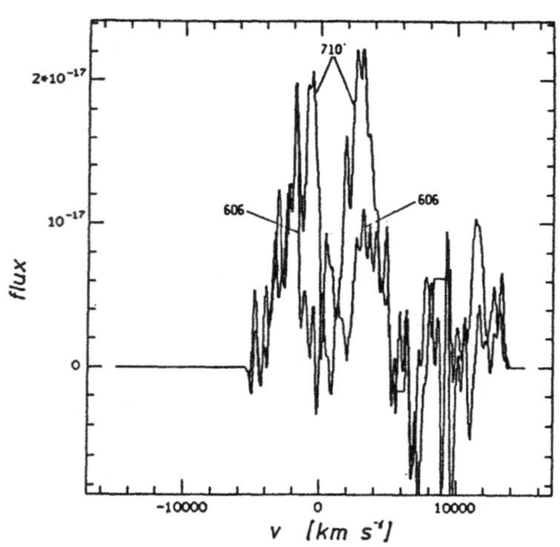

Fig. 5c. Hβ difference profiles for the epochs JD 244 7606 and 7710 showing only the blue and red component

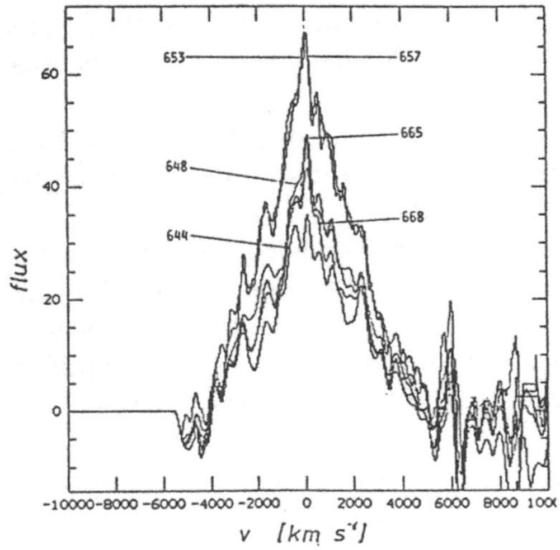

Fig. 6. The Variation of the Hβ difference profiles for the second increasing and decreasing period from JD 2447644 until JD 2447668 after subtraction of the blue component (JD 2447509)

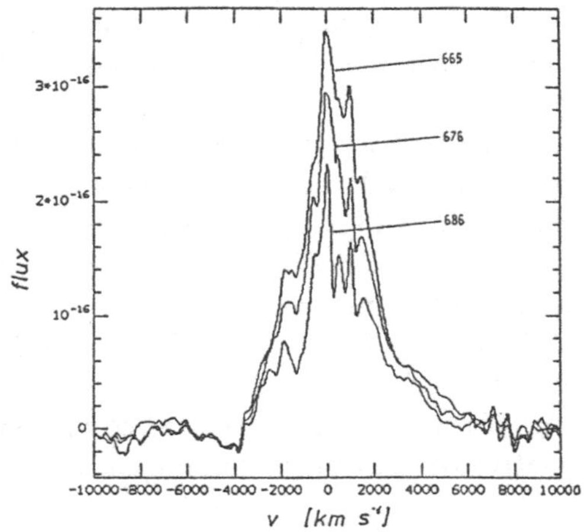

Fig. 7. Variation of the Hα difference profile relative to JD 2447749 for the second decreasing period JD 2447665 until JD 2447686 after subtraction of the blue component (JD 2447509)

4. Discussion

The variations of the central component can be compared with theoretical considerations (Ulrich et al.,1984) of the response of the broad line profiles to an outburst of the ionizing source. The dominant motion of the region where the central component originates must be chaotic motion. The emission line profiles have the same width at zero intensity early in a continuum outburst, at maximum state and in the decreasing state. Rotation and radial motions can be excluded for the central component.

The blue and the red components are not correlated with the continuum variations in the same way as the central component is. Therefore they must originate in a different region. The blue component is centered at a radial velocity of about $-1800.$ km s^{-1} and the red component at 2800. km s^{-1}.

Both components are smaller than the central one and have a full width at half maximum of $\sim$2000. km s^{-1}. The blue component was first detected by Peterson (1987) in the Hβ line, while the red component was first found by Stirpe et al. (1988) in the Hα line. The existence of a red component in the Hβ profile was confirmed by Peterson et al. (1990).

The following models for the origin of the blue and red component exist:

It was first proposed that the blue and the central component are the broad line regions of a supermassive binary. However, an additional red component was later found.

Then it was proposed by Stirpe et al. (1988) that the double peaked structure of the red and blue components is due to a rotating disk. However, the two components do not vary together (Peterson et al.,1990).

A further possiblity is that they originate from a two sided jet. There seem to be some similarities with the L_1 and L_2 lines on either side of the CIV line at 1550 Å in NGC 4151 (Ulrich et al.,1985). In both cases the two components vary independently. They are best visible during low active states. The profiles are narrower than the central component, and in both cases the radial velocities relative to the central component are different.

5. Conclusion

The broad line profiles of Hβ and Hα underwent strong variations during the optical monitoring campaign. Three different components are found in the broad Hβ and Hα profiles. The intensity of the strong central component is correlated with the UV continuum. The difference profiles are nearly symmetric after subtraction of the blue component. During the variability events the emission line profiles have a constant width at zero intensity. Therefore chaotic motion is the dominant motion for the part of the broad line region where they originate. The blue and the red components vary in different ways from each other and have narrower line widths than the central component; therefore they must originate in a physically distinct region. One possibility is that the double peaked structure arises in a disk-like structure. One argument against the accretion disk model is the independent variation of the two components. Another possibility is that the blue and the red component originate in a two sided jet. Many of the arguments for a two sided jet in NGC 4151 are also valid for NGC 5548.

Acknowledgements: This work has partly been supported by grant Ko 857/6-1, Bi191/6-2 of the Deutsche Forschungsgemeinschaft.

References

Anderson, K. : Astrophys.J.,**169**, 469 (1971)
Clavel, J., et al. :Astrophys,.J., in press (1990)
Peterson, B.M.: Astrophys.J., **312**, 79 (1987)
Peterson, B.M., Reichert, G., Korista, K., Wagner, M. : Astrophys.J. **352** 68 (1990)
Peterson, B.M., et al. :Astrophys.J., in press (1991)
Stirpe, G.M., de Bruyn, A.G., van Groningen, E.: Astron.Astrophys., **200**, 9 (1988)
Ulrich, M.H. et al.: Mon.Not.R.Astr.Soc., **206**, 221(1984)
Ulrich, M.H. et al.: Nature, **313**, 747 (1985)

Continuum Variability in NGC 5548: Implications for Theoretical Models

Silvano Molendi [1] [2], Laura Maraschi [1] and L. Stella [3]

[1]Dipartimento di Fisica, Università di Milano, Italy
[2]Max–Planck–Institut für Extraterrestrische Physik, Garching, Germany
[3]Osservatorio Astronomico di Brera, Milano, Italy

Abstract: The recent campaign on NGC 5548 has shown the continuum to vary simultaneously at ultraviolet and optical frequencies, with an upper limit to the delay of less than 4 days. We examine this result in the context of accretion disk theory. In the first part the standard α-disk model is discussed. The effects connected with heating of the outer regions by radiation from the inner regions are also considered. Neither scenario lends itself to a simple explanation of the observed data. A more successful model involves a geometrically thin outer disk heated by a central extended source of X-ray radiation.

1. Introduction

During a recent campaign the Seyfert galaxy NGC 5548 was observed with the IUE satellite every four days for a period of eight months. The continuum was measured at three different wavelengths 1350, 1840 and 2670 Å. In the optical band the source was observed with a number of telescopes and the continuum was measured at 4870 Å. Variability was clearly present at all wavelengths. Inspection of the four lightcurves and cross-correlation analysis (Clavel et al. 1990; Peterson et al. 1990) show that the lightcurves at different wavelengths are strongly correlated with an upper limit to any delay of less than 4 days.

These results pose serious difficulties for the standard disk models which have been employed to explain the UV bump of Quasars and Seyfert galaxies.

2. The data set

We consider continuum flux measurements at two different epochs, the first from day 2447582 (JD) corresponding to a low state of the source, the second from day 2447622 (JD) corresponding to a high state. The ultraviolet data are from Clavel et al. (1990) and the optical data are from Peterson et al. (1991). We did not correct the ultraviolet data for contribution from line blends and Balmer continuum, since the entity of the contribution is rather uncertain. We corrected the optical data for starlight contribution as specified in Peterson et al. (1990). All fluxes were converted to luminosities using a redshift of z = 0.0166 and H_0 = 50 km s^{-1} Mpc^{-1}. No corrections were made for reddening.

3. The accretion disk model

We compared the data from the low state with an accretion disk spectrum. The standard model for a disk around a Schwarzschild black hole was used (Shakura and Sunyaev 1973). We assumed a local blackbody spectrum and a face on disk. The free parameters are the mass of the black hole M and the accretion rate $\dot{m}$ which we measure in units of $M_{\rm Edd}$, $M_{\rm Edd} = L_{\rm Edd}/\eta c^2$ where $\eta = 0.057$. The spectrum and the model show agreement for $\dot{m} = 3.3 \times 10^{-3}$ and M = 2.5 x $10^8 M_\odot$, cfr. fig. 1. The outer radius of the disk is set at r = 70r$_g$ where r$_g$ = 2GM/c^2. We tried to match the high state fluxes with an accretion disk spectrum for the same value of the mass and a different accretion rate. The best result is obtained for $\dot{m} = 7.5 \times 10^{-3}$, cfr. fig. 2.

The timescale on which $\dot{m}$ variations are expected to propagate is the radial drift time, $t_d = \frac{1}{\alpha}\,\frac{r}{h}\,t_k$, where t_k is the keplerian timescale. Even at r = 20r$_g$, for a maximal value of $\alpha = 1$ this is of the order of several years in our case, thus we cannot interpret our result in terms of a variation in the $\dot{m}$ parameter.

An inner radiation pressure dominated region exists even at the subcritical rates derived above. The radius r_{ab} at which the transition between the gas pressure dominated outer region of the accretion disk and the radiation pressure dominated inner region occurs is r = 20r$_g$ for $\alpha = 10^{-1}$. The thermal timescale at this transition radius is 78 days for $\alpha = 10^{-1}$. This is interestingly similar to the quasi period in the optical and ultraviolet lightcurves. It is therefore tempting to interpret the variability in terms of some instability that starts at the radius r_{ab}, grows on the thermal timescale and propagates to the inner region, where the ultraviolet radiation is emitted, at the speed of sound. For the inner region $v_s \simeq c/\sqrt{3}$ so that the propagation time is of the order of 2 days. There are however a number of problems. First, it is not clear how an instability that propagates in the form of a sound wave can produce a 100% increase in the luminosity. Second, a considerable fraction of the optical radiation is emitted at radii larger than 20. This can be clearly seen in fig. 1 and 2 where the 2 lower continuous lines indicate the contribution to the total disk spectrum respectively from radii larger and smaller than 20.

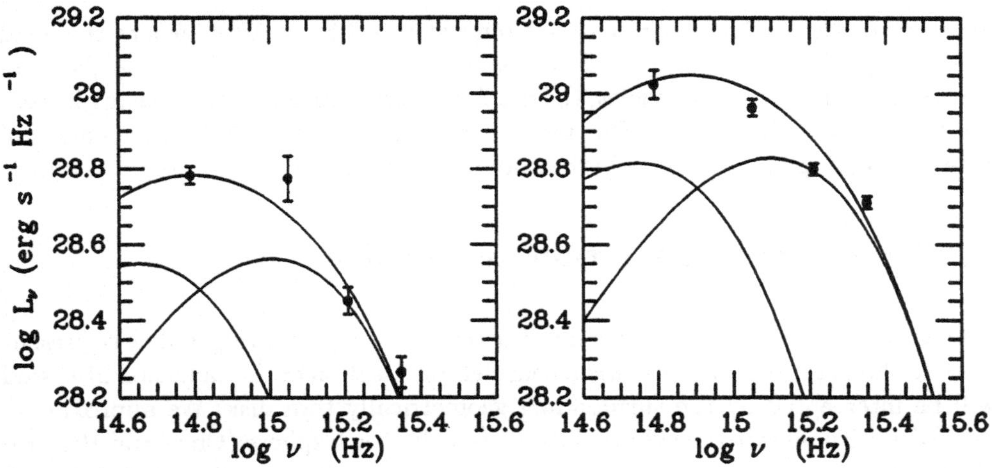

Fig. 1,2. Luminosity versus frequency. The data points and errors are respectively from day 2447582, low state, and day 2447622, high state, of the campaign on NGC 5548. The upper continuous line is the spectrum from the whole disk, the two lower continuous lines are the contributions from radii larger and smaller than 20 r_g. See the text for parameters.

4. The self-irradiating disk model

An alternative explanation of the lack of delays between the optical and ultraviolet lightcurves is that the emission originates from reprocessing of the flux emitted by the innermost regions.

A large h/r ratio in the outer regions of the disk is required, in order to intercept a significant fraction of the light emitted by the innermost region. Since h/r scales with $(\frac{r}{r_g})^{1/8}$, it is convenient to go to smaller masses of the central object. The integrated luminosity of the disk scales like the product of the mass times the accretion rate, thus if we reduce the mass by a certain amount, we have to increase $\dot{m}$ by approximatively the same amount. However, the fraction of reradiated flux to locally produced flux

$$\frac{F_r}{F_l} = 6.2 \times 10^{-5}\, \alpha^{-1/10} \dot{m}^{3/20} m^{-1/10} \left(\frac{r}{r_g}\right)^{9/8}$$

for $\dot{m} = 1$, $M = 10^6 M_\odot$ $\alpha = 1$ is > 1 only at $r > 1.88 \times 10^4 r_g$ where the local spectrum peaks in the infrar ed, $T = 1.3x10^3\ ^\circ K$, at smaller radii, where the local spectrum peaks in the ultraviolet, the dominant contribution to the heating is from the local viscous dissipation. This result is not changed by including the effects of external heating on the vertical structure of the disk.

5. The central bulge + thin disk model

The situation is drastically different if a central X-ray emitting bulge is present close to the black hole. In this case a considerable fraction of the radiation liberated will be intercepted by the surrounding geometrically thin disk. We approximate the bulge with an isotropically radiating optically thick sphere (Chen and Halpern 1989). We have fixed the mass of the central object at $m = 2.5 \times 10^6$. Figure 3 and 4 show the best fits for the low and high state of the source. The inferred value for the bulge radius is $r = 800 r_g$. The luminosity radiated by the bulge is 1.14×10^{44} for the high state and 0.55×10^{44} for the low state. In a thin accretion disk context this would correspond to $\dot{m} = 0.35$ and $\dot{m} = 0.17$. The fraction of the total luminosity emitted at 4870 Å coming from a region of the disk 4 lightdays in radius is larger than 98%. This is consistent with the simultaneous variability in the optical and ultraviolet lightcurves. The radiation from the bulge may be emitted in the hard X-ray band, cfr. Maraschi and Molendi (1990), or in the soft X-ray band below 1 Kev. The α_{ox} parameter for NGC 5548 is 1.2, and therefore a strong medium X-ray component (2 - 10 Kev) can be ruled out.

The timescale on which variations in $\dot{m}$ are expected to propagate in the bulge, i.e. the radial drift time $t_d = \frac{1}{\alpha}\frac{r}{h} t_k$, where t_k is the keplerian timescale at the outer radius of the bulge, and the thermal timescale at the outer radius of the bulge $t_{th} = \frac{1}{\alpha} t_k$, which for $\frac{h}{r} \simeq 1$, coincides with t_d, are of 58 days for $\alpha = 1$ and $r = 800 r_g$. This compares well with the two quasi periods of 40 and 80 days seen in the light curves.

6. Conclusions

Simple accretion disk models have been used in the past few years by a number of authors (Malkan (1983), Sun and Malkan (1989), Laor and Netzer (1990)) to model the optical–UV bump of Quasars and Seyfert galaxies. Simultaneous variations in the optical and ultraviolet continuum of NGC 5548 are not easily explained in the framework of standard accretion disk models. The self–irradiating disk model, which gives a very natural explanation to the simultaneous variations, does not work because the fraction of energy intercepted by the outer regions of the disk is too small. A more sophisticated model consisting of an X-ray emitting central bulge and a thin outer disk yields better results.

Acknowledgements: Silvano Molendi acknowledges the financial support of an "Anglo Della Riccia" scholarship.

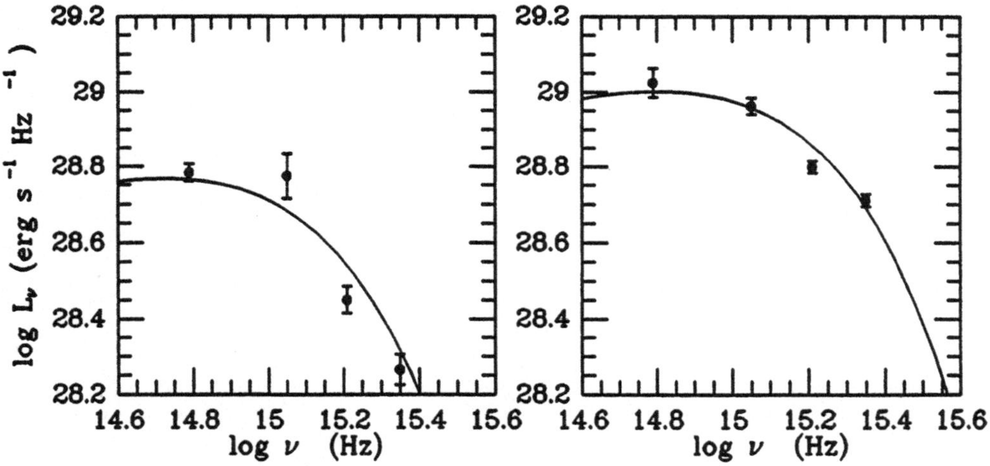

Fig. 3,4. Luminosity versus frequency. The data points and errors are respectively from day 2447582, low state, and day 2447622, high state, of the campaign on NGC 5548. The continuous line is the spectrum from the geometrically thin disk illuminated by the thick disk. See the text for parameters.

References

Chen, K. and Halpern, J.P.: 1989, *Astrophys. J.*, **344**, 115

Clavel, J. et al.: 1990 *Astrophys. J.*, in press..

Laor, A. and Netzer H.: 1989, *Mon. Not. R. astr. Soc.,* **238**, 897

Maraschi, L. and Molendi, S.: 1990, *Astrophys. J.*, **352**, 452

Malkan, M.A.: 1983, *Astrophys. J.*, **268**, 582

Peterson, B.M., Reichert, G.A., Korista, K.T., and Wagner, R.M.: 1990, *Astrophys. J.*, **352**, 68

Peterson, B.M. et al.: 1991, *Astrophys. J.*, in press

Shakura, N.I. and Sunyaev, R.A.: 1973, *Astron. Astrophys.*, **24**, 337

Sun, W.- H. and Malkan, M.A.: 1989, *Astrophys. J.*, **346**, 68

Variability in Markarian 279

Gioavanna M. Stirpe

Osservatorio Astronomico di Bologna, Via Zamboni 33, 40126 Bologna, Italy

1. Introduction

Markarian 279, a spiral galaxy with an inclination of about 40°, was immediately identified as a Seyfert 1 galaxy upon its discovery (Markarian 1969). All its characteristics make it an 'average' medium-bright object of its kind (see Table 1). Its strong broad emission lines present both a median shift and an asymmetry towards red wavelengths: neither of these characteristics is extreme within the Seyfert 1 population (Stirpe 1989). In this paper I will review what is known about the broad line variability of Markarian 279 and present some recent results.

Table 1. Main characteristics of Markarian 279

$\alpha,\delta\,(1950)$	135153.6+695108
z	0.0303
m_V	14.5
L_{bol}	$10^{45}\,\mathrm{erg\,s^{-1}}$
$\mathrm{FWHM_{broad}}$	$\sim 6000\,\mathrm{km\,s^{-1}}$
$W_\lambda(\mathrm{H}\beta)$	~ 100 Å

2. Early observations

High quality spectra of Hα and Hβ were published by Osterbrock and Shuder (1982), who suggested that some line variation may have occurred between May 1980 and March 1981. The first clear detection of emission line variation in Markarian 279 was obtained between 9 March and 2 May 1981 by Peterson et al. (1982), who observed Hβ to decrease at negative velocities. After this, the object disappeared from the optical literature for several years, except for Peterson et al.

(1985) reporting that no further profile variations had been detected. UV continuum variations were observed by Chapman et al. (1985), who also had some indication that the UV lines are variable. No further evidence for this has been published.

In May 1985, high quality Hα and Hβ profiles of the source were obtained by Stirpe (1990): at the time, the Hβ profile looked very similar to the March 1981 profile by Peterson et al. (1982). A lower quality unpublished Hβ spectrum taken in July 1986 showed little variation with respect to the previous year. On the whole, no spectacular variations were detected in this source in the early to mid eighties.

3. Results from monitoring campaigns

By the time Stirpe and de Bruyn (1990) started monitoring Markarian 279 with the 2.5m Isaac Newton Telescope (INT) on La Palma in the summer of 1987, the flux of both continuum and emission lines had dropped by a factor 2, making this the largest variation observed up to then. During the two-month campaign both continuum and line fluxes underwent a further decrease of 30% (optical continuum), 20% (broad Hα) and 40% (broad Hβ). The decrease of Hα was clearly delayed with respect to that of the continuum, and a comparison between the light curves indicates that most of the line flux is emitted within a radius of $\sim$ 20 light days. All line variations always occurred both at negative and positive velocities: on time scales as short as those of the campaign, which resolve the response of the BLR to the continuum variations, this is a strong indication that the predominant motions are *not* radial, otherwise we would observe delay effects between the two sides of the lines.

Two monitoring efforts were conducted on Markarian 279 in 1988. One (Stirpe and de Bruyn, in preparation) was a follow-up of the 1987 campaign on La Palma, and involved obtaining high resolution ($\sim$ 2 Å), high signal-to-noise ($\sim$ 50 : 1) spectra of the Hα and Hβ regions. The major drawback of this campaign was the scarcity of epochs awarded (8 distributed irregularly over a period of 4 months). The other campaign was conducted at Wise Observatory (Maoz et al. 1990), and yielded low resolution spectra covering both Hα and Hβ, taken at average intervals of 4 days over a period of 5 months. While the time coverage of this data-base is very good, the low resolution and signal-to-noise severely limit any conclusion on BLR kinematics.

The results from the 1988 season can be resumed as follows. A gradual increase of 30% in the continuum was observed at Wise Observatory between early May and late June, before and after which no significant variations were observed. Hα and Hβ had higher mean fluxes ($\sim$ 15% and $\sim$ 40% respectively) at the end of the campaign than at the beginning. The uncertainties on the flux values do not allow to follow the detailed response of the lines to the continuum variation, but in any case the observations confirm that the BLR is smaller than one light-month: the formal lag (12 $\pm$ 3 light-days) obtained from the cross-correlation of the line and continuum light curves can be considered consistent with the size obtained by Stirpe and de Bruyn (1990; see above), given the considerable uncertainties affecting both results.

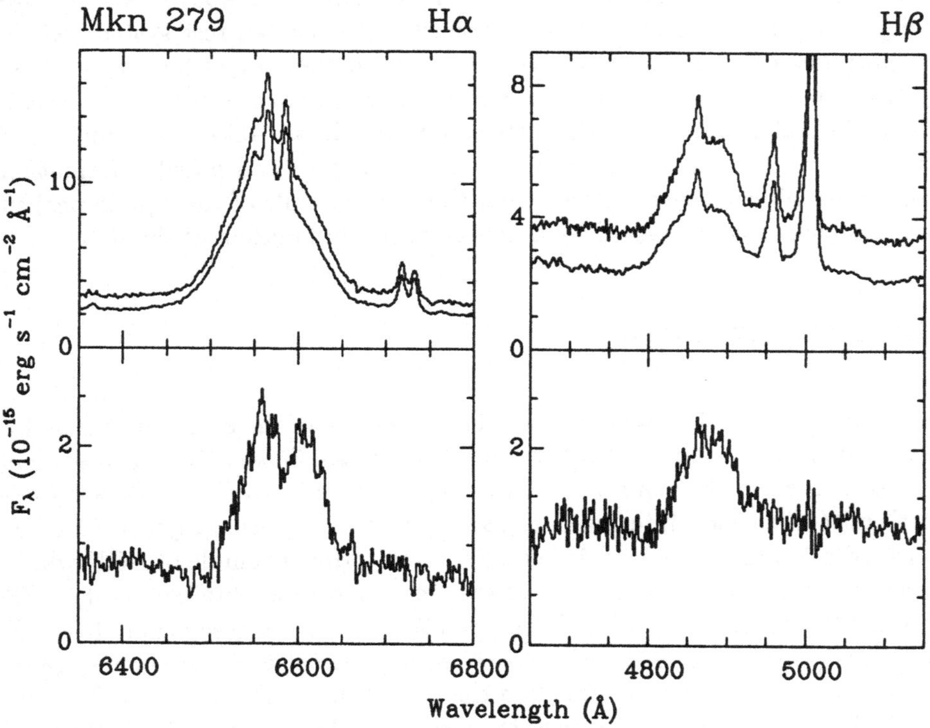

Fig. 1. The top panels show the Hα and Hβ spectra of Markarian 279 taken at the INT at La Palma on 4 April (lower) and 3 August 1988 (higher). The difference spectra are displayed in the bottom panels. The wavelength scale has been corrected for the redshift of the narrow lines (Table 1), and the spectra have been re-binned in 1.5 Å intervals. The flux scale is subject to uncertainties in the absolute calibration, but the spectra have been set on a consistent scale by normalizing the flux of the narrow lines (assumed to be non-variable on short time scales)

The Wise Observatory results are consistent with those obtained during the La Palma campaign conducted that same year. Figure 1 shows the Hα and Hβ spectra from the beginning and end of this campaign (4 April and 3 August), and the respective difference spectra. As well as the continuum increase at both wavelenghts, an increase of the broad Hα and Hβ fluxes of 15% and 50% respectively is clearly visible in the difference spectra. From intermediate spectra it appears that most of the variation occurred before 21 June. As observed previously in NGC 5548 (Stirpe et al. 1988), the two difference spectra are not alike in shape: while the broad Hβ profile does not appear to have changed, the Hα difference spectrum displays the double-peaked appearance already observed in other objects (Stirpe et al. 1988; Alloin et al. 1988). This profile is characteristic of an accretion disk, and has led to the suggestion that the velocity field of the BLR is dominated by rotation. Notice, however, that one of the two peaks, rather than the dip between them, is centered on $v_r \sim 0$ km s^{-1} (where v_r is the observed radial velocity with respect to the system defined by the narrow lines). If the profile is indeed produced by an accretion disk with isotropic emissivity, the disk itself would have to have a velocity of at least 1000 km s^{-1} with respect to the chosen rest system, which for our purposes can be considered to coincide with the rest system of the host galaxy (the narrow lines are usually shifted by no more than $\sim$ 200 km s^{-1} with respect to the stellar lines when the latter are visible in an AGN spectrum). This situation requires careful consideration about its stability and possible formation mechanism. An alternative solution is that the velocity shift of the double peaked profile is a spurious effect caused by a non-uniform illumination of the disk.

4. The LAG campaign

The latest monitoring effort was conducted at La Palma in the first half of 1990 by the Lovers of Active Galaxies (**LAG**) collaboration. **LAG** was formed to exploit the 5% fraction of the observing time at the Canary Islands observatories, assigned by the Comité Cientifico International to major international projects. Most of the allocated observing time was dedicated to monitoring a sample of AGN which included Markarian 279. CCD spectra of the Hα region were obtained at the INT and at the 4.2m William Herschel Telescope (WHT), with a resolution of $\sim$ 2 Å and $\sim$ 1 Å respectively. Spectra of Hβ were also obtained at some epochs. CCD images in several bands were obtained on the same time scales at the 1m Jacobus Kapteyn Telescope, in order to monitor the continuum light curve. This is the best quality data set available up to now for Markarian 279, if we consider the time coverage (the average interval between spectroscopic observations is 6.1 days), resolution, signal-to-noise ratio (typically $\sim 70:1$ in the Hα continuum), and the availability of CCD photometry. The data are still being analysed at present, and only some preliminary results are presented here.

Firstly, a major increase was noticed in the line and continuum flux with respect to the 1987 and 1988 observations (Figs.2 and 3, compare with Fig.1). Hα was at least as bright as in 1985 (Stirpe 1989), and Hβ was at the highest level ever observed. Throughout the campaign Markarian 279 remained brighter than observed in 1987 or 1988. The fastest variation in Hα was observed at the beginning of the campaign, in an interval of 5 days (Fig.2), when the broad component increased by $\sim$ 15%. Contrary to what was observed in 1988 (Fig.1), the profile of the difference spectrum is very similar to that of the individual spectra. After 24 January, the source underwent a slow, steady decrease which was still continuing at the end of the campaign. Figure 3, for instance, shows how the broad Hβ flux decreased by 25% in an interval of 3 months: intermediate Hα spectra show that no distinct features are present in the light curve, apart from the gradual decrease. Notice also the strong decrease of the He II λ4686 line, and the bump in the difference spectrum at ~ 5020 Å, due presumably to the variation of Fe II λ5018. By the end of the campaign Hβ had become weaker by about 40%, and Hα by ≥ 20%. In general, these observations confirm that Markarian 279 tends to vary slowly, i.e. on time scales of several weeks.

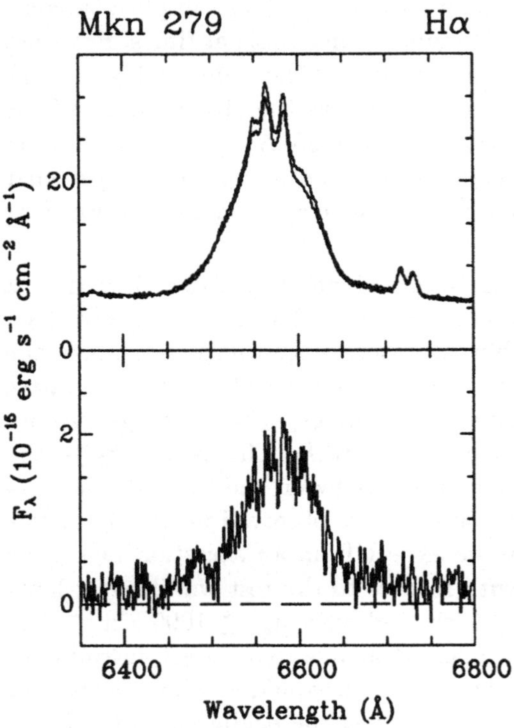

Fig. 2. Top panel: Hα spectra of Markarian 279 obtained at the WHT on 19 (lower) and 24 January 1990 (higher). The spectra are treated as those in Fig.1, except for a different bin width (1 Å). Bottom panel: the difference between the two spectra

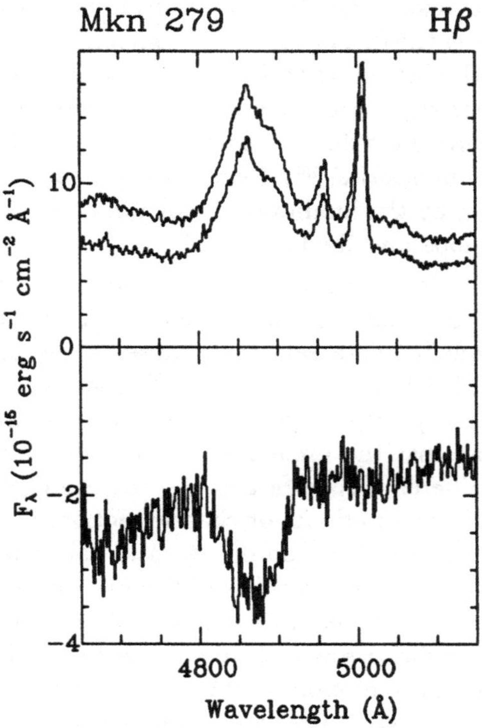

Fig. 3. Top panel: Hβ spectra obtained at the INT on 11 February (higher) and 14 April 1990 (lower). The spectra are treated as those in Fig.1. Bottom panel: the difference between the two spectra

5. Conclusions

Figure 4 summarizes the extreme states in which Markarian 279 has been observed up to now: highest in June 1985 and more recently in January 1990, lowest in Summer 1987 to May 1988. In June 1988 and June 1990 the object was at an intermediate level between those shown here. The data collected in the last 5 years show that variations in this object can have a very large amplitude, causing the flux of the broad component in Hβ to more than double at times. However, the largest variations also occur on *long* time scales (of the order of 1 year): the strongest variation within a single observing season was the 50% increase of the broad Hβ flux in an interval of 2 months, observed in 1988. As well as the range of luminosities, Fig.4 evidences the fact that the optical continuum becomes steeper when the luminosity is higher, and that there is a strong correlation between its slope and the luminosity.

It is by now well established that the BLR in Markarian 279 is less than a light-month in size: a more accurate determination awaits more complete, high quality data sets than those as yet published. More uncertainty surrounds the BLR kinematics. In all difference spectra, variations are visible both at negative and positive velocities, even on the shortest time scales, which strongly argues against the predominance of radial motions in the BLR. The symmetry in the difference spectra and the absence of delay effects between the two sides favours either rotational or chaotic motion. The double peaks observed in the Hα difference spectrum on one occasion are what we expect from an accretion disk. Due to the different shifts of the two peaks with respect to the rest wavelength, however, a model of this kind would require a fairly high velocity (≥ 1000 km s^{-1}) of the disk itself with respect to the NLR, unless dishomogeneities are introduced. Chaotic motion would not produce double peaks, but would explain the lack of profile variation observed on a very short time scale at the beginning of the **LAG** campaign. We should ask ourselves why double-peaked or flat-topped difference profiles are seen on some occasions (see also Stirpe and de Bruyn 1990), while on most occasions the difference profiles do not differ from the individual ones, and bear in mind that disk profiles are not necessarily double-peaked (Dumont and Collin-Souffrin 1990). A more complete analysis of the **LAG** data set, and simulations such as those performed by Robinson et al. (1990) will allow to set constraints on the kinematics of the BLR. For the moment, the only strong statement that can be made about the prevailing type of motion in this source is that it is almost certainly not radial. The lack of major variations in the line profiles of Markarian 279 is in contrast with what is common occurrence in other Seyferts, such as NGC 5548 and Akn 120. This different behaviour suggests that, although Markarian 279 and NGC 5548 have comparable intrinsic luminosities, the BLRs of the two sources may present different characteristics, as is already suggested in fact by their considerably different line profiles. As well as studying the luminosity dependence of AGN variability, therefore, it would be worthwile to monitor the behaviour of sources with widely differing line profiles: this would help us to de-

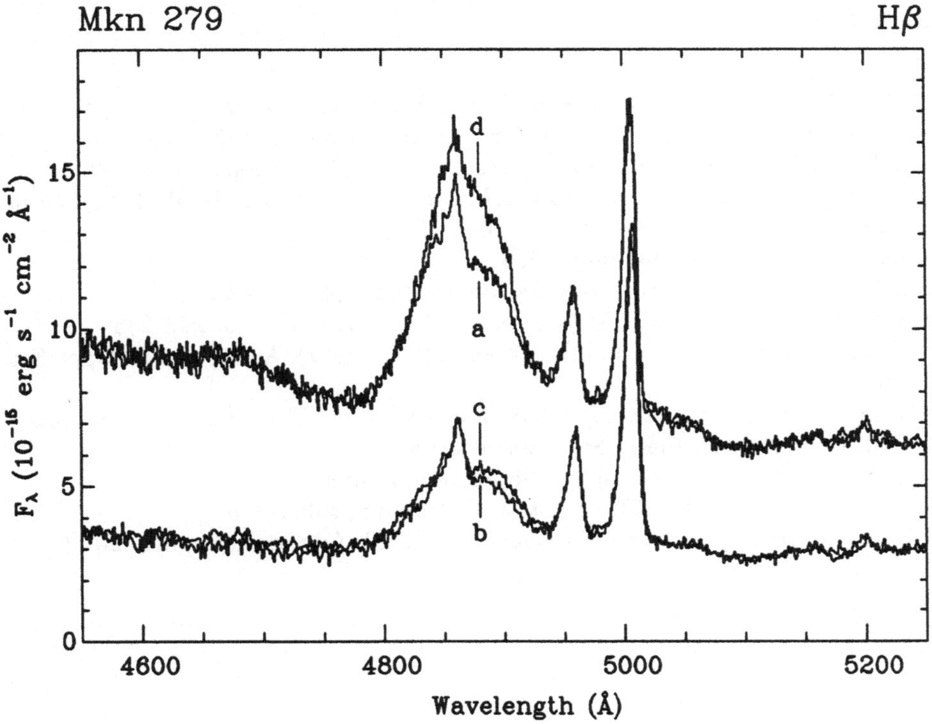

Fig. 4. Hβ spectra of Markarian 279 taken in 4 different years, according to the following key: a – June 1985, b – June 1987, c – April 1988, d – January 1990. The spectra have been treated as those in Fig.1, but are binned in 1 Å intervals

termine which factors (different velocity fields, projection angles, etc.) cause the diversity observed in AGN emission lines.

Acknowledgements: I am grateful to the other members of **LAG** for allowing the use in this paper of the data taken by the collaboration.

The William Herschel Telescope and the Isaac Newton Telescope are operated on the island of La Palma by the Royal Greenwich Observatory at the Spanish Observatorio del Roque de los Muchachos of the Instituto de Astrofisica de Canarias.

References

Alloin, D., Boisson, C., Pelat, D. (1988): Astron. Astrophys. **200** 17
Chapman, G.N.F., Geller, M.J., Huchra, J.P. (1985): Astrophys. J. **297** 151
Dumont, A.-M., Collin-Souffrin, S. (1990): Astron. Astrophys. Suppl. Ser. **83** 71
Maoz, D., Netzer, H., Leibowitz, E., Brosch, N., Laor, A., Mendelson, H., Beck, S., Mazeh, T., Almoznino, E. (1990): Astrophys. J. **351** 75
Markarian, B.E. (1969): Astrophysics **5** 286
Osterbrock, D.E., Shuder, J.M. (1982): Astrophys. J. Suppl. Ser. **49** 149
Peterson, B.M., Crenshaw, D.M., Meyers, K.A. (1985): Astrophys. J. **298** 283
Peterson, B.M., Foltz, C.B., Byard, P.L., Wagner, R.M. (1982): Astrophys. J. Suppl. Ser. **49** 469
Robinson, A., Pérez, E., Binette, L. (1990): Monthly Notices Roy. Astr. Soc., in press
Stirpe, G.M. (1989): Ph.D. thesis, Sterrewacht Leiden
Stirpe, G.M. (1990): Astron. Astrophys. Suppl. Ser., in press
Stirpe, G.M., de Bruyn, A.G. (1990): Astron. Astrophys., submitted
Stirpe, G.M., de Bruyn, A.G., van Groningen, E. (1988): Astron. Astrophys. **200** 9

A Search for Variability in PKS 1302-102

Neil Jackson

NRAL, Jodrell Bank, Macclesfield, Cheshire SK11 9DL, England

In recent years it has become clear that line variability is not confined to lower luminosity objects such as Seyfert galaxies. Higher luminosity objects such as radio quasars have been found to contain variable line emission. The timescales of the variations are generally longer than those seen in Seyfert galaxies, but are still shorter those expected from photoionization models. For example, Penston, Pérez and Moles (1989) find variations in the Lyα and CIV lines of PKS2134+004, in the CIII] and CIV lines of 3C 445, and in the core of the MgII line of 2344+092, all on timescales of months. In addition, variations of 20% -40% have been found in the Balmer lines of 0736+017, 2135-147 and 2141+175 (Zheng and Burbidge 1986; Zheng *et al.* 1987).

The LAG (Lovers of Active Galaxies) collaboration was set up to bid for the CCI international 5% of time on the La Palma telescopes for 1989-1990. Part of the LAG programme involved the monitoring of line and continuum variability, mostly in Seyfert galaxies but also two quasars. 1302-102 was chosen as one of these quasars because it is one of the brightest radio core dominated quasars in the appropriate part of the sky. (Most variable quasars mentioned above are also core dominated: for a possible explanation of this see Pérez *et al.* 1989). Spectra were obtained on the William Herschel and Isaac Netwon telescopes (INT and WHT) on La Palma.

Figure 1 shows the preliminary results, in the form of spectra obtained over the six months of the campaign. These spectra have been extracted according to an algorithm which weights spatial columns according to the excess of counts in them over the sky background. It appears that, to an accuracy of $\leq$ 10% , no variation took place in the first five months. A possible small variation can be seen in the last two spectra, in the shape of either a continuum rise or a decrease in Hβ flux, by comparison with the [OIII] lines which are assumed to be constant. Jacobus Kapteyn Telescope dara, which were also obtained during the six months, will resolve this ambiguity.

The coadded spectrum of the best WHT epochs (Figure 2) is of very high quality. The small narrow component can be clearly seen: note that the centroid of the broad Hβ line has a small but definite shift to the red with respect to the narrow Hβ line which sits on top of it. The contaminating FeII lines (at 4924Å and 5018Å) can also be seen and should allow a more unambiguous deconvolution of the FeII contamination from the Hβ profile than has hitherto been possible.

Acknowledgements: The INT, WHT and JKT are operated by the Royal Greenwich Observatory at the Spanish Observatorio del Roque de los Muchachos of the Instituto de Astrofisica de Canarias.

References

Zheng, W., Burbidge, E.M., 1986. *Astrophys. J.*, **306**, L67.
Zheng, W., Burbidge, E.M., Smith, H.E., Cohen, R.D., Bradley, S.E., 1987. *Astrophys. J.*, **322**, 164.
Pérez, E., Penston, M.V., Moles, M., 1989. *Mon. Not. R. astr. Soc.*, **239**, 55.

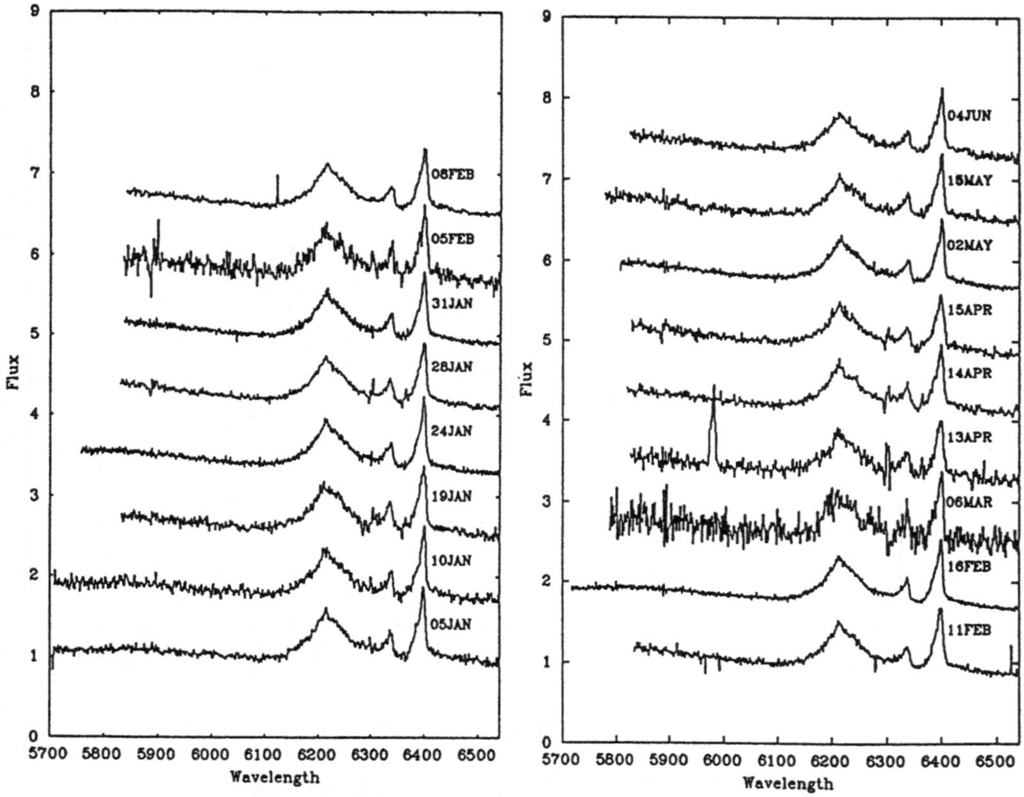

Fig. 1. Spectra of the Hβ and [OIII] region in the quasar 1302-102 over 17 of the epochs. There is a slight residual in many of the spectra from subtraction of the [OI] skyline at 6300 Å. Cosmic rays remain to the blue of Hβ in the spectra of February 8th and April 13th.

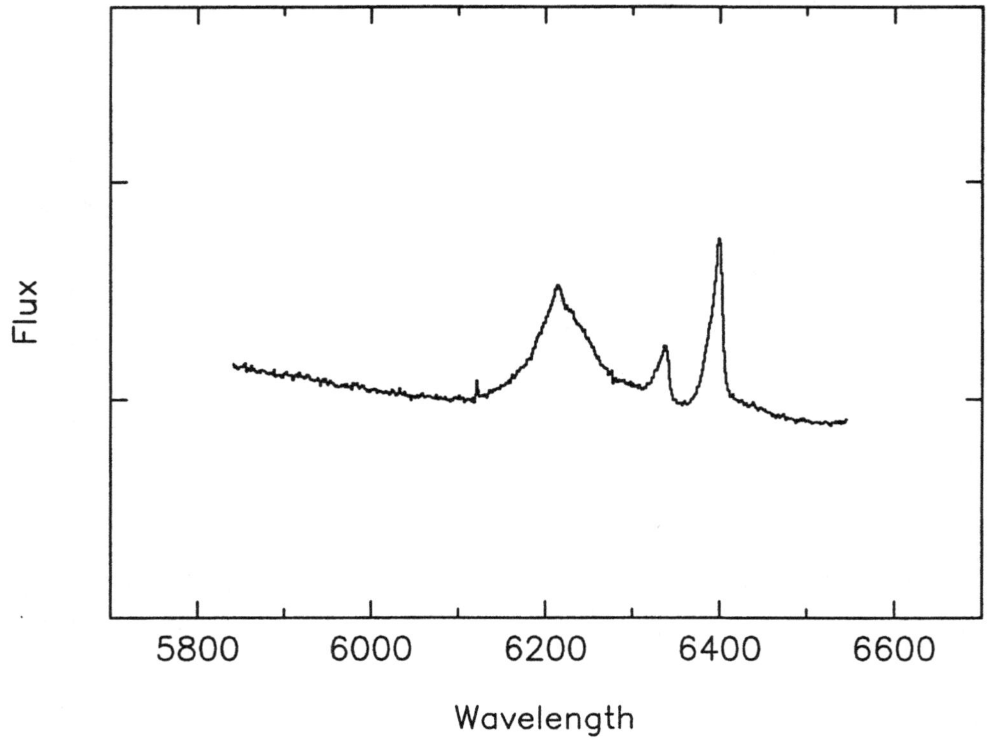

Fig. 2. The best five WHT spectra added together.

Meteorology of the Broad Line Region

Andrew Robinson [1], Enrique Pérez [2] and Luis de la Fuente [2]

[1]Institute of Astronomy, Madingley Rd., Cambridge CB3 0HA, U.K.
[2]Instituto de Astrofísica de Canarias, 38200 La Laguna, Tenerife, Spain

Abstract: We discuss the response of the broad emission lines to ionizing continuum variations in the context of the discrete cloud model. We consider cases in which the cloud system has a thick, spherical geometry and discuss several reverberation phenomena which may be relevant to the interpretation of observed light curves. We find that the lag obtained from cross-correlation analyses does not give a reliable indication of the effective "size" of the BLR; this can instead be obtained directly from the centroid of the transfer function. A notable property of this type of model is the relationship between the variability behaviour of an emission line and the shape of its profile. Computer simulations are presented which illustrate the evolution of the line profile in response to a 'realistic' continuum light curve.

1. Introduction

It is generally believed that the broad emission lines of active galactic nuclei are produced by photo-ionization of the emitting gas (the broad line region or BLR) by a central continuum source. The observed line intensity variations are then naturally explained (though this is not the only possibility) by fluctuations in the ionizing luminosity of the source. If, furthermore, the BLR is spatially extended, light travel delays modify the line response in such a way that it "lags" and is "smoothed" with respect to the continuum light curve. In principle, therefore, various important properties of the BLR—systemic geometry, effective size, emissivity distribution—can be inferred from a comparison of the line and continuum light curves. Similarly, the velocity field can be obtained from the evolution of the line profile. Thus, emission line variations can be used to 'map' the structure and kinematics of the BLR. However, although this possibility has been recognised for many years, such an analysis requires high quality data and regular sampling. It is only recently, with the advent of intensive monitoring campaigns using both the IUE satellite and ground-based telescopes (see contributions of Clavel, Peterson and Netzer for details), that data-sets meeting these criteria have been obtained.

Nevertheless, interpreting the observations in terms of models for the structure and dynamics of the BLR is by no means straightforward. In this paper we consider the response of the broad line region (BLR) to ionizing continuum variations in terms of the discrete cloud model. We discuss, in particular, several properties of the model which may be relevant to the interpretation of variability data. The work summarized here and also by Pérez *et. al.*, in this volume forms part of a more general study of the properties of discrete cloud models of the BLR which is described in a series of recent or forthcoming papers (Robinson & Pérez 1990 [RP]; Robinson, Pérez & Binette 1990 [RPB]; Robinson 1990; Pérez, Robinson & de la Fuente 1990 [PRF]; Robinson, Pérez, Binette & Rodríguez-Espinosa, in preparation and Pérez, Robinson & de la Fuente, in preparation.

2. The discrete cloud model

We will investigate the response behaviour of the broad emission lines in the context of the discrete cloud model for the BLR. In this model, the BLR consists of an ensemble of small clouds which are photoionized by the radiation field emitted by a central, point continuum source of ionizing photon luminosity ($Q(t)$). Physical conditions in the BLR are generally believed to be such that line emission from any given cloud responds effectively instantaneously to changes in the local ionizing photon flux. The response of the cloud system as a whole is therefore dominated by light travel delays introduced by differences in photon flight paths between the continuum source, the various re-processing clouds and the observer. As a result of this 'reverberation' effect, it will appear to a distant observer that at time t, all clouds lying on the parabolic surface of constant delay (or 'light front') defined by $ct = r(1 - \cos\theta)$ (where r is the radial distance from the source and θ is the angle to the line- of-sight) experience the ionizing continuum in the same state. Formally, the integrated line emission from the entire cloud ensemble can be represented (Blandford & McKee 1982) by

$$L(t) = \int T(t')L_c(t - t')dt' \qquad (1)$$

where $L_c(t)$ is the ionizing continuum luminosity and $T(t)$ is a 'transfer function' which depends on the BLR geometry and the distribution and reprocessing efficiency of the line-emitting material. In the following, we will refer to this scenario as the 'reverberation model'.

We will consider only the case in which the individual clouds are distributed within a spherical shell of "thickness ratio" $y = r_s/r_0$, where r_0 and r_s are the inner and outer radii, respectively. Transfer functions for other types of systemic geometry are discussed by Pérez *et. al.*, elsewhere in this volume. For simpliciy, we assume that the clouds are linear reprocessors of ionizing radiation i.e., the emitted power, p_c, in a given line is simply proportional to the rate at which the cloud absorbs ionizing photons. Thus, for a cloud of cross- section A_c, $p_c \propto A_cF$, where F, the incident ionizing flux, is given by $F(t) = Q(t)/4\pi r^2$. The volume emissivity distribution in line is, in turn, given by $\epsilon_V = \rho p_c(t)$, where ρ is the cloud number density. In general, ρ, p_c and hence ϵ_V will depend on distance from the continuum source and we will assume that this dependence can be represented by the power law $\epsilon_V \propto r^{\beta}$. Similarly, in discussing the line profile formed by the system, we assume that cloud velocities are given by $v \propto r^p$.

It is obvious that the "size" of the BLR will play an important role in determining the response behaviour of the emission lines. The "size" of an extended system (i.e., $y \gg 1$), however, is not a precisely defined quantity. Perhaps the most appropriate choice in this case is to measure "size" in terms of the "emissivity radius", which we define as follows:

$$|R| = \frac{\int r \epsilon_{V0}(r) dV}{\int \epsilon_{V0}(r) dV} \tag{2}$$

where ϵ_{V0} represents the steady state volume emissivity (i.e., that corresponding to the time-averaged ionizing continuum).

In discussing the response properties of this model, we will assume that the that the transfer function is not itself time-dependent or, in other words, that the structure and velocity field of the system are maintained in a steady state over timescales of interest. In addition, we must also suppose that the ionizing continuum light curve is known. In practice, of course, the ionizing ultra-violet cannot be observed directly and in analyzing real data it is necessarily assumed that the ionizing continuum fluctuates in phase with the optical or UV continuum underlying the emission lines and with the same amplitude. These and other assumptions implicit in analyses of emission line variability based on the reverberation model are critically discussed by Penston (1990).

3. Analysis of response light curve

It is convenient to discuss the response of the spherical cloud ensemble in terms of the dimensionless time $\tau = t/t_s$, where $t_s = 2r_s/c$ is the light crossing time at the outer shell boundary. In these units, the transfer function (1-DTF) for the model is given by

$$\begin{aligned} T(\tau) &\propto y^{\beta+2} - 1 & \tau < 1/y \\ &\propto y^{\beta+2}(1 - \tau^{\beta+2}) & \tau > 1/y \end{aligned} \tag{3}$$

The function is flat-topped until $\tau = 1/y$, the light crossing time at the inner shell boundary, and then decays more or less steeply according to the value of β (see Fig. 2). In order to gain an insight into the importance of the continuum variability timescale, we consider the line response to a series of box-car continuum pulses of duration $\Delta\tau$ and amplitude $A-1$. Note that $\Delta\tau$ measures the continuum timescale relative to the shell light crossing time and if the pulse duration in absolute units is fixed, parameterizes the size scale of the BLR.

3.1. The response function

The line response functions, $F_s(\tau)$, are discussed in detail by RP, here we will simply list their more important properties. Some examples are shown in Fig. 1. It is clear that models characterized by a highly peaked transfer function, such as thick shells (e.g, $y \geq 10$) with relatively steep emissivity distributions (e.g., $\beta = -4$), will respond to a given continuum event more rapidly and with a greater amplitude than those characterized by flatter ones such as thin shells ($y \sim 1$) or models with relatively flat emissivity distributions (e.g., $\beta = -2$). However, the response function also depends on the continuum pulse duration, $\Delta\tau$. As was originally pointed out by Bahcall, Kozlovsky & Salpeter (1972), the line response reaches its maximum amplitude when $\tau = \Delta\tau$. Thus, the rise time is entirely determined by the continuum timescale. The subsequent decay is largely governed by the transfer function and it is this which determines the length of the "lag" between continuum and emission line light curves. The lag is longer for models in which a large fraction of the reprocessed energy is emitted at times $\tau > \Delta\tau$, as is the case for small y (see below). Both the form and in particular, the amplitude, of the response function depend on $\Delta\tau$. As $\Delta\tau$ is decreased, the BLR responds less 'coherently' (smaller fractions of the cloud ensemble will be illuminated by the continuum pulse at any given time) and the response amplitude therefore decreases. Nevertheless, for large y, a significant emission line response (say, with an amplitude relative to the continuum of $\geq 10\%$) can be still be obtained for continuum pulses of duration $\sim 1\%$ or so of the light crossing time of the shell.

3.2. Cross-correlation function

Until recently, analyses of emission line variability have concentrated on determining the lag between the continuum and emission line light curves, usually by calculating the cross-correlation function (e.g., Gaskell & Sparke 1986; Gaskell & Koratkar 1989). However, the lag does not necessarily have a straightforward interpretation in terms of the size and structure of the BLR. Computing the CCF's for the line response models shows, for example, that the lag depends on $\Delta\tau$, a point emphasized by Netzer (this volume). This is not surprising given that, as shown by Penston (1990), the CCF is the convolution of the continuum auto-correlation function with the 1-DTF. The lag also depends on the volume emissivity distribution but it is particularly sensitive to the shell thickness ratio, y (RP). It has been previously pointed out (Gaskell & Sparke 1986; Edelson & Krolik 1988) that we should expect the lag to be biased towards the inner regions of the BLR, since they respond more coherently. In fact, for $\Delta\tau \ll 1$, the CCF lag for our models is $\approx 1/y$, the light crossing time at the inner shell radius.

The inner radius will be similar to the effective emission radius as represented by R only if the shell is thin ($y \approx 1$). It follows that not only is the lag affected by the continuum light curve but, for thick shells, it underestimates the emission radius (by up to an order of magnitude in some models). This may partly explain the discrepancy, which has been reported for several objects (notably Akn120, Gaskell & Peterson [1987] and NGC5548, Netzer *et. al.*, [1990]), between the "size"

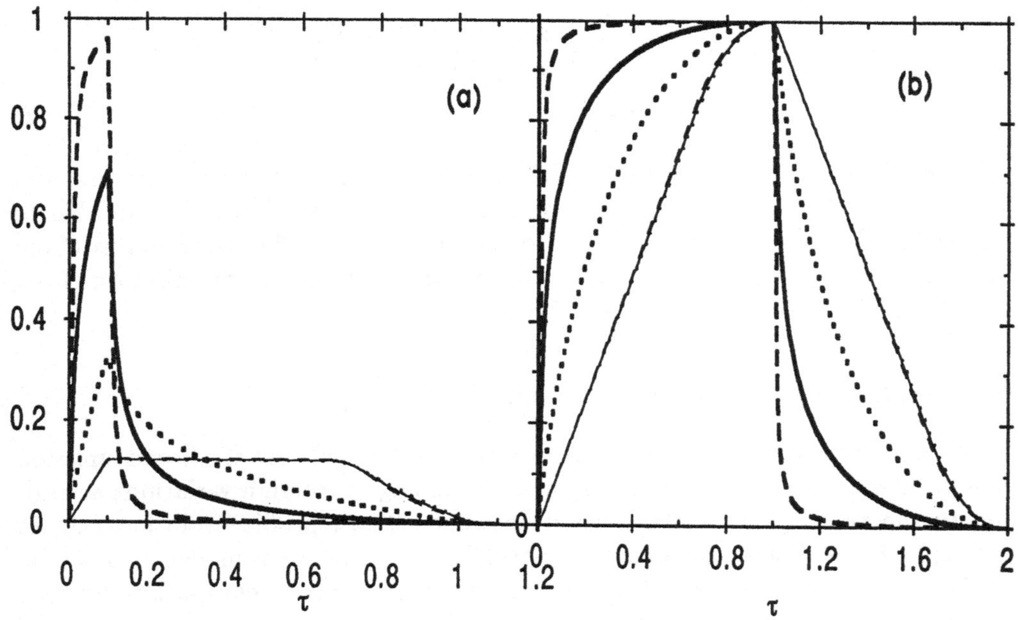

Fig. 1. Line response light curves for a box-car continuum pulse. The response amplitude relative to that of the continuum pulse $(A-1)$ is plotted against time in units of the shell light crossing time: **(a)** response curves for a pulse duration $\Delta\tau = 0.1$, shell thickness ratios $y = 100$ (thick lines) and $y = 1.5$ (thin lines) and volume emissivity distributions $\epsilon_V \propto r^{\beta}$, where $\beta = -2$ (dotted lines), -3 (full lines) and -4 (dashed lines); **(b)** as (a) but for $\Delta\tau = 1.0$.

of the BLR as estimated from the CCF and that deduced from the standard 'single cloud' photoionization model: we might expect that the BLR "size" deduced from the standard model will be close to R, since the calculation depends on inferring the ionization parameter from the relative intensities of various emission lines, whereas the CCF lags actually indicate the smallest radius at which line emission is produced.

3.3. The "size" of the BLR

It is evident that the lag does not give a straightforward indication of the "size" of the BLR, even given that our assumption of a spherical systemic geometry is correct. On the other hand, it is easy to show that the emission radius, R, is given *directly* by the centroid of the 1-DTF:

$$\tau_{cen} = \frac{\int_0^1 \tau T(\tau) d\tau}{\int_0^1 T(\tau) d\tau}$$

$$= \frac{1}{2y}\left(\frac{\beta+3}{\beta+4}\right)\left(\frac{y^{\beta+4}-1}{y^{\beta+3}-1}\right)$$
$$\equiv \tfrac{1}{2}\tfrac{R}{r_s} \qquad (4)$$

(c.f., RP). Of course, if the 1-DTF could be reconstructed from the line and continuum light curves to sufficiently high accuracy, we would also be able to determine the inner radius (from the break at $\tau = 1/y$), the outer radius (from the value of τ at which $T(\tau) = 0$) and the volume emissivity distribution (from the shape of the decay curve at $\tau > 1/y$).

3.4. Profile shapes and the 1-D transfer function

It is a well known consequence of the reverberation model that the broad emission line profiles evolve differently in response to ionizing continuum variations according to the velocity field of the cloud ensemble. Thus, profile evolution can, in principle, be used to distinguish between different kinematical models (e.g. radial outflow or Kepler rotation). However, we may also expect a relationship between the response light curve of a given emission line and the shape of its time-averaged profile. This arises because both the transfer function and the line profile depend on the volume emissivity distribution. Thus, for the type of model outlined in section 2, the profile shape is, in general †, given by

$$L_\lambda = constant\left[1 - \left(\frac{\max(c\Delta\lambda/\lambda_0, v_{min})}{v_{max}}\right)^{\eta}\right] \qquad 5$$

where $\eta = (\beta+3)/p - 1$, $c\Delta\lambda/\lambda_0$ is the velocity shift and v_{max} and v_{min} are, respectively, the maximum and minimum radial velocities (Robinson 1990). For a given velocity gradient, therefore, the profile wing is concave, linear or convex, depending on the radial distribution of ϵ_V (for $p = -0.5$, $\beta > -4, \beta = 0$ or $\beta < -4$, respectively). Examples of profiles for different values of β can be compared with the corresponding 1-DTF's (equation 3) in Fig. 2. Thus, models characterized by steep emissivity distributions (say, $\beta < -4$) produce broad, convex profiles and have highly peaked transfer functions which would result in a light curve showing a large amplitude, relatively rapid response to continuum variations. On the other hand, models characterized by flatter emissivity distributions (say $\beta > -4$) produce narrower, concave profiles and have flatter transfer functions implying a relatively slow response of small amplitude.

This relationship suggests an explanation for correlations between profile width and variability behaviour in some objects, notably NGC4151 (Ulrich *et. al.*, 1984; Clavel *et. al.*, 1987), which does not involve assigning the various lines to physically distinct components of emitting gas. The efficiency with which lines of different species, say CIVλ1550 and MgIIλ2800, are produced will be more or less sensitive, as the case may be, to the local physical conditions (radiation flux, gas density etc.) at any point in the cloud ensemble. In effect, therefore, each line has a unique

† The solution is logarithmic for $(\beta+3)/p = 1$, c.f., Capriotti *et. al.*, (1980)

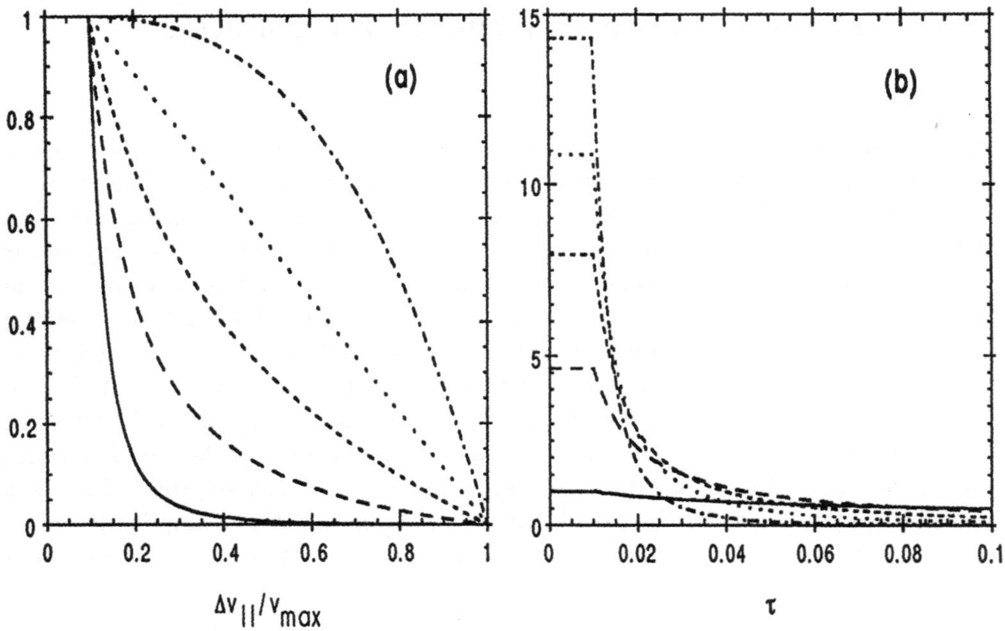

Fig. 2. Comparison of line profile and 1-D transfer function. (**a**) Line profiles for $p = -0.5$ and, in order of decreasing concavity, $\beta = -2, -3, -3.5, -4$ and -5. (**b**) Corresponding 1-D transfer functions (the steepest decay corresponds to $\beta = -5$).

volume emissivity distribution and its profile and variability behaviour will be correspondingly different from those of other lines. We might expect, for example, that CIV will have a relatively steep emissivity distribution compared to MgII, since it is particularly sensitive to the ionization parameter (RPB). It then follows that CIV would have a broader profile and exhibit stronger variability than MgII, as is observed.

We can envisage a similar relationship holding statistically for a given line within samples of AGN of similar luminosity. This would be the case, for example, if the radial distribution of emitting clouds differs widely from object to object. The slope of the volume emissivity distribution is then an important parameter in determining profile width and we would expect a statistical relationship between FWHM and variability behaviour; objects with broader lines would also show larger amplitude variability.

4. Line response simulations and profile evolution

Real continuum light curves are, of course, much more complex than the box-car functions discussed in the previous section. For example, short term 'flares' ($\sim$ 10 days) are commonly superposed on longer term ($\sim$ 1 year) transitions between 'high', and 'low' states. It is also possible that several flare events may occur within a BLR light crossing time so that at any given time, different regions of the BLR will be responding to different events, probably with different characteristic timescales. We have run computer simulations to study both the total emission line response and the evolution of the line profile in such a case (RPB; PRF). The input continuum light curve is based on the observed optical light curve of NGC4151 (Gill *et. al.*, 1984), which we suppose to be representative of ionizing continuum variability in a typical AGN. In fact, we use the square of the light curve to crudely allow for the possibility of a greater variability amplitude in the EUV (c.f., Penston 1986). The simulations are based on the discrete cloud model outlined in Section 2 and contain 3– 5×10^4 clouds.

4.1. Response light curves

The response simulations discussed here all have $y = 20$, $\beta = -3$ and the same radial outflow velocity field ($p = -0.77$, giving a factor 10 decrease in radial velocity over the shell) but differ in scale size: r_0 and r_s are adjusted to give values of R = 25, 50, 100, 200, 400 and 800 *ld*. Fig. 3 shows the corresponding response light curves during a 1000 day period (note that in each case, the response calculation takes into account the continuum 'history' for at least one shell light crossing time prior to the epoch at which the simulation starts). As expected, the line response decreases in amplitude and becomes 'smoother' as R is increased. The fall in amplitude can be understood in terms of the $\Delta\tau$ parameter of the response functions discussed in Section 3: as R (and hence r_s) increases $\Delta\tau$ decreases since the continuum light curve is fixed. An additional effect, which further reduces the response amplitude, is that as the BLR becomes larger, the total line luminosity at any epoch includes the responses to a greater number of different continuum events.

The CCF, when calculated for the 1000-day light curves shown in fig 3, exhibits a lag approximately equal to the light crossing time at the inner radius for each model (c.f., Section 3). However, since the CCF is influenced by the continuum light curve, the actual value of the lag depends on the length and position of the data 'window' from which the portions of light curves used in the calculation are taken (PRF90).

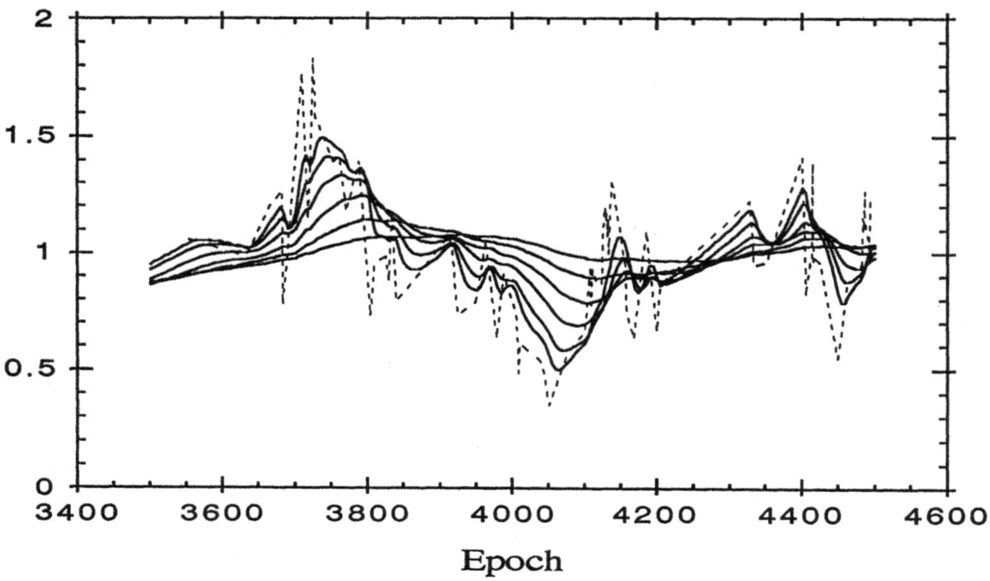

Fig. 3. Line response light curves from computer simulations. The input ionizing continuum is shown by the dashed line. The full lines are the response light curves for a series of models of increasing "size". In order of decreasing amplitude, R = 25, 50, 100, 200, 400 and 800 ld.

4.2. Reconstructing the 1-D transfer function

As already noted, much more information can be obtained from the 1- DTF than from the CCF. Blandford & McKee (1982) have shown that, in principle, the 1-DTF can be derived by Fourier transforming the line and continuum light curves. However, this method requires particularly good sampling and consequently has rarely been applied in practise (but see Moaz *et. al.*, 1990). To test its reliability with 'perfect' data (i.e., regularly sampled, infinite signal:noise) we have applied the method to our response simulation for the $R = 200\,ld$ model (Fig. 4). The result is disappointing. The Fourier transform method clearly fails to retrieve the model transfer function accurately. It would not be possible to gain a clear idea of the emissivity distribution or, indeed, distinguish unambiguously between a spherical cloud shell and a disk system. The maximum entropy deconvolution technique discussed by Horne appears to offer a better alternative. In any case, response simulations can be used to check such deconvolution techniques, assess the significance of the results and ultimately, model the reconstructed transfer functions.

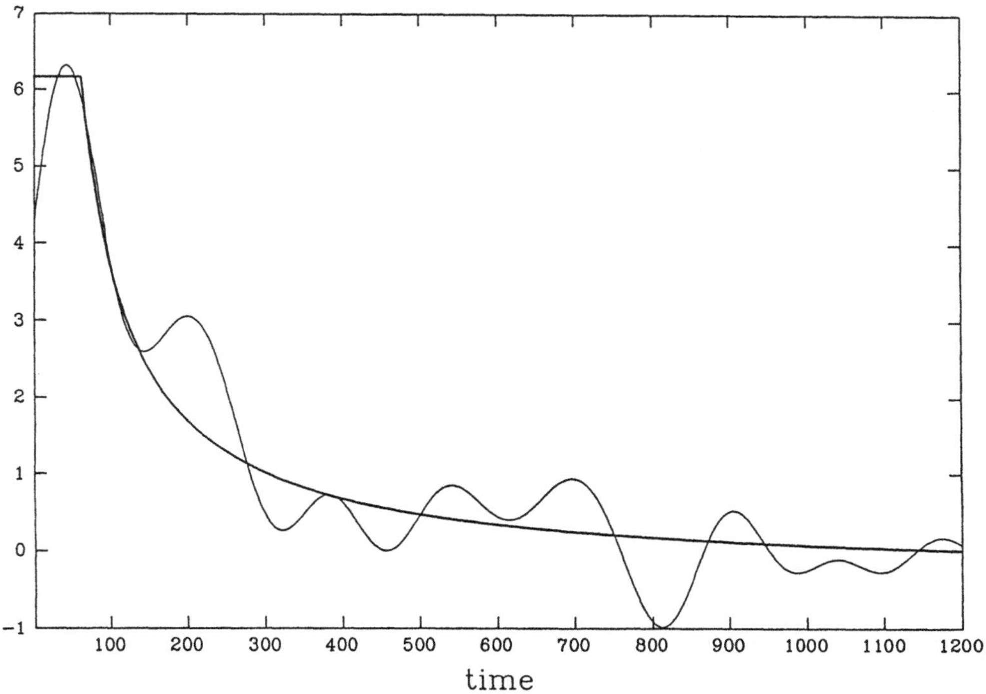

Fig. 4. 1-D transfer function recovered from the response light curve of the $R = 200\,ld$ model using the Blandford & McKee Fourier transform method. The analytical expression (eq. 3) is overplotted as the thick line.

4.3. Line–continuum correlations

The very simplest BLR photoionization model predicts a linear relationship between the luminosity of the continuum source and the resulting emission line luminosity from the surrounding gas ($L_{line} \propto L_s^\gamma$, where $\gamma \approx 1$). Statistically, this does in fact appear to be very nearly the case (Yee 1980; Shuder 1981), although in samples of radio loud quasars the CIV luminosity appears to be almost independent of the continuum (the so-called 'Baldwin Effect'; Baldwin *et. al.*, 1978). However, variability in individual AGN apparently produces a non-linear relationship between the line and continuum luminosities. This was first noted by De Bruyn (1980) who found that, typically $\gamma \approx 1/3$ for Hα. Although there are several possible interpretations, we note that departures from linearity are to be expected in the reverberation model due to light travel delays (Fig 5): depending on the scales involved, γ could lie anywhere between 0 (no line response; $\Delta t \ll R/c$) and 1 (instant line response; $\Delta t \gg R/c$). As an example, de Bruyn's result ($\gamma = 1/3$) can be accounted for purely in terms of reverberation in the $R = 100\,ld$ model. We would also expect that reverberation effects may contribute significantly to the statistical scatter in line-continuum correlations.

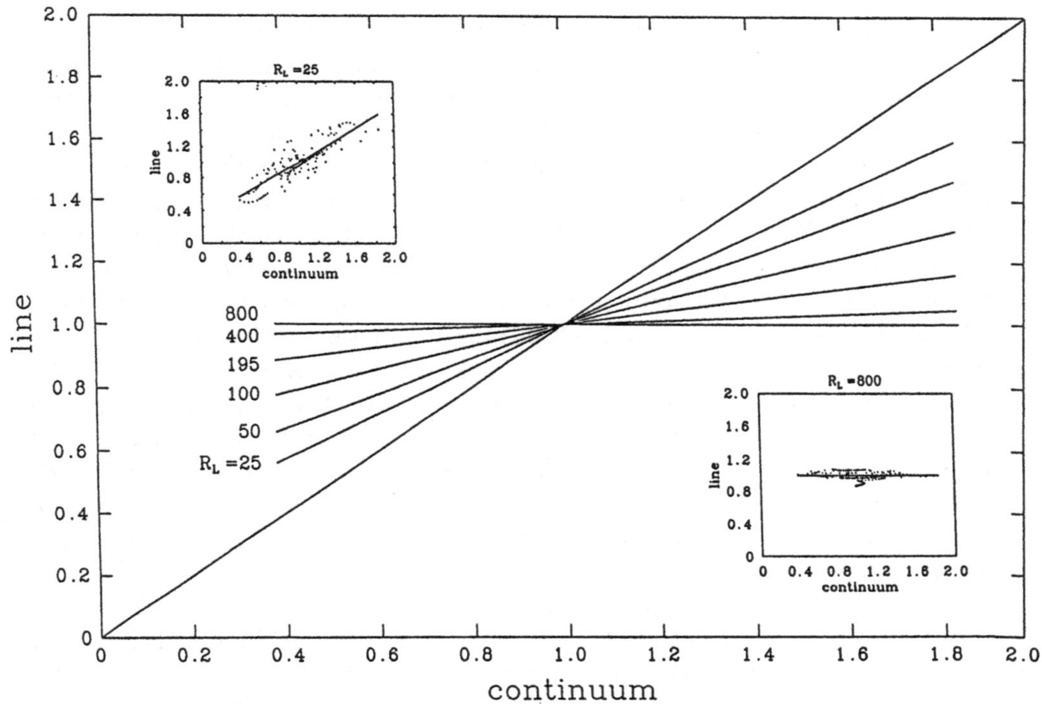

Fig. 5. Relationship between line and continuum fluxes for the series of response simulations where R increases from 25 to 800 ld. The lines are fits to the distribution of single-epoch points produced by each simulation. The insets show the distributions for the smallest and largest models, respectively.

4.4. Profile evolution

The evolution of the line profile in response to ionizing continuum variations is shown in Fig. 6 as an intensity map in the projected velocity–time (v-t) plane. Two models models are shown, both having radial outflow velocity fields but with $R = 25$ and $200\,ld$, respectively. The way in which the profile evolves can be understood by noting that for an infinitesimally thin shell of radius r the response propagates along the projected velocity axis from $-v(r)$ (blue wing) through to $+v(r)$ (red wing) at the constant rate $1 - ct/r$, thus forming a diagonal line in the v-t plane. The corresponding two- dimensional transfer function (2-DTF) is then simply a series of such lines corresponding to different radii (see contributions of Pérez *et. al.*,, this volume). The 2-DTF for a case similar to that considered here (i.e. spherical cloud system undergoing radial outflow) is presented by Pérez *et. al.*,. A comparison with Fig. 6 shows that the main features of the 2-DTF can be discerned in the underlying pattern of the profile evolution in the v-t plane.

It is evident from Fig. 6 that profile evolution in the radial outflow model is marked by several distinctive characteristics (these are discussed in more detail by RPB90 and PRF90). The most obvious is that the blue wing responds more rapidly and with greater amplitude than the red wing. The decrease in response amplitude from blue to red arises because the separation between any two light fronts corresponding to a fixed time interval decreases with time (i.e., as they 'propagate' through the shell; RP90; RPB90). A well known consequence of the greater variability of the blue wing (of course, if the direction of the flow is reversed—inflow instead of outflow— the red wing will be the more variable) is the appearance of transient profile asymmetries. In particular, since the blue wing is more closely correlated with the continuum, the profile tends to be blue asymmetric when the continuum is in a high state but red asymmetric when it is in a low state. Statistically, however, the distribution of asymmetries along the light curve tends to be skewed to the blue, a property which may be relevant to interpreting profile asymmetry distributions for statistical samples of AGN. Another characteristic of this model is that, as was first shown by Capriotti *et. al.*, (1982), continuum flares gives rise to much localized sub-structure in the profile (i.e., 'spikes' and 'bumps'). This sub-structure is naturally transient but in larger models (high R) features in the core of the profile (which is mainly produced by clouds close to the outer boundary of the shell) can be quite persistent, giving it a double-peaked appearance (see Fig. 6).

However, while many AGN do have asymmetric profiles, sometimes with prominent sub-structure, asymmetric *changes* in the profile are apparently rarely seen or are very weak. Since such events would provide an unambiguous indication of a radial velocity field, the fact that they rarely occur argues against models in which radial flow dominates. Nevertheless, it should be noted that the prominence of the evolutionary phenomena disucussed above depends on BLR size (R). Whereas the profile produced by the $R = 200\,ld$ model may remain asymmetric over several sampling epochs (separated by 50 day intervals), strong asymmetries are very short-lived in the $R = 25\,ld$ model and are thus rarely sampled, even with an interval of 10 days. Furthermore, partly as a result of undersampling the profile wing variations and partly because of the speed with which the entire system responds, the statistical distribution of asymmetries for small R models tends to be more symmetric than for large R models and may even be skewed to the red.

5. Conclusion

The calculations discussed in this paper show that the response of the broad emission lines to ionizing continuum fluctuations can be quite complex, even in the basic reverberation model. Nevertheless, there is little doubt that more complicated models will have to be considered. The most obvious next step is to calculate the response for non-spherical geometries, such as disks or bi-polar flows, and transfer functions for these cases are presented by Pérez *et. al.*, in this volume. A further complication is that, under the physical conditions prevailing in the BLR, most lines are not linear reprocessors and future response models should therefore incorporate a more sophisticated treatment of the line reprocessing efficiencies. However, it should not be forgotten that point-source reverberation may itself be an inadequate approximation. Diffusion of the ionizing radiation field by scatter-

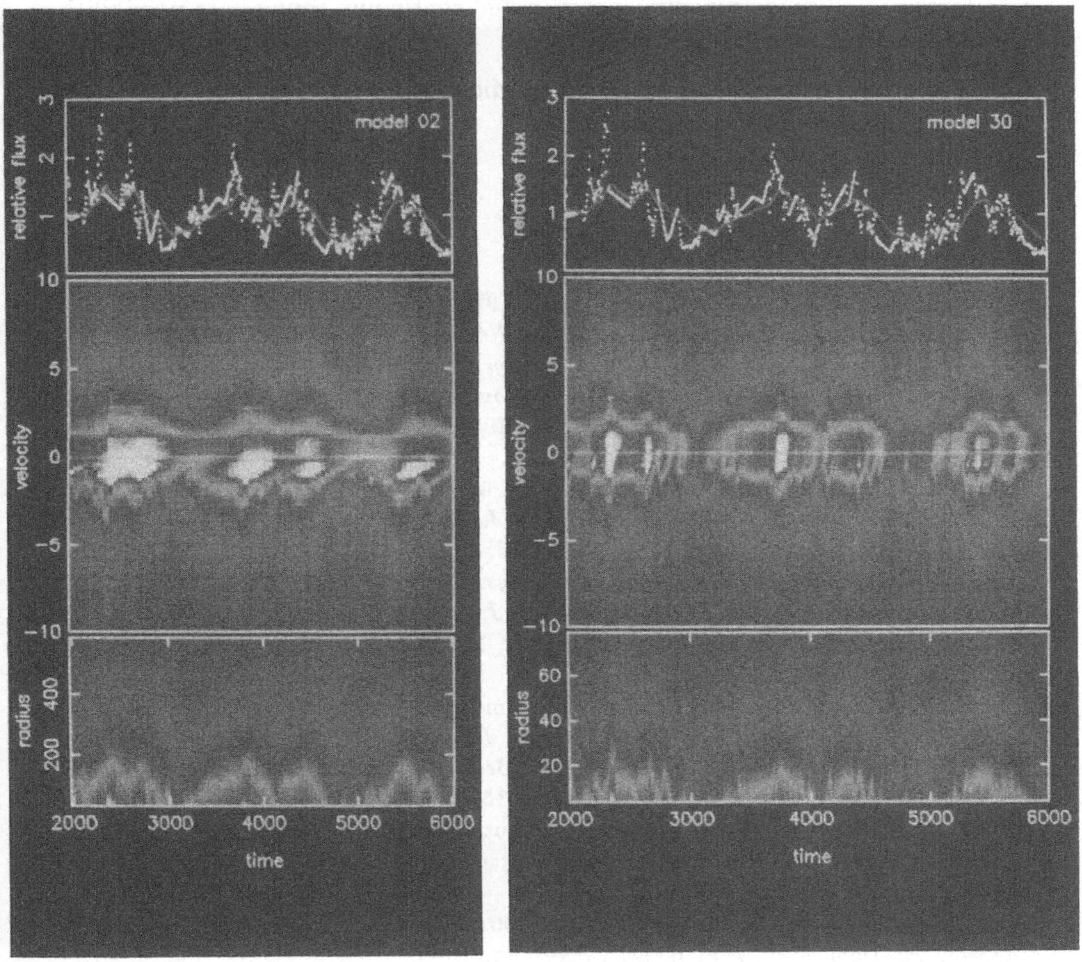

Fig. 6. Profile evolution for spherical cloud ensemble characterized by a radial outflow velocity field. **(a)** Model with $R = 200\,ld$. The profile is shown in the middle panel as an intensity map in the projected velocity–time plane over a 6000 day period (with a sampling interval of 50 days). The corresponding line (full line) and continuum (dotted line) light curves are shown in the top panel. The lower panel shows the radial luminosity distribution $(\Delta L/\Delta r)$ of the cloud ensemble. **(b)** As (a) but for the model with $R = 25\,ld$.

ing processes and models involving distributed continuum sources are possibilities which should not be ignored.

Acknowledgements: We are grateful to Michael Penston for many stimulating discussions on this and many other topics.

References

Bahcall, J.N., Kozlovsky, B. & Salpeter, E.E., 1972. *Astrophys. J.*, **171**, 467.
Baldwin, J.A., Burke, W.L., Gaskell, C.M., & Wampler, E.J., 1978. *Nature*, **273**, 431.
Blandford, R.D. & McKee, C.F., 1982. *Astrophys. J.*, **255**, 419.
Capriotti, E., Foltz, C. & Byard, P., 1980. *Astrophys. J.*, **241**, 903.
Capriotti, E.R., Foltz, C.B. & Peterson, B.M., 1982. *Astrophys. J.*, **261**, 35.
Clavel, J., *et. al.*, 1987. *Astrophys. J.*, **321**, 251.
de Bruyn, A.G., 1980. *Highlights in Astronomy*, **5**, 631.
Edelson, R.A. & Krolik, J.H., 1988. *Astrophys. J.*, **333**, 646.
Gaskell, C.M. & Koratkar A.P., 1989. *Astrophys. J.*, **345**, 637.
Gaskell, C.M. & Peterson, B.M., 1987. *Astrophys. J. Suppl.*, **65**, 1.
Gaskell, C.M. & Sparke, L.S., 1986. *Astrophys. J.*, **305**, 175.
Gill, T.R., Lloyd, C., Penston, M.V. & Snijders, M.A.J., 1984. *Mon. Not. R. astr. Soc.*, **211**, 31.
Maoz, D., Netzer, H., Mazeh, T., Beck, S., Almoznino, E., Leibowitz, E., Brosch, N., Mendelson, H. & Laor, A., 1990. *Astrophys. J.*, in press.
Netzer, H., Maoz, D., Laor, A., Mendelson, H., Brosch, N., Leibowitz, E., Almoznino, E., Beck, S. & Mazeh, T., 1990. *Astrophys. J.*, **353**, 108.
Penston, M.V., 1986. In *Physics of Accretion onto Compact Objects*, eds. Mason, K.O., Watson, M.G. & White, M.E., Springer Verlag.
Penston, M.V., 1990. RGO preprint.
Pérez, E., Robinson, A., & de la Fuente, L., 1990. *Mon. Not. R. astr. Soc.*, submitted.
Robinson, A., 1990. *Mon. Not. R. astr. Soc.*, submitted.
Robinson, A. & Pérez, E., 1990. *Mon. Not. R. astr. Soc.*, **244**, 138 .
Robinson, A., Pérez, E. & Binette, L., 1990. *Mon. Not. R. astr. Soc.*, **246**, 349.
Shuder, J.M., 1981. *Astrophys. J.*, **244**, 12.
Ulrich, M.H., *et. al.*, 1984. *Mon. Not. R. astr. Soc.*, **206**, 221.
Yee, H.K.C., 1980. *Astrophys. J.*, **241**, 894.

An Atlas of 2-dimensional Transfer Functions of the Broad Line Region

Enrique Pérez [1], Andrew Robinson [2] and Luis de la Fuente [1]

[1]Instituto de Astrofísica de Canarias and Depto. de Astrofísica, Universidad de La Laguna, 38200 La Laguna, Tenerife, Spain
[2]Institute of Astronomy, Madingley Rd., Cambridge CB3 0HA, U.K.

1. Introduction

The time dependent reprocessing of ionizing continuum photons into emission line photons in the Broad Line Region (BLR) is generally assumed to be well represented by an equation of the form

$$L(t) = \int_0^t T(t-\tau)C(\tau)d\tau \tag{1}$$

where $L(t)$ is the integrated intensity in an emission line at time t, $C(\tau)$ is the continuum at some previous epoch $\tau \leq t$ and T is the transfer function, which maps C into L. The transfer function T incorporates the information about the geometrical and kinematical properties of the BLR which we want to find out. Given the observables $L(t)$ and $C(t)$[1], the problem becomes one of inverting eq.(1) to find T. This so called "reverberation mapping" of the BLR was initially proposed by Blandford and McKee (1982), which used standard Fourier analysis to invert the transfer equation.

It is only during the last couple of years that light curves of sufficient quality have become available (e.g. Maoz et al., 1990 and references therein to the Wise Observatory monitoring programme; Clavel et al., 1990; Peterson et al., 1990) to carry out a one dimensional inversion to obtain $T(t)$ (Maoz et al., 1990). But the 1D transfer function (TF) does not contain enough information about the BLR; for example, different types of geometry can project onto similar 1D TF and, most importantly, it does not provide any clues about the kinematics. It is thus necessary to gather long trains of high signal/noise, high spectral resolution line profiles to invert the two-dimensional (velocity,time) problem.

[1] We do not really observe the ionizing continuum, but we have to assume that the observed continuum varies with time in a similar fashion that the ionizing continuum.

We note that when $C(t)$ takes the form of a pulse $\delta(t)$, equation (1) simplifies to give $L(t) = T(t)$. In this way, when modelling transfer functions we can circumvent the inversion problem and directly produce the TF by using a single continuum pulse propagating through the BLR.

For the last two years we have been engaged on a theoretical project to investigate the structure and dynamics of the Broad Line Region (Robinson and Pérez 1990; Robinson, Pérez and Binette 1990; Pérez, Robinson and de la Fuente 1990a; Pérez, Robinson and de la Fuente 1990b; Robinson 1990).

As part of this work, we present an atlas of 2-D TF for simple combinations of geometry and velocity field, which can be used to compare with the data now starting to become available. We will show here how the geometry, size and kinematics of the BLR can be obtained in a straightforward way from an analysis of the 2-D TF.

This contribution should be regarded as a progress report. A more comprehensive presentation and analysis will be given elsewhere.

2. The atlas

To generate the TF we have used the computer code OBFRAM (Robinson 1990; Pérez, Robinson and de la Fuente 1990a). The BLR is described by an ensemble of clouds occupying a global geometry with an inner, r_o, and an outer radius, r_s, and characterized by a radially dependent velocity field. The radii are the same for all the models, $r_o = 10$ light-days (ld) and $r_s = 100$ ld. The volume emissivity is given by $\epsilon(r) = \epsilon_o(r/r_o)^\beta$, where ϵ_o is the power emitted by a cloud at the inner radius. In the models we present here $\beta = -3$. The minimum velocity for each model is always chosen to give a maximum velocity of 10^4 km s^{-1}, the radial dependence being of the form

$$v(r) = v_o \left(\frac{r}{r_o}\right)^p \tag{2}$$

Three different global geometries are considered: spherical, discoidal and biconical. At a given time, the spherically symmetrical light front in the BLR rest frame is seen as a parabolic light front by a distant observer, given by the equation

$$t = r\,(1 - cos\theta) \tag{3}$$

where we have taken the speed of light c=1. Figure 1 shows an overview with four different light fronts as seen by an observer looking from below.

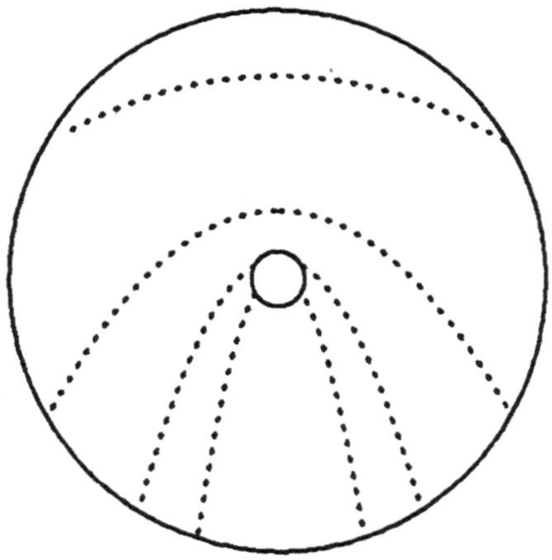

Fig. 1. Four parabolic light fronts as seen by a distant observer looking from below

2.1. Spherical geometry

We consider six different velocity fields of the form given by eq.(2), with the values of $p = -2, -1, -0.5, 0.5, 1, 2$. The corresponding TF are shown in figure 2. It is very clear that the envelopes which define the 2-D TF change with the radial velocity dependence. We now discuss the functional forms of these envelopes and how they relate to the BLR parameters.

The velocity component projected to the line of sight, $v^{\|}$, is given by

$$v^{\|}(r) = v(r)\, cos\theta = v_o \left(\frac{r}{r_o}\right)^p cos\theta \tag{4}$$

where θ is the angle between the radius vector of a given point and the line of sight. Substituting eq.(3) into eq.(4) we have

$$v^{\|}(r) = v_o \left(\frac{r}{r_o}\right)^p - \frac{v_o}{r}\left(\frac{r}{r_o}\right)^p t \tag{5}$$

For a fixed value of the radius, $v^{\|}$ is a linear function of time; then the evolution of the light front at a given radius will project onto a straight line starting at $t = 0$, $v^{\|} = -v_o \left(\frac{r}{r_o}\right)^p$ and ending at $t = 2r$, $v^{\|} = v_o \left(\frac{r}{r_o}\right)^p$, ($2r$ being the light travel time for the radius r. Negative velocities are approaching to and positive receding from the observer). All these lines will be enclosed by the two lines projecting the evolution of the light front at the inner and outer radii.

Note that from $t = 2r_o$ to $t = 2r_s$ the end point of these straight lines is given by the leading point of the light front, which evolves exactly as the radial dependence of the velocity field. Thus the envelope formed by these points, i.e. the evolution of the red wing of the emission line profile after the light crossing time of the inner radius, gives the precise kinematics of the BLR.

For $p \geq 0$, the lower envelope is given by the evolution of the light front at the outer radius, for $0 \leq t \leq 2r_s$; while the upper envelope is given by the

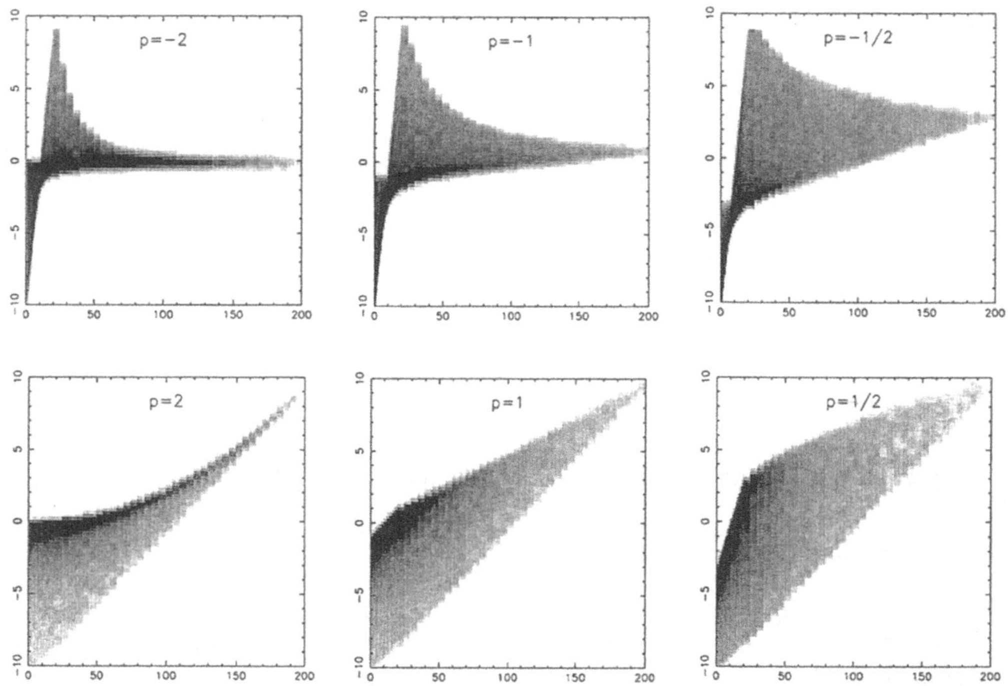

Fig. 2. 2-D Transfer Functions for six different radial velocity fields in a spherical geometry.

evolution at the inner radius, for $0 \leq t \leq 2r_o$, and the form of the velocity field for $2r_o \leq t \leq 2r_s$. For $p < 0$ the upper envelope is as for the previous cases, but now the maximum approaching velocity is not always given by the outer radius, as the sense of the velocity field is reversed.

2.2. Disc geometry

In figure 3 we present six models with a flattened configuration. For all the cases the velocity field is Keplerian rotation, $v(r) = v_o(r_o/r)^{1/2} sin\, i$, where the inclination i is the angle of the disc axis to the line of sight. The top three models correspond to a disc half height of 5 ld and inclinations of 0°, 45° and 90°, respectively. The bottom three models are as above but for a thicker disc, of half height 20 light days. For the sake of simplicity, in the following we will consider an edge on disc, $i = 90°$.

As the velocity vector is now tangential to the radius vector, the projected velocity is now given by

$$v^{\parallel}(r) = v(r) sin\theta = v_o \left(\frac{r_o}{r}\right)^{1/2} sin\theta \tag{6}$$

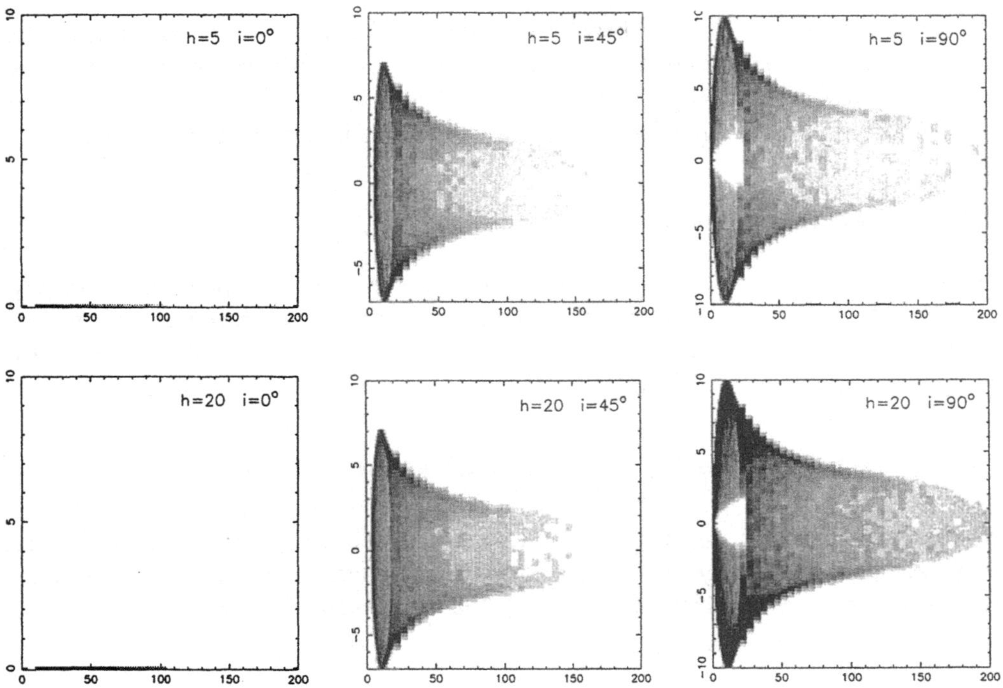

Fig. 3. 2-D Transfer Functions in a disc geometry.

Clearly, for the face on discs $v^{\|} = 0$. By substituting $sin\theta$ from eq. (3) and re-arranging it is easy to see that, for a fixed radius (i.e. for a ring BLR), the TF is an ellipse in $v^{\|}$ and time. A disc can be regarded as a succession of rings, and thus its 2-D TF can be formed by a succession of ellipses with their centres in $(v^{\|}, t) = (0, r)$.

The envelope in fig. 3 is defined by three functions. Up to $t \simeq 13.3$, $v^{\|}_{max}$ is given by $v^{\|}(r_o)$. We obtain the equation of the envelope by substituting for $sin\theta$ from eq. (3) and finding the maximum of the resulting function with respect to r, at a given time. This gives the radius at which the projected velocity is maximum, $r_{v^{\|}_{max}} = 0.75t$, which substituted into eq. (6) gives

$$v^{\|}_{max}(t) = v_o\left(\frac{32r_o}{27t}\right) \tag{7}$$

This is valid until $t = 133.3$. From there to $t = 200$, $v^{\|}_{max}$ is given by the velocity at the outer radius, $v^{\|}(r_s)$.

The inclination of the disc acts on the 2-D TF as a 'reduction factor'. The ellipse corresponding to each radius has both semi-axes reduced by a factor $\propto sin\,i$. We can see this in fig. 3, for $i = 45°$, where the ellipse corresponding to r_o now has $v^{\|}_{max}(r_o) = v_o\,sin45 = \pm 7.07$, while it starts at $t = 0.5\,r_o sin45 = 3.53$. This is the reason why the 'clapper' in the bell shaped TF for $i = 90°$ disappears for $i = 45°$, as the response of the ring representing the outer radius now starts at $t = 0.5 r_s sin45 = 35.3$.

We also note that the effect of the disc thickness is not readily noticeable.

2.3. Biconical geometry

We can regard the biconical geometry as a limited case of spherical geometry, where we 'cut' thru the sphere with some opening angle 2α. In general, we will be looking at this structure from an angle i with respect to the axis of the cone. Now the projected velocity is given by

$$v^{\parallel}(r) = v_o \left(\frac{r}{r_o}\right)^p (cos\, i\; cos\theta + sin\, i\; sin\theta\; sin\phi) \tag{8}$$

where i has a fixed value, ϕ is the azimuthal angle about the cone axis and θ can only take a limited range of values in two ranges, given by $i - \alpha \leq \theta \leq i + \alpha$ for the near side cone and $180 + i - \alpha \leq \theta \leq 180 + i + \alpha$ for the far side cone.

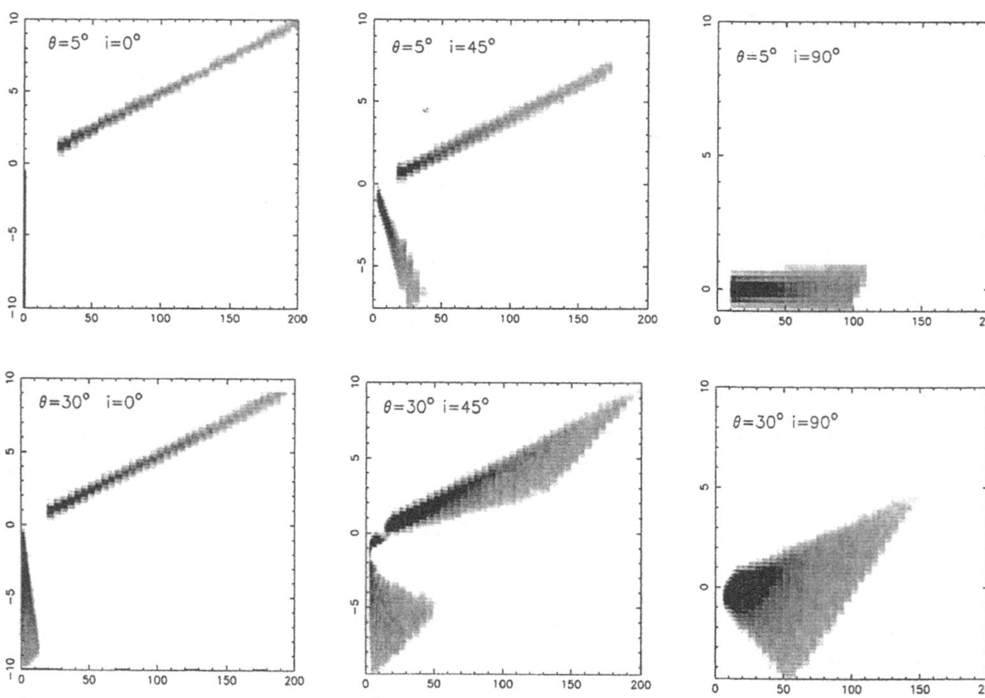

Fig. 4. 2-D Transfer Functions in a bi-conical geometry with a radial velocity field.

In figure 4 we show six different transfer functions for a biconical geometry with a radial velocity field $p = 1$. The three in the top row correspond to a cone semi-opening angle $\alpha = 5°$ and viewing inclination angles of $i = 0°, 45°$ and $90°$. The bottom row is similar but for a wider cone of semi-opening angle 30°. Note that these cases correspond to different cuts thru the spherical BLR of $p = 1$ in figure 2 above.

2.4. Randomly oriented orbits

We consider here the case of a spherical geometry in which the matter travels in Keplerian orbits whose orientation to the line of sight are randomly uniformly distributed. This is illustrated in figure 5. This case exhibits properties of both spherical geometry with radial velocity and disc geometry with Keplerian velocity. In fact, for an orbit at radius r the projected velocity is given again by eq.(6) above, with an extra $sin\, i$ for the inclination of the orbit. But the evolution of $v^{\|}_{max}$ is now different.

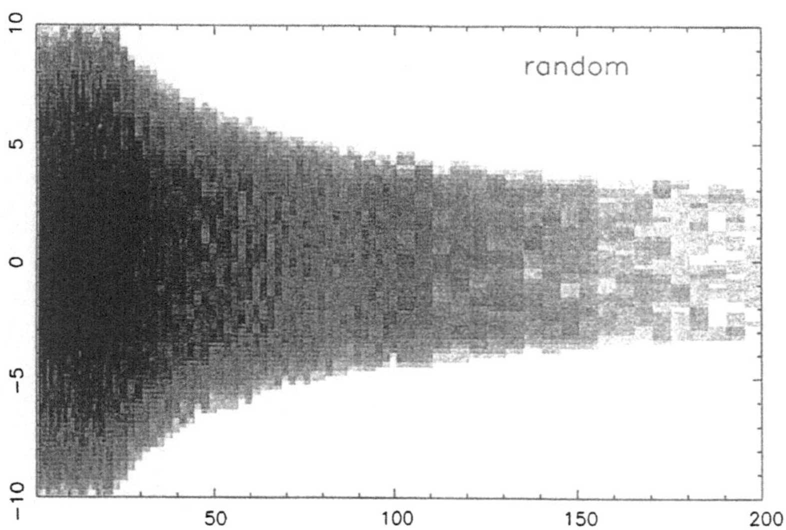

Fig. 5. 2-D Transfer Function for randomly oriented orbits in a spherical geometry.

Until the pulse has traveled for the light crossing time of the inner radius, $t_{r_o} = 2r_o$, we see components of the projected velocity at all possible velocities ($t = 20$ in fig. 5). Beyond this point the maximum projected velocity is sampled by the leading point of the light front, just as in the case discussed above for the upper envelopes in fig. 2. But now the radial dependence of the velocity field is Keplerian and thus $v^{\|}_{max} \propto (t/(1 - cos\theta))^{1/2}$. This is just what we can see as the envelope in fig. 5.

References

Blandford, R. and Mckee, ., 1982,*Astrophys. J.*, **255**, 419.
Clavel, J., *et al.* (56 co-authors), 1990, *Astrophys. J.*, in press.
Maoz, D., Netzer, H., Mazeh, T., Beck, S., Almoznino, E., Leibowitz E., Brosch, N., Mendelson, H., and Laor, A., 1990, *Astrophys. J.*, in press.
Pérez E., Robinson, A., and de la Fuente, L., 1990a, *Mon. Not. R. astr. Soc.*, submitted.
Pérez E., Robinson, A., and de la Fuente, L., 1990b, *Mon. Not. R. astr. Soc.*, submitted.
Peterson, B.M., *et al.* (63 co-authors), 1990. *Astrophys. J.*, in press.
Robinson, A., 1990, *Mon. Not. R. astr. Soc.*, submitted.
Robinson, A. and Pérez, E., 1990, *Mon. Not. R. astr. Soc.*, **244**, 138.
Robinson, A., Pérez, E., and Binette, L., 1990, *Mon. Not. R. astr. Soc.*, **246**, 349.

Limits on the Expansion of the Radio Source OQ 208 in Mkn 668

A. Ger de Bruyn

NFRA, Dwingeloo, The Netherlands

The radio source OQ208 (1404+28) is well-known amongst VLBI radioastronomers because its compactness makes it suitable as a calibrator. I have for a long time had an interest in this source because of its peculiar giga-hertz peaked spectrum, the possible detection of stimulated radio recombination lines and the fact that it is also a Seyfert galaxy (Mkn 668, z=0.077, 16th magnitude) with complex very broad optical emission lines. It could therefore learn us something about the relationship and possible interaction between the radio source and the broad emission line region.

Most compact radio sources show variability in their flux density, which is often linked with motion and/or expansion. OQ208, however, had not shown any variations in the 6-year period that it was monitored with the Algonquin telescope soon after its discovery as bright radio source in the OHIO surveys (1967-1973). I therefore began to systematically monitor the flux density of the radio source with the WSRT at a frequency of 5 GHz in 1979 at intervals of about 3-6 months. In the last decade I have also gathered flux data at frequencies of 1.4, 0.6 and 0.3 GHz using the WSRT. Recently I have mapped the source with high dynamic range at frequencies of 5, 1.4 and 0.3 GHz. The 6 cm map show a faint halo with a total flux density of 6 mJy and an angular size of about 20" (corresponding to about 30 kpc at the assumed distance of 300 Mpc (H=75)). If this halo is optically thin emission with a spectral index of α = -1.0 the halo is responsible for about 50% of the observed flux density at 92 cm. On the basis of this assumption I have constructed the true core spectrum.

The striking, and very interesting, property of the spectrum of OQ208 is that the core has a very steep falloff at both low frequencies, down to at least 0.3 GHz, and on the optically thin side of the spectrum. The peak lies around 5 GHz. The steep falloff at high frequencies could imply that there is no fresh injection of relativistic plasma in the core of this object. Furthermore, the steepness in the optically thick part of the spectrum – which is almost certainly due to synchrotron self-absorption because the spectrum appears to be straight and the brightness temperature is about 10^{11} K – suggests that the source is sharply confined. That it, it is not likely to consist of a long jet or a series of components peaking at different frequencies. The VLBI data barely resolve this source as stated earlier.

In interpreting the (lack of) variability data assembled on OQ208 one has to make a distinction between frequencies below and above the peak which is at about 5 GHz. The low-frequency data have shown no variation, the best limits being set by WSRT data at 1.4 GHz over a 11 year period from 1977 through 1988. Adopting a conservative limit of 2% on the accuracy of the flux determinations at 1.4 GHz I conclude that at 1.4 GHz the source varies less than 0.2% per year. Limits at 0.3 and 0.6 GHz are poorer but consistent with no variations. At 5 GHz, where the source becomes optically thin, the source has decreased by about 10% in the last 10 years, most of it in the last 5 years.

I interpret the absence of any flux increase at frequencies below the peak as evidence that the solid angle subtended by the radio source has not increased at all. The flux density at the optically thick frequency of 1.4 GHz must come from the outer skin of the radio source. If the solid angle of the source scales with radius^2 the lack of any increase of the source flux at 1.4 GHz ($dS/S < 0.002$/year) can be used to set a limit on the fractional increase of radius ($dr/r < 0.001$/year). This value depends only slightly on the change of effective value of the magnetic field in the region from where most of the radiation comes from. At the distance of 300 Mpc the observed dimension of the source is of the order of 1 parsec or less. The limit on the expansion then translates to a velocity of less than 1000 km/sec! This is a limit much better than can be measured using VLBI proper motion measurements. If we apply a similar reasoning to the radio nucleus of 3C84, which in the past 10 years has increased its flux density by 5% per year at 1.4 GHz, where the source is very optically thick, one predicts an expansion velocity of about 0.25c, close to the observed value using VLBI studies.

The low value of the expansion velocity of the radio source OQ208 in the Seyfert galaxy Mkn668 is in striking contrast with the very high velocities (up to 10000 km/sec) inferred for the gas in the broad line region. The thermal pressure in the BLR, as well as the dynamic pressure of clouds moving around in the BLR, is much higher than that in the radio source, assuming equipartition, hence the radio source could not confine the BLR gas, nor prevent its outward motion. This therefore is an argument against spherical outflow of the BLR gas. The same argument would naturally apply to a hot intercloud medium often invoked to confine the BLR clouds. In fact, it is hard to see how, at least in OQ208, there could be a confining medium in the BLR. Solutions to the problem of the coexistence of the BLR and the radio source must therefore be found in the geometrical configurations and relative disposition of these two components.

A full account of this work, with supporting material and relevant references will be submitted to Astron. & Astrophysics.

BLR Models and the L-M Relation for AGNs

Hagai Netzer

School of Physics and Astronomy and The Wise Observatory, Tel Aviv University and Astronomy Department, Ohio State University, Columbus

Abstract: This paper discusses the L-M relation for AGNs, in view of the currently available data and the present understanding of the theory of active galactic nuclei. I review all data sets that are, in my opinion, reliable enough to deduce a *cross correlation size* for the BLR (only 5 objects) and discuss some important information about the gas distribution that can be extracted from them. An apparent contradiction between different observations of NGC 5548, and the transfer function for the NGC 4151 $H\beta$ line, are the best evidences for the existence of a thick geometry in those objects. The complications due to the non-linear response of several emission lines are also discussed, and a preliminary L-M relation is suggested.

1. Existing data sets

In my opinion, there are only 5 AGNs with good enough line and continuum light curves to measure a statistically significant lag from the peak of the the Cross Correlation Function (CCF). Four of those are the result of extensive monitoring programs, from the ground and space, with good sampling and absolute flux calibration. The fifth, Akn 120, is somewhat different, since the time lag, in this case, was obtained by adding up the results of several observing seasons, each with a marginal time sampling. Most of the work on this source is by Peterson and collaborators (see Peterson 1988) who produced some impressive results and demonstrated the importance of long monitoring efforts. This galaxy will not be discussed here any further.

1.1. The NGC 5548 data set (optical and UV)

This is, no doubt, the best data set obtained so far. The IUE campaign, in 1989 (see J. Clavel contribution) was extremely successful, resulting in reliable lag measurements for several lines. The strong ultraviolet lines of $L\alpha$ and $CIV\lambda1549$ lag the ultraviolet continuum by about 10 days, and there is a clear tendency for lower ionization lines to show longer lags. This is described in Clavel et al. (1990). The parallel optical monitoring (Peterson et al 1991) was as successful, giving a very significant time lag between the $H\beta$ and the optical continuum, of 18 ± 2 days (the uncertainty is due to the determination of the CCF peak). This is very different from the time lag obtained for $L\alpha$ and $CIV\lambda1549$. It is also in conflict with the Wise Observatory result for this source (see below) of 7 ± 3 days, obtained a year earlier. This apparent contradiction is a key for the understanding of the BLR geometry, as explained in section 2.2 below.

1.2. The NGC 4151 data set (UV)

This is the least luminous AGN among the five considered here. Despite of the vast amount of telescope time invested in the monitoring of this source, it is only recently that reliable measurements of the time lags of the ultraviolet lines, with respect to the ultraviolet continuum, have been measured (Clavel et al. 1991). Most UV emission lines lag the UV continuum by about 4 days, and there is no clear evidence for a dependence on the degree of ionization. The Wise observatory data for $H\alpha$ and $H\beta$ are discussed below.

1.3. The F-9 data set (UV)

This is the most luminous of the 5 objects for which the time lag of the ultraviolet lines is well known (Clavel et al. 1989). The object has faded by a very large factor (30!) over the observing period, and a reliable lag measurement (about 150 days for the strong UV lines) was possible, despite of the poor sampling, because of the simple continuum behavior over that period.

1.4. The Wise Observatory data set (optical)

The "Wise observatory project" was an extensive observational effort, during 1987-1988, to monitor 14 AGNs from the ground, using the Wise Observatory 1m telescope. The campaign lasted for 8 months and took some 80% of the telescope time during that period. All the observers in Israel (5 faculty and 4 graduate student: E. Almoznino, S. Beck, N. Brosch, A. Laor, E. Leibowitz, H. Mendelson, D. Maoz, T. Mazeh and H. Netzer) took part in this effort. It resulted in some 600 spectra, with an unprecedented temporal resolution of 4-5 days for most objects.

The observational method and the reduction procedure for the project are described in Maoz et al. (1990). They involve simultaneous observations of near-by stars, some 4-8 arc-min away from the AGN under study. The stars, used later as local standards, were observed through the same long-slit aperture as the target source. This is possible due to the large scale of the telescope (30 arcsec/mm) at its f/7 spectroscopic mode. The wide slit (20 arcsec), linear detector (RCA CCD) and special reduction routine, enabled absolute flux calibration, over the spectral range of 4500-7000Å, during times of full moon and bad weather conditions.

Three out of the 14 AGNs monitored showed clear signs of line and continuum variations. These are discussed below.

1.4.1. Mkn 279

This source is somewhat more luminous than NGC 5548. Its 1988 monitoring at the Wise Observatory showed a clear correlation between the $H\alpha$ and $H\beta$ light curves and the optical continuum, with a best measured lag of 12 days. More details are given in Maoz et al. (1990).

1.4.2. NGC 5548

This source was constant in the beginning of the campaign, but later showed a small amplitude continuum pulse of about 40 days in duration. The $H\alpha$ and $H\beta$ response to this pulse have been analyzed and gave a significant CCF peak, at 7 days, for $H\beta$ but inconclusive results for $H\alpha$. The uncertainty in the $H\beta$ lag, assuming a thin-shell geometry for the BLR (see Maoz and Netzer 1989 for the analysis) is about 3 days. More details can be found in Netzer et al. (1990).

The short lag found for $H\beta$ in 1988, seems to be in conflict with the much longer lag found by Peterson et al (1991) a year later. This discrepancy is discussed below.

1.4.3. NGC 4151 and the transfer function

NGC 4151 is the third source found to vary during the Wise Observatory campaign. The observations and results are described in Maoz et al. (1991).

The Wise data for NGC 4151 are good enough to attempt, for the first time, a full deconvolution of the Balmer line light curves, in order to derive their transfer functions. Fig. 1 shows the transfer function for $H\alpha$ and a model to the line light curve.

The shape of the transfer function reveals several important facts about the distribution of the $H\alpha$ emitting gas in this source. First, there is some BLR material along the line of sight to the source, as is evident from the strong peak of the transfer function at zero time. Second, the gas is distributed in a thick shell, extending from a very small radius up to about 15 light days. Other thick geometries, such as an edge-on thick disk, are also consistent with the above transfer function. The cross correlation of the $H\alpha$ and $H\beta$ light curves with the optical continuum light curves give a very significant peak at a lag of 9 days. The interpretation of this result, in view of the thick geometry of the BLR, is discussed below.

The Wise Observatory data for NGC 4151 is good enough to test the idea of radial gas motion in the BLR, by correlating the light curves of the blue and the red wings of the Balmer lines. The two are found to be identical, within the

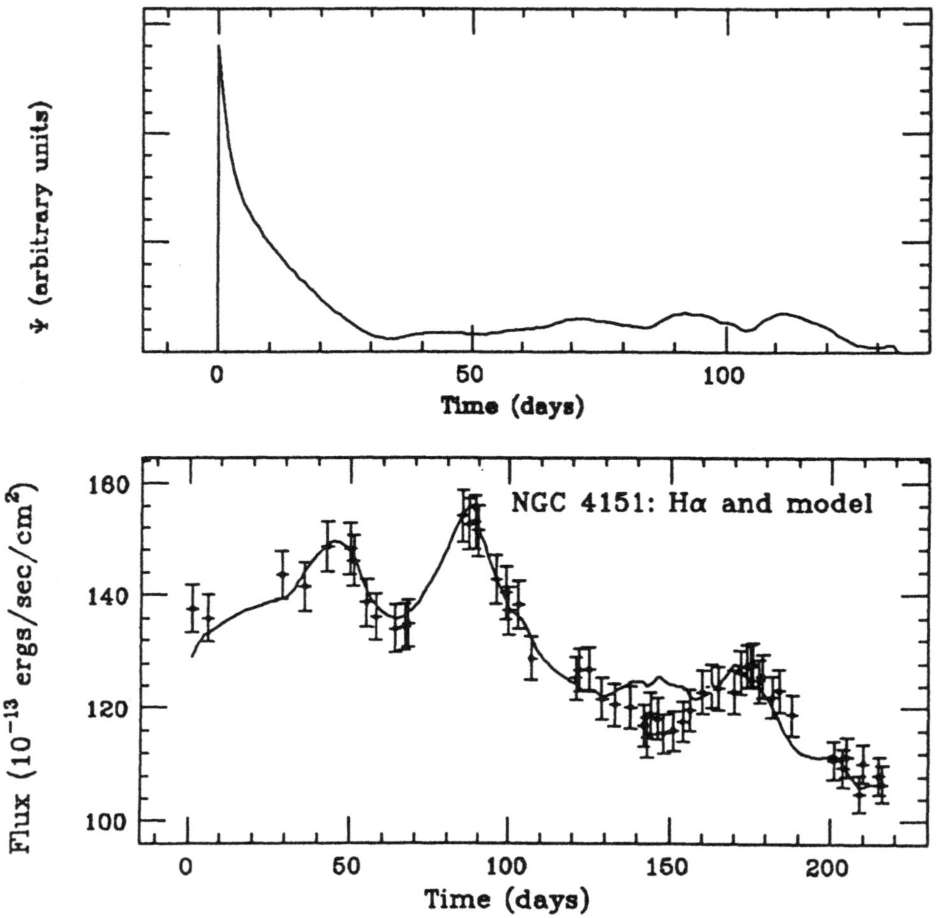

Fig. 1. Top: Transfer function for NGC 4151, obtained from the Wise Observatory data using a maximum entropy deconvolution. Bottom: A model emission line light curve obtained from the transfer function, on top of the $H\alpha$ light curve

observational uncertainties, indicating no radial motion of the broad line gas. For more details see Maoz et al. (1991)

2. Modelling the BLR

The minimal requirement for a successful monitoring campaign is the ability to estimate the "size" of the BLR. There are, however, several different ways to define this "size". Perhaps the most meaningful one is the *emissivity-weighted radius*, defined, for each line, by

$$r_{\mathrm{av}} = \frac{\int_{r_{\mathrm{in}}}^{r_{\mathrm{out}}} r dE(r)}{\int_{r_{\mathrm{in}}}^{r_{\mathrm{out}}} dE(r)} , \tag{1}$$

where $E(r)$ is the cumulative line emission up to distance r. Another "size" is obtained from cross-correlating the line and continuum light curves. Unfortunately, this *cross-correlation size*, which is the one easiest to measure, may be rather different from r_{av}, especially for thick geometries. Moreover, it depends on the nature of the continuum variability, as explained below.

Another potential difficulty is the relatively short crossing time of the broad line clouds. This *dynamical time* is given approximately by

$$t_{dyn} = \frac{r_{av}}{< v >} \simeq 30L_{46}^{1/2} \ years, \tag{2}$$

where L_{46} is the ionizing luminosity in units of 10^{46} $erg\ s^{-1}$. For NGC 5548 it is about 3 years and for NGC 4151 about one year.

We shall now assume that the dynamical time is long enough, and the recombination time short enough, not to affect the interpretation, and proceed to examine the consequences of different pattern of continuum activity.

2.1. Thin-shell, thick-shell and the case of NGC 5548

The apparent contradiction between the cross-correlation sizes obtained for $H\beta$ in 1988 (Wise Observatory data, 7 light-days, see Netzer et al 1990) and 1989 (18 ligh-days, Peterson et al. 1991) is one of the most important clues for understanding the gas distribution in this source. A similar apparent contradiction has recently pointed out by Netzer and Maoz (1991) who have examined the ultraviolet observations obtained by Clavel et al. (1990). The ultraviolet continuum light curve, during 1989, shows three distinct "events" (pulses), lasting for 70, 86 and 52 days, respectively. A careful examination of the lags of the strong lines of $L\alpha$ and $CIV\lambda1549$, with respect for the continuum, reveal a clear dependence of the lag on the duration of the pulse, being larger for longer events. This is clearly demonstrated in Fig. 2, that shows the light curves of the UV continuum and $L\alpha$ obtained by Clavel et al. Inspection of the data, and cross-correlation analysis, confirm that the cross-correlation sizes deduced for the three events are 6, 16 and 3 light-days, respectively. The lag quoted by Clavel et al. (12 days) is simply a weighted mean of the three. Examination of the Peterson et al. (1991) $H\beta$ results reveals a similar effect, although the S/N in the last (shortest) event is not good enough to measure the delay.

Turning now to the Wise Observatory data, the continuum pulse of 1988 is of about the same duration as the *shortest* event in 1989. The expected lag is therefore much shorter than the one found by Peterson et al. (1991), which is a weighted mean of the three events.

The above behavior is, in fact, expected for a very large class of thick geometries. As is well known, the CCF is the convolution of the continuum Autocorrelation function with the transfer function. The transfer function of a thick geometry is peaked at zero time and drops to zero after about $2r_{out}/c$, where r_{out} is the outermost radius (see Fig. 1). Thus a longer continuum event must result in longer lag as measured from the peak of the CCF. This is demonstrated in Fig. 3 that shows the response of a 5-50 light-days shell to triangular shaped continuum pulses of different duration. As expected, the delay between the line and continuum light curves is longer for longer continuum pulses, and the effect "saturates"

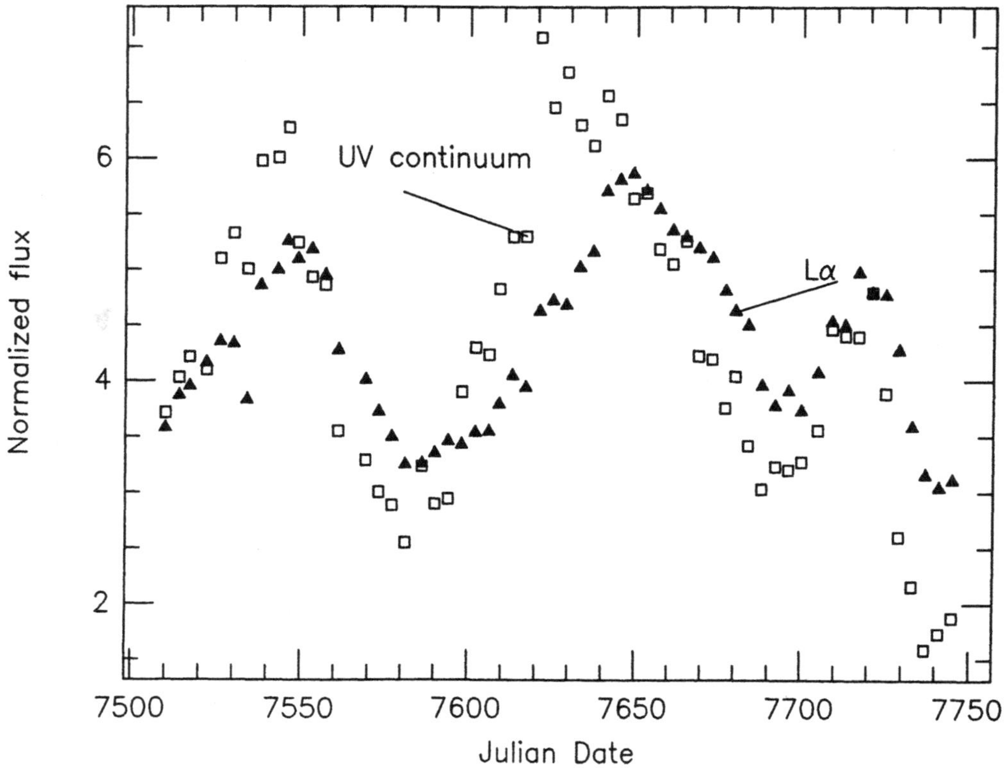

Fig. 2. The Clavel et al. (1990) light curves for the ultraviolet continuum and $L\alpha$ (normalized to the continuum luminosity) in NGC 5548. Note the different lags between the line and continuum light curves for the three continuum events

when the duration of the continuum pulse exceeds the light crossing time of the shell. Thus, the different lags found for the different continuum events in NGC 5548 suggest the existence of a thick BLR (i.e. $r_{\rm out}/r_{\rm in} \gg 1$).

A simulation using the observed ultraviolet continuum of NGC 5548 in 1989, with an assumed thick-shell BLR of 30 light-days in radius, confirm that this is indeed a good first order approximation for the gas distribution in this source. The position of the three peaks is recovered (Fig. 4) but not their relative amplitudes. This raises the question of the line response to continuum variation, which is discussed in the following section.

2.2. Photoionization calculations

A complete photoionization model for the BLR requires two separate sets of calculations:

2.2.1. Isolated clouds

Several models for isolated clouds must be calculated, under a variety of conditions, changing the fundamental input parameters such as the shape of the continuum, the gas density and column density and the ionization parameter. The observed line ratios serve as an initial guess for the model parameters and the calculations are rather complex (e.g. Kwan and Krolik 1981, Rees et al. 1989 and references therein).

2.2.2. A system of clouds

The second stage of the calculations involves some assumptions about the gas distribution in the source. The run of the (hypothetical) confining pressure, or the density dependence on distance, are the fundamental input parameters. They determines the dependence of the covering factor, and the cloud emissivity, on radius. Line intensity observations cannot, by themselves, provide enough constrains on the gas distribution, and variability data, as well as line profile information, is needed to narrow down the range of possibilities. Several examples of such complete models are described in Rees et al. (1989) and Netzer (1989).

The line and continuum light curves of NGC 5548, and the different lags found for the different events, suggest a thick geometry BLR, thus future photoionization models for this source must take this into account. The disagreement of the simple model results with the observed amplitudes of $L\alpha$ and other emission lines (Fig. 4) suggest that one other effects must be considered. The non-linear response of the emission lines is perhaps the most important.

2.2.3. Non-linear response

Fig. 5 shows a set of calculations for an isolated BLR cloud, where the variable parameter is the input flux of ionizing photons (or the ionization parameter). The diagram shows that some lines, like $L\alpha$, respond linearly to changes in ionizing flux, up to the point at which the gas becomes transparent at the Lyman continuum. Further increase of the ionizing flux does not increase the line emission. The HeII lines are similar in this respect, except that they become non-linear at a higher flux level, due to the different optical depth structure. $CIV\lambda1549$ is very different from the above two lines, increasing more rapidly than linear at low flux levels (temperature increase) and dropping very rapidly beyond a certain point, due to the almost complete ionization of C^{+3}.

The new observations of NGC 5548 suggest that non-linear effect are important for several broad lines. Future modelling of this source, in particular the attempts to construct transfer functions for the lines, must take this into account.

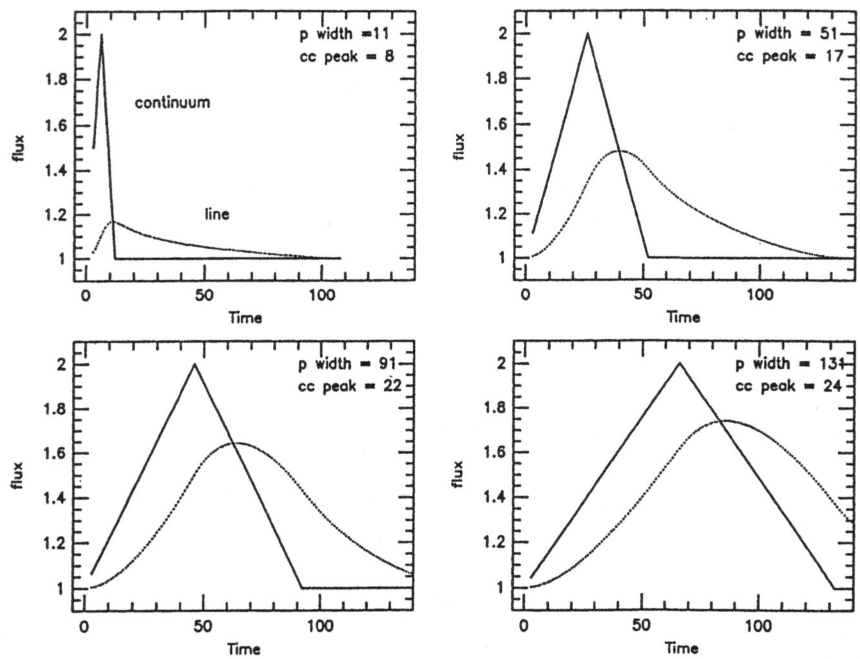

Fig. 3. The response of a 5-50 light-days shell, with constant emissivity per unit radius (dotted lines) to triangular shaped continuum pulses (solid lines) of different duration. The duration of the pulses, and the peaks of the CCF, are marked on the diagrams

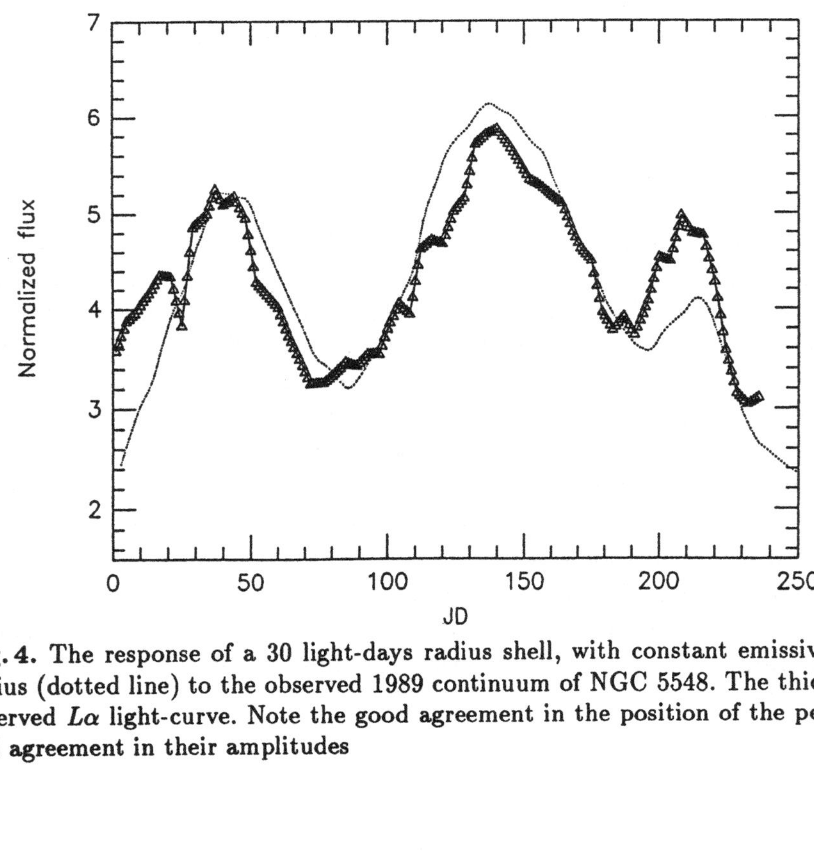

Fig. 4. The response of a 30 light-days radius shell, with constant emissivity per unit radius (dotted line) to the observed 1989 continuum of NGC 5548. The thick line is the observed *Lα* light-curve. Note the good agreement in the position of the peaks and the bad agreement in their amplitudes

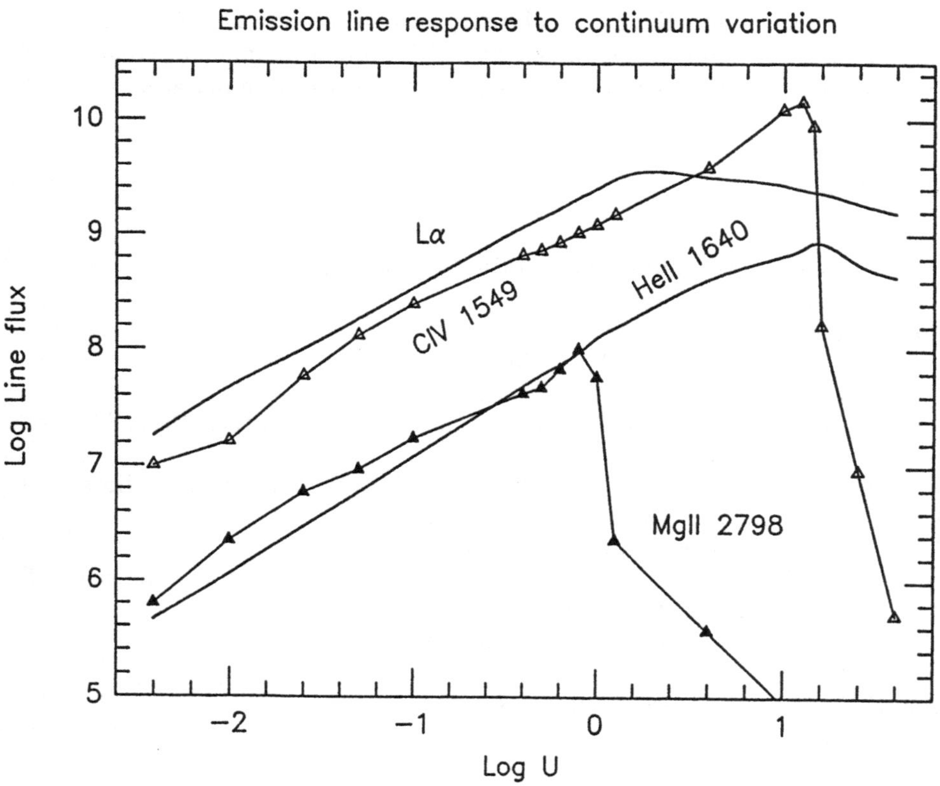

Fig. 5. The response of different emission lines to changes in ionization parameter, for an isolated cloud model with constant density of $10^{10}\ cm^{-3}$, $N_{\rm col}= 10^{23.5}\ cm^{-2}$ and central continuum as in Rees et al. (1989).

3. L-M for AGNs

The available observational information is far from enough to investigate the general L-M relation for AGNs. It is tempting, however, to try and calibrate the relation, at the low luminosity end, using the hand-full of measurements discussed above.

Assume that $r_{\rm av}$ for the above sources is twice the cross-correlation size, and the observed line width are due to Keplerian velocities. We find the following approximate relation between L and M:

$$M = M_1 L_{46}^{1/2} \ , \tag{3}$$

where $M_1 \sim 10^9 M_\odot$. As for the more general relation, the dependence of the ionization parameter and the line width on L must be better understood before the above expression can be extended to more luminous AGNs (Netzer 1989).

Acknowledgements: The Wise Observatory project is supported by grant no. 85/00085 from the United States-Israel Binational Science Foundation, the Israel

Academy of Sciences and the Israel Ministry of Science and Development. I am grateful to the Astronomy Department at OSU Columbus, and its chairman E. Capriotti, for their generous hospitality and much help in this research.

References

Clavel, J., Wamsteker, W., and Glass, I., 1989 Ap.J., **337**, 236.
Clavel et al. 1990, Ap.J., (in press)
Clavel et al., 1991, MNRAS, (in press).
Kwan, J., and Krolik, J.H., 1981, Ap.J., **250**, 478.
Maoz, D., and Netzer, H., 1989, MNRAS, **236**, 21.
Maoz, D., Netzer, H., Leibowitz, E., Brosch, N., Laor, A., Mendelson, H., Beck, S., Mazeh, T., and Almoznino, E., 1990, Ap.J., **351**, 75.
Maoz, D., Netzer, H., Mazeh, T., Beck, S., Almoznino, E., Leibowitz, E., Brosch, N., Mendelson, H., and Laor, A., 1991, Ap.J., (in press)
Netzer, H., 1989, Comments on Astrophysics, **14**, 137.
Netzer, H., Maoz, D., Laor, A., Mendelson, H., Brosch, N., Leibowitz, E., Almoznino, E., Beck, S., and Mazeh, T., 1990, Ap.J., **353**, 108.
Netzer, H., and Maoz, D., 1991, Ap.J.Letts (in press).
Peterson, B.M., 1988, PASP, **100**, 18.
Peterson, B.M., et al., 1991, Ap.J., (in press).
Rees, M., Netzer,H. and Ferland, 1989, Ap.J., **347**, 640.

Variability of BL Lac Objects in the Radio Regime

Esko Valtaoja

Metsähovi Radio Research Station, Helsinki University of Technology, Otakaari 5A, SF-02150 Espoo, Finland
and Turku University Observatory, Tuorla, SF-21500 Piikkiö, Finland

Abstract: Results from the Finnish quasar monitoring program are discussed. Both the spectra and the radio variability indicate that both the basic structure and the mechanism for variability are the same for all AGN. The basic radio properties of low and high polarization quasars are in agreement with orientation-dependent scenarios. The contradictory properties of BL Lac objects can be understood if they in reality consist of two physically different classes. One class, consisting of highly boosted, strongly variable, distant BL Lacs with quasarlike spectra, represents the extreme end of the FR II - LPQ - HPQ population, sources aligned to a few degrees with our line of sight. The other BL Lac population consists of much less boosted, moderately variable, nearby FR I radio galaxies beamed towards us.

1. Introduction

Active galactic nuclei have been divided into ever-increasing numbers of different classes: using only the radio continuum properties, one finds (to give just a few examples) variable, non-variable, flat-spectrum, steep-spectrum, superluminal, radio-loud, radio-quiet, empty field, compact and extended sources. Optical properties give other classifications, apparently without much connection to radio properties: galaxies, high and low polarization quasars, optically violent variables, blazars, BL Lac objects, and so on *ad infinitum.* What physical reality underlies these observational classifications? Is the basic mechanism for activity the same for all sources? How are the different types related to each other? How much can be explained by variations of a single parameter, such as the viewing angle?

To answer these questions one requires large and well-defined samples of different objects for statistical studies, and extensive data sets for detailed investigations of the source properties. The Finnish quasar monitoring program, a joint effort between the Helsinki University of Technology and the University of Turku, was initiated with the central aim of obtaining such data on high radio frequencies.

The backbone of the monitoring program is intensive 22 and 37 GHz monitoring of bright, compact extragalactic radio sources using the Metsähovi 13.7-m radio telescope near Helsinki. The number of objects included is presently about 60. In addition, we have used the SEST telescope in Chile to obtain 100 and 230 GHz data of the monitoring sources and of various other samples, and collaborated with mm-telescopes in Crimea and Brazil for better coverage in both time and frequency.

Our over 11000 Metsähovi observations since 1980 form by far the largest data set of AGN flux variability on radio frequencies above 15 GHz. Since lower frequencies are usually dominated by remnants of old outbursts and other more extended structure, and the core and the new outbursts start to dominate only around 20-40 GHz, higher frequency observations are a necessity for studying the actual sites of activity. Our data is especially useful in shedding new light on the spectra and variability of the core components and on the relationships between different classes of radio sources. In this paper I will mainly discuss the about 3000 published 22 and 37 GHz observations up to 1986 (Salonen et al. 1987), since the newer data has not yet been fully reduced and analyzed. I will first consider what can be said about the physics of AGN cores, their structure and their variations, and then proceed to compare the results of statistical studies with orientation-dependent ("unified") scenarios. Finally, I will discuss the place of BL Lac objects in the overall scheme.

2. The sample

Our monitoring sample of 26 sources is not well-defined, since we originally selected the most active sources for investigating the characteristics of high frequency radio variations. The list of individual sources can be found in Valtaoja et al (1988). However, it is a quite representative subsample of bright, flat-spectrum sources, as can be seen from Figure 1. Of the 36 Northern sources in the Kühr catalogue (Kühr et al. 1981) with S(>10 GHz) > 2 Jy and α(2.7-5 GHz) > –0.3, 16 are included in our sample. The other sources in our sample form a random group of sources which are slightly weaker or not included in the Kühr catalogue, Southern sources and sources with steeper 2.7-5 GHz spectra (3C 84 and 3C 345). Only the steep-spectrum ($\alpha < -0.5$) sources, most of which are 3C radio galaxies or "normal" (low polarization) quasars, are excluded as a class. Our sample also includes 9 of the 26 BL Lacs with S(>10GHz) > 1 Jy in the Véron-Cetty and Véron (1985) catalogue, and 9 of the 24 known superluminals in the list of Zensus (1989).

We have divided the sources into three basic categories, which seem to be the most suitable: low polarization quasars (LPQ), high polarization quasars (HPQ) and BL Lac objects (BL Lacs). LPQs are "normal" quasars, and HPQs (or OVVs, which to all purposes is an identical definition) are blazars. We have also studied the LPQ/blazar and QSO/BL Lac divisions. In addition, there are two galaxies (3C 84, sometimes also classified as a BL Lac object, and 3C 120), and one empty field source, NRAO 150.

3. Spectra and variability: the physics of the core

In Valtaoja et al. (1988) we derived separate quiescent and flare spectra for the sources in the monitoring program. We found an overwhelming similarity in both the quiescent (i.e., outside flares) spectra and the evolution of outbursts in all the sources. The quiescent spectrum of a source could be explained as a combination of extended nonvariable components and remnants of old ejected components with slow secular decay, dominant on low frequencies, and of an apparently constant inhomogeneous jet (also called the milliarcsecond or the core jet), which is the main contributor to high frequency quiescent flux. While continuum spectra alone cannot reveal the nature of the underlying components, the similarity of the quiescent spectra strongly suggests that all compact radio sources have the same basic building blocks: extended emission, old knots, and an inhomogeneous mas jet ("the core").

Within the observational accuracy, all the flare spectra we could derive had the same shape, that of an simple, homogeneous synchrotron source with $\alpha_{thick} \approx$ +2.5 and $\alpha_{thin} \approx$ -0.2, implying that the "universal" electron energy exponent γ = $(2\alpha+1) \approx 1.4$, and not 2.5 as often assumed (cf. also Hughes et al. 1989, 1990). During the last stages the spectra appeared to steepen due to radiating losses. The spectral evolution of the flares also seemed to be similar, and in accordance with shock models (Marscher and Gear 1985). From this we concluded that most, if not all, intrinsic radio variability has the same origin, growing and decaying shocks in a relativistic jet. In fact, the similarity in the radio continuum properties is so great that it is not possible to reliably classify a given source purely on the basis of its radio properties. This result agrees well with unification schemes: all sources have the same basic structure, and the same basic mechanism for variability.

Conversely, this puts strong constraints on the existence of alternative causes for variability. For example, if a significant fraction of BL Lacs were gravitationally lensed objects, as has been suggested (e.g., Ostriker and Vietri 1985; Stickel, these Proceedings), it is difficult to understand just why microlensing would produce radio variations so similar to shock-induced variations. A case in point is AO 0235+164, a favored candidate for microlensing. The optical spikes and the radio flares are almost simultaneous, and they must thus have a common origin: if the optical event is caused by microlensing, then so is the radio flare. Yet the radio outbursts in AO 0235+164 behave just as shocks would, with delays in the maxima and smaller amplitudes towards low frequencies (Valtaoja et al. 1988) and with no indications of a different origin for the variability.

4. A generalized shock model

The shock model of Marscher and Gear (1985) predicts approximately correctly the general behavior of the shock, but fails in some details, most notably in predicting too steep optically thin spectra (Valtaoja et al. 1988). The shock code of Hughes et al. (1989, 1990) gives impressive fits to lower frequency flux, polarization and VLBI data, but seems to become progressively worse on higher frequencies, probably because the growth phase of the shock is not explicitly included in the description. Generally, the last decay stages of shocks seem to be rather well understood, but a realistic physical theory for the formation and early evolution of shocks can only be formulated when we have more observations from the relevant millimeter-submillimeter regime.

A generalized description of shock behavior can nevertheless be given. Every shock must have a growth stage when, according to both observations and general physical arguments, the turnover flux will rise rapidly while the turnover frequency decreases. At some stage energy gains and energy losses will balance, and the shock reaches its maximum development stage. Finally, the shock must decay, becoming more transparent and radiating less and less. If the spectral shape of the shock remains constant until the last stages, as the data indicates, the evolution of the shock is described by the evolutionary track of the turnover peak, schematically plotted in Figure 2.

While all the shocks evolve similarly, the *observed* behavior depends crucially on the frequency ν_{max} where the shock reaches its maximum development, and on the observing frequencies ν_1 and ν_2 (Fig. 2). From observations, ν_{max} in shocks can vary from below 10 GHz to over 100 GHz, and possibly over a still wider range. If the observing frequencies are well below ν_{max}, as is the case for most low frequency observations, the shock will be in its last decaying stages by the time it reaches the observing frequencies. These "high-peaking" flares will approximately follow the old van der Laan (1966) model, with time delays between frequencies, transition from optically thick to optically thin at burst maximum, and burst amplitudes dependent on the slope β of the turnover track: $\Delta\ S(\nu) \propto\ \nu^{\beta}$. If, on the other hand, the observing frequencies are above ν_{max}, the shock is observed while still in its growth stages, and its observed behavior is very different. Such a "low-peaking" flare will have lightcurves tracking each other closely and with no delays, since the shock on the observing frequencies is optically thin most of the time, and α_{thin} is close to zero. The observed burst amplitude is now dependent on the spectral index of the shock: $\Delta S(\nu) \propto\ \nu^{\alpha(thin)}$. Courvoisier et al. (1990) show examples of both high-peaking and low-peaking flares in 3C 273.

We have calculated the maximum amplitude of variations during 1980-1985 for our sources at 4.8, 8 and 14.5 GHz from the observations of Aller et al. (1985), and at 22.2 and 36.8 GHz from our own data. For each source we then identify the frequency at which ΔS is largest, and plot the other $\Delta S(\nu)$ normalized by $\Delta S(\nu_{max}) = 1$. The result is shown in Figure 3. Straight line fits through ν_{max} gives for frequencies below the maximum $\Delta S \propto\ \nu^{+0.55}$, in accordance with several earlier results from low-frequency observations (cf. Altschuler 1989). The new result is the dependence for higher frequencies: $\Delta S \propto \nu^{-0.25}$, close to the value $\alpha_{thin} \approx -0.2$ independently deduced from shock spectra, and thus in good agreement with the prediction of the generalized shock model.

In summary, the generalized shock model can provide a simple unified description for a number of different observed properties of radio outbursts and associated VLBI and optical phenomena. In the context of unified models, basically similar shocks seem to provide a good description of radio variability for both quasars and BL Lacs.

5. Variability analysis

Even if the basic mechanisms are the same, there still should be some differences in the average radio properties of different classes of sources, detectable by statistical analysis. We have analyzed our monitoring data to find answers to three questions:

1) Are there differences between different classes of sources?
2) Are there correlations between different radio properties?
3) How does it all fit with orientation-dependent scenarios?

For our LPQ / HPQ / BL Lac classification the relevant unified model is that presented by Barthel (1989). Going from the assumed parent population of strong (FR II) radio galaxies to normal (=LPQ) quasars and to blazar-QSOs (=HPQ), the viewing angle decreases, resulting in increased amount of Doppler boosting in the jet and, it is hoped, the observed classification. BL Lacs were earlier assumed to be the quasars most closely beamed towards us, but presently the most common view seems to be that they have another parent population, FR I radio galaxies (cf. Browne 1989, Ulrich 1989).

We have used a number of different indexes in the analysis. For the spectra, the best index was found to be the quiescent spectral index between 5 and 37 GHz, determined from the data in Valtaoja et al. (1988). This index describes the overall spectral slope in the radio regime. To describe the variability, we calculated the average fractional variability $V = \Delta S/\langle S \rangle$ as defined in Seielstad et al (1983) for each source. This index is basically a measure of the amplitude of variations, with no account of the time structure of the variations. These are described by the average variability on different timescales, $T(n) = N^{-1} \Sigma |S_i - S_j|/(\sigma_i^2 + \sigma_j^2)^{1/2}$, where the summation is done over all the N pairs $(S,\sigma)_{i,j}$ separated by a given time interval Δt_n, for example over all pairs of flux measurements separated by not less than 16 and not more than 25 days (Wardle et al. 1981). In the absence of variability, $T(n) = (2/\pi)^{1/2}$.

5.1. Spectrum vs. classification

In our sample we find an increasing overall flatness in the spectra, when proceeding from LPQs to HPQs and BL Lacs (Figure 4). The difference between low and high polarization quasars is in reality much more pronounced, since we have excluded the steep-spectrum sources, most of which are radio galaxies or LPQs, from our sample.

5.2. Classification vs. variability

In general, whatever variability indicator is used, we find a general tendency for the HPQs to be somewhat more variable than the LPQs, while the BL Lacs do not seem to occupy any definite place in the variability scale. The variability indexes of BL Lacs span the whole observed range of variability; some are among the least variable in the sample, while others are the most variable by any index. Figure 5 shows an example, V(37 GHz).

5.3. Variability vs. spectrum

The wide range of variability properties of BL Lacs is also reflected in the correlation analysis: no dependence between any spectral and variability properties is found. Looking only at the class of quasars (both LPQ and HPQ), we find no correlation between the average fractional variability V and the overall spectral index (5-37). However, there is a clear correlation for the *rapidity* of variations, measured by T(n) over timescales of weeks or months: for quasars, the rapidity of variations increases with spectral flatness. One example, T(16-25 days) vs. (5-37) is shown in Figure 6.

6. Orientation, spectra and variability: a simple model

The correlation between spectra and rapidity of variations in quasars, and the lack of correlation with average variability can be understood in terms of orientation effects. To demonstrate this, we consider a simple model, similar to that of Eckart et al. (1989).

On low frequencies the flux is dominated by extended components and older ejected VLBI knots, which are either unboosted or less boosted than the inner components. On high frequencies the core-jet and the shocks start to dominate the total flux. In the simplest approximation we may assume that the core and the shocks are boosted by the same amount, and S(shock)/S(core) is in the average constant which depends on the details of the shock and jet physics. (From our high frequency SEST observations, this ratio seems to be about 0.4). If all the quasars have the same intrinsic Lorentz factor Γ, the amount of boosting and the prominence of the core relative to the less boosted low frequency components increases with decreasing viewing angle Θ.

At 5 GHz, most of the flux in a typical QSO comes from the extended components, while at 37 GHz the core starts to dominate. The 5-37 GHz spectral index therefore measures the relative strengths of the outer and the inner components and thus, in the model, the viewing angle: the smaller Θ, the flatter the overall spectrum. The average variability V measures the ratio S(var)/S(non-variable) = S(shock)/[S(extended)+S(core)]. On low frequencies, where $S_{ext} \gg S_{core}$, the average variability will increase with increased amount of boosting, i.e. with decreasing viewing angles, as has been found to be the case by Eckart et al. (1989). On high frequencies this dependence should vanish: now $S_{core} > S_{ext}$, and the average variability depends on the ratio S_{shock}/S_{core}, which, assuming equal boosting factors, does not depend on Θ. This agrees with our findings. However, although the *average* variability does not depend on Θ at high frequencies, the *rapidity* of variations still does: the more Doppler boosting, the more rapid the variations appear. Thus, we expect to find a correlation between T(small timescales) and spectral flatness, as indeed is the case.

Thus, although the present data does not compel one to accept orientation dependence as the underlying reason for the correlation between spectral flatness and rapidity of variations in quasars, it certainly is in accordance with such a hypothesis.

7. The trouble with BL Lacs

As already indicated, the results of the variability analysis for BL Lacs show no clear patterns. One thing that *can* be stated is that the characteristics of variability, both the details of outbursts in individual sources, and the overall statistical properties of the whole class, are so similar to those of quasars that a common origin for both seems certain. However, the orientation hypothesis which works for quasars seems to fail for BL Lacs. No clear correlations are found, and the results point to two opposite directions.

BL Lacs have flatter spectra than quasars, which could indicate more core boosting and smaller viewing angles, as would also their general lack of extended structure. Some BL Lacs are also among the most active and rapidly variable of all known radio sources; their derived brightness temperatures exceed the 10^{12} K limit by orders of magnitude, indicating large Doppler boosting and small viewing angles. An extreme example is the outburst in OJ 287 in the middle of 1989. The fluxes increased by a factor of 3 in less than two months, giving T_B up to 10^{16} K and Doppler boosting factors 15-20. The lightcurves at different frequencies show clearly that the burst was intrinsic to the source. This and other similar flares in OJ 287 are the most extreme radio events for which the intrinsic shock nature of the variations can be securely established. D > 15 means that OJ 287 *must* have a viewing angle less than 4 degrees (cf. Cohen and Unwin 1984), and the intrinsic Lorentz factor Γ of the flow must be > 8.

But equally strong evidence indicates that BL Lacs do *not* have especially small viewing angles. Several BL Lacs are among the least variable in our whole sample, and as a class BL Lacs are not more variable than quasars (cf. also Altschuler 1982, 1983 and Fiedler et al. 1987). The lack of large apparent superluminal motions in BL Lacs (Mutel, 1990) also argues against the small viewing angle hypothesis. Finally, the detailed model fits to the flux and polarization data of 1749+096 (OT 081) and BL Lac itself by Hughes et al. (1989, 1990) require viewing angles of about 40 degrees.

8. Two classes of BL Lac objects?

How can these contradictory results, also reflected in the large spread of variability indicators and lack of correlations, be explained? We propose the following solution: ***the sources classified as "BL Lac objects" in reality consist of two physically quite different populations.*** One population represents the extreme end of the quasar population: they are distant, strong FR II radio galaxies beamed most closely towards us, and thus belong to Barthel's unification scheme. The other population is made up of weaker, nearby FR I radio galaxies beamed towards us.

The classification of Burbidge and Hewitt (1989) seems to reflect this dichotomy. Burbidge and Hewitt divided BL Lacs into "BL Lac objects in galaxies", mainly nearby $z < 0.3$ sources, and "BL Lac objects with QSO spectra", all with $z > 0.3$. We have calculated the Doppler boosting factors from our full 10-year variability data for the most rapid outburst in each quasar, "distant" $z > 0.3$ BL Lacs and "nearby" $z < 0.3$ BL Lacs according to the classification of Burbidge and Hewitt. The results are shown in Figure 7. The distant BL Lacs are as much or more boosted than quasars, as expected if they really are quasars with typical intrinsic QSO Lorentz factors but smaller viewing angles. The nearby BL Lacs are hardly boosted at all, clearly less than quasars: they must have a different parent population, which may be identified with FR I radio galaxies. The lack of boosting and strong variability indicates both larger viewing angles and smaller intrinsic Lorentz factors for these sources as compared with QSOs / distant BL Lacs.

The space distribution of BL Lacs also supports the existence of two different classes. It has been known for some time that BL Lacs are oddly distributed in space. This point has been well represented by Browne (1989): in complete samples of radio sources, there is a large number of BL Lacs with $z < 0.1$, while quasars have a completely different redshift distribution (cf. Figure 1 in Browne, 1989). On the other hand, the redshift distribution of these nearby BL Lacs is similar to that of FR I radio galaxies, encouraging the view that they are the parent population of nearby BL Lacs. But there is also another peak in the redshift distribution presented by Browne: the distant BL Lacs, which do seem to be distributed similarly to flat spectrum quasars, and could therefore have the same parent population.

The hypothesis of two separate classes of "BL Lac objects" can be further tested in several ways. Deep optical images should reveal FR I galaxies around nearby BL Lacs and FR II galaxies around distant BL Lacs. If nearby BL Lacs are less aligned and less relativistic than distant BL Lacs, this difference should be reflected in their radio properties (as has been indicated here for a small sample), and the dichotomy should also extend to optical and X-ray properties. The viewing angles could be directly estimated by measuring both the apparent speeds v/c (from VLBI) and the Doppler factors (from variability timescales or X-ray data) for a larger number of sources. For nearby BL Lacs the distribution of viewing angles should be clearly different from that of quasars, while the distant BL Lacs should join smoothly with the QSO distribution.

9. Conclusions

We find that the basic radio structure and the mechanism for radio variability are the same both for quasars and for BL Lacs (and possibly for radio galaxies also, although this conclusion is based on only two objects). All features of the radio variability, with the possible exception of intraday variability, seem to be explainable with shocked jet models, and no class of AGN is dominated by gravitational lensing or other exotic mechanisms. The "universal" electron energy exponent is ≈ 1.4, not 2.5. The differences in the radio spectra and variability between LPQs and HPQs can be explained by different viewing angles.

The objects classified as "BL Lacs" consist in reality of two different classes of sources with different parent populations and different intrinsic properties. The "nearby" / "in galaxies" BL Lacs are FR I radio galaxies beamed towards us, and the "distant" / "with QSO-like spectra" BL Lacs are the extreme end of FR II - LPQ - HPQ population, quasars aligned within a few degrees with our line of sight.

References

Aller, H.D *et al.*: 1985, *Astrophys. J. Suppl.* **59**, 513
Altschuler, D.R.: 1982, *Astron. J.* **87**, 387
Altschuler, D.R.: 1983, *Astron. J.* **88**, 16
Altschuler, D.R.: 1989, *Fundamentals of Cosmic Physics* **14**, 37
Barthel, P.D.: 1989, *Astrophys. J.* **336**, 606
Browne, I.W.A.: 1989, in *BL Lac Objects*, eds. L. Maraschi, T. Maccacaro, M.-H. Ulrich, Springer-Verlag, Berlin, p. 401
Burbidge, G., Hewitt, A.: 1989, in *BL Lac Objects*, eds. L. Maraschi, T. Maccacaro, M.-H. Ulrich, Springer-Verlag, Berlin, p. 412
Cohen, M.H., Unwin, S.C.: 1984, in *VLBI And Compact Radio Sources*, eds. R. Fanti, K. Kellerman, G. Setti, Reidel, Dordrecht, p. 95
Courvoisier, T.J.-L. *et al.*: 1990, *Astron. Astrophys.* **234**, 73
Eckart, A., Hummel, C.A., Witzel, A.: 1989, *M.N.R.A.S.* **239**, 381
Hughes, P.A., Aller, H.D., Aller, M.F.: 1989, *Astrophys. J.* **341**, 68
Hughes, P.A., Aller, H.D., Aller, M.F.: 1990 preprint
Kühr, H., *et al.*: 1981, *Astron. Astrophys. Suppl.* **45**, 367
Marscher, A.P., Gear, W.K.: 1985, *Astrophys. J.* **298**, 114

Mutel, R.L.: 1990, preprint
Ostriker, J.P., Vietri, M.: 1985, *Nature* **318**, 446
Salonen, E. *et al.*: 1987, *Astron. Astrophys. Suppl.* **70**, 409
Seielstad, G.A., Pearson, T.J., Readhead, A.C.S.: 1983, *P.A.S.P.* **95**, 842
Stickel, M.: 1990, these proceedings
Ulrich, M.-H.: 1989, in *BL Lac Objects*, eds. L. Maraschi, T. Maccacaro, M.-H. Ulrich, Springer-Verlag, Berlin, p. 45
Valtaoja, E. *et al.*: 1988, *Astron. Astrophys.* **203**, 1
van der Laan, H.: 1966, *Nature* **211**, 1131
Veron-Cetty, M.-P., Veron, P.: 1985, *ESO Scientific Report* **4**
Wardle, J.F.C., Bridle, A.H., Kesteven, M.J.L.: 1981, *Astron. J.* **86**, 848

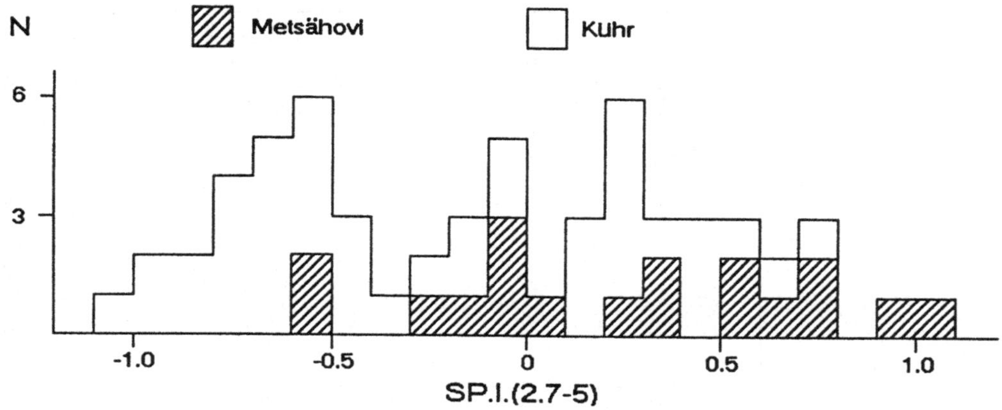

Fig. 1. The fraction of bright Northern sources in the Kühr catalogue included in the Metsähovi monitoring sample.

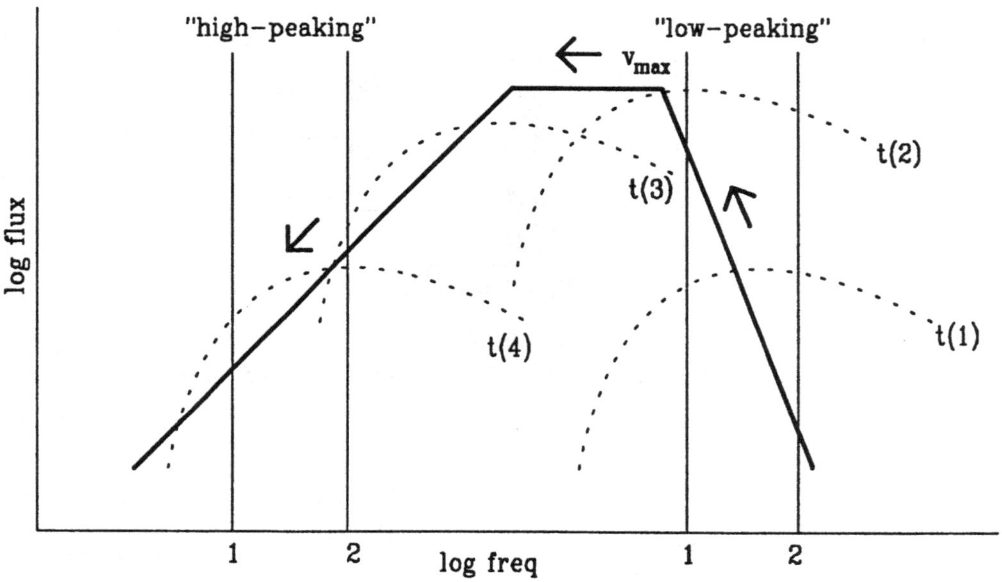

Fig. 2. A schematic model for the evolution of a shock. The thick line is the evolutionary track of the shock spectrum's turnover peak. The vertical lines show two neighboring observing frequencies (1,2) in the case of a high-peaking flare (left) and in the case of a low-peaking flare (right).

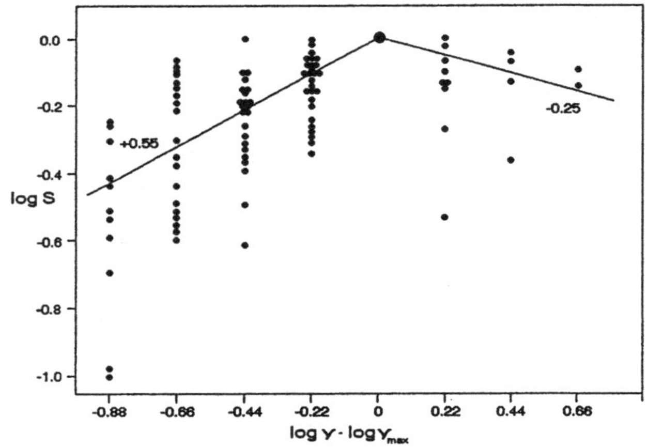

Fig. 3. The maximum amplitude of outbursts on frequencies below and above the maximum development frequency, normalized to $\Delta S(\nu_{max}) = 1$. The straight line fits for ΔS vs. ν -dependence are also shown.

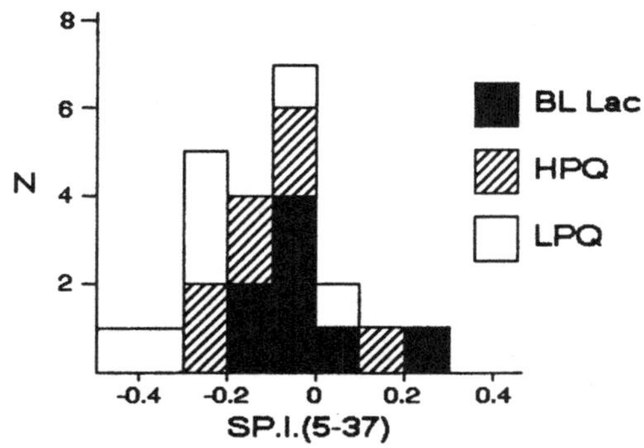

Fig. 4. The distribution of the quiescent 5-37 GHz spectral index.

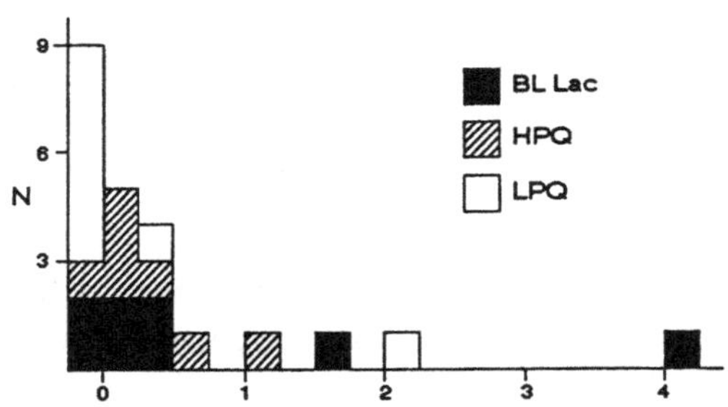

Fig. 5. The distribution of the average fractional variability V at 37 GHz.

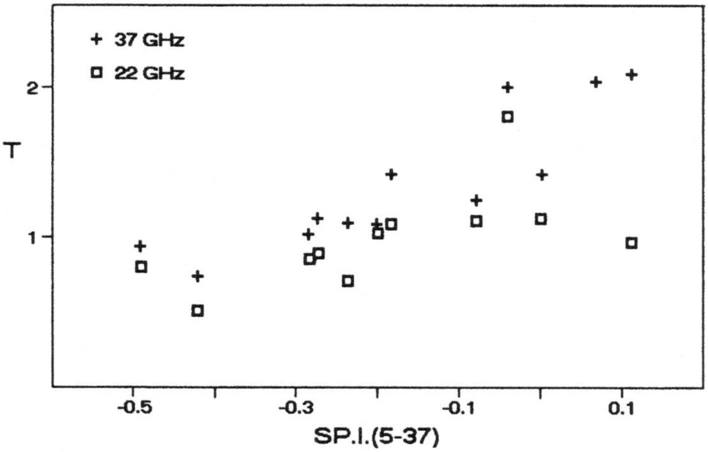

Fig. 6. The average variability of quasars on timescales 16-25 days as a function of the quiescent 5-37 GHz spectral index.

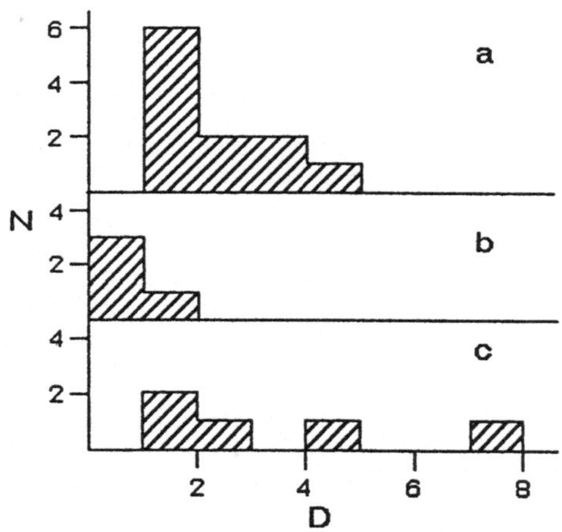

Fig. 7. The distribution of the Doppler boosting factor $D = (T_B/10^{12}\ K)^{1/3}$ for quasars (a), nearby (z¡0.3) BL Lacs (b), and distant (z>0.3) BL Lacs (c).

Intraday Radio Variability of Quasars and BL Lac Objects

Andreas Quirrenbach

Max-Planck-Institut für Radioastronomie, Auf dem Hügel 69,
5300 Bonn 1, Germany

Abstract: Light curves of 49 flat spectrum radio sources with time resolutions of 1 to 4 hours have been obtained with the Effelsberg 100 m telescope and the VLA at cm wavelengths. Intraday variations with amplitudes up to $\sim 30\%$ were found in several sources; in the quasar 0917+624 they are correlated with variations of the polarized flux density by a factor ~ 3. In the BL Lac object 0716+714, coordinated radio and optical observations have revealed correlated variations in both wavelength regimes with typical time scales between ~ 1 and ~ 7 days. The latter findings indicate that large amplitude intraday variations are probably intrinsic to the sources; a statistical analysis suggests that that additional contibutions of interstellar scintillation are present at a lower level.

1. Introduction

The variability of extragalactic radio sources has been the subject of many investigations ever since its first detection by Sholomitskii (1965). For numerous sources, the basic characteristics of variations with typical time scales of several months could be explained by a simple model based on adiabatically expanding clouds emitting synchrotron radiation (Pauliny-Toth and Kellermann 1966). More sophisticated models that attribute variations of total intensity and polarization to a shock propagating in an underlying jet have been successfully used to interpret observations of blazars such as BL Lac, 0735+178, and 3C279 (e.g. Hughes et al. 1990). It was recognized by Shapirovskaya (1987) that at long wavelengths – mainly in the decimetric regime – refractive scattering in the interstellar medium (RISS) can produce additional variability of compact sources. Because of its dependence on galactic latitude, small-amplitude variability on time scales of a few days found in a large sample of flat spectrum radio sources at 9 cm wavelength was also attributed to this effect (Heeschen 1984, Heeschen and Rickett 1987).

In an attempt to trace these variations down to even shorter time scales, Witzel et al. (1986) and Heeschen et al. (1987) detected considerable intraday variations in several high declination sources. This effect had not been anticipated, since rapid intrinsic variations require apparent brightness temperatures so high that it is difficult to reconcile them with the inverse Compton limit even if Doppler boosting is taken into account, and since very small (~ 0.1 AU) ionized structures in the ISM are needed to produce intraday variations by RISS.

Here I will report on observations intended to reveal the frequency of occurrence as well as the main characteristics of intraday radio variations of quasars and BL Lac objects, and I will try to give some clues to the question whether they are intrinsic to the source or due to propagation effects.

2. Observations and calibration

The investigation of intraday radio variability began in 1985, when the 100 m telescope of the Max-Planck-Institute for Radio Astronomy in Effelsberg was used to monitor a complete sample of 15 flat spectrum sources ($\alpha_6^{11} \geq -0.5$, $S_\nu \propto \nu^\alpha$) north of $\delta = 60°$ with a time resolution of 2 to 4 hours. Later, quasi-simultaneous dual frequency and polarization observations were performed with the same telescope, and the NRAO's Very Large Array (VLA)[1]. was added in May 1989 to obtain simultaneous light curves at five wavelengths. A second VLA campaign in February 1990 was intended to give longer time coverage and simultaneous optical and radio data. Finally, in April 1990 a second complete sample consisting of all 18 flat spectrum sources stronger than 1.3 Jy in declination range $35° \leq \delta \leq 50°$ was observed at 6 cm to obtain additional statistical data, and to verify that intraday radio variability is not confined to a certain region in the sky. A summary of the observations is given in Tab. 1.

For the Effelsberg measurements, we used correlation receivers with circularly polarized feeds at 6 and 11 cm wavelength equipped with IF-polarimeters. Since the accuracy of our measurements was not limited by thermal noise, we performed cross scans consisting of 4 to 12 subscans on the source positions (rather than ON-OFFs), so that we were able to correct for pointing offsets. By including almost 50% steep spectrum sources (which were all confirmed to be not variable within the measurement accuracy) in our schedules we could derive elevation dependent gain curves and time variations of the sensitivity due to weather or day/night effects internally from our observations. After corrections for these effects, which were typically of order $\sim 1\%$, we could achieve an r.m.s. accuracy of order 0.3% at 11 cm and in the range 0.3% to 0.5% at 6 cm wavelength. The parameters describing the instrumental polarization (including crosstalks of the Stokes I, Q, and U channels and ellipticity of the polarimeter) were determined by frequent measurements of the strongly polarized sources 3C286 and 0836+710.

The VLA was operated at 2, 3.5, 6, and 20 cm wavelength. In May 1989 we used the full array in B configuration to obtain snapshot maps of all sources, afterwards the VLA was split into two subarrays to get faster sampling; in February 1990 we

[1] The National Radio Astronomy Observatory is operated by Associated Universities, Inc., under cooperative agreement with the National Science Foundation of the USA.

observed with a subarray of four antennas. After standard initial calibration of the visibility data, the snapshot maps were used to subtract the effects of extended structures and confusion; afterwards the VLA data were calibrated in the same way as the Effelsberg observations.

A more complete description of the observational procedures is given by Witzel and Quirrenbach (1990); parallel optical observations are discussed by S. Wagner in this volume.

Table 1. Summary of the observations

Date	Duration	No. of Sources	100 m telescope u_0 (6 cm)	u_0 (11 cm)	Pol	VLA	optical
May 85	50 hrs	17	-	0.27%	-	-	-
Aug 85	87 hrs	31	-	0.26%	-	-	-
Dec 85	100 hrs	31	-	0.27%	-	-	-
Mar 88	112 hrs	17	0.39%	0.27%	-	-	-
Jun 88	104 hrs	19	0.45%	-	+	-	-
Dec 88	147 hrs	14	0.29%	0.18%	+	-	-
May 89	127 hrs	14	0.38%	0.46%	+	+	+
Feb 90	449 hrs	4	-	-	-	+	+
Apr 90	108 hrs	48	0.71%	-	-	-	-

3. The occurrence of rapid variations in flat spectrum radio sources

Up to now, a total of 49 flat spectrum and 32 steep spectrum sources (with the threshold between steep and flat set to $\alpha = -0.5$) have been observed at one or more epochs. To analyze the observed variations, we have computed fluctuation indices $u = 100 \cdot \sigma/\langle S\rangle$, with σ being the r.m.s. scatter of the measurements of a source and $\langle S\rangle$ being the mean flux density. Since none of the steep spectrum sources have shown significant variability in our observations, the mean fluctuation index u_0 of all steep spectrum sources observed at a given epoch is used as an estimate of the total measurement error; values of u_0 are listed in Tab. 1.

A standard χ^2-test applied to our data shows that at any given epoch more than one third of all flat spectrum sources are significantly variable. The frequency of occurrence of short time scale variability can also be derived from Fig. 1, which shows the distributions of the variability amplitudes $Y = 3\sqrt{u^2 - u_0^2}$ of steep spectrum and flat spectrum objects measured at 6 cm wavelength. (Whenever u was smaller than u_0, Y was set to 0.) Whereas the distribution for the steep spectrum sources peaks sharply in the 0% to 1% bin, the maximum in the distribution for the flat spectrum objects occurs in the 1% to 2% bin; the median value of the distribution is 2.3%. The deficiency of flat spectrum sources in the first bin is even more pronounced, if only sources with compact VLBI stuctures – these are sources with at least 80% of their total flux density at 6 cm contained in a unresolved VLBI core – are taken into account (non-hatched area in Fig. 1(a)), i.e. if sources with a prominent VLBI jet like 0836+710 or 1928+738 are disregarded. This result indicates that virtually all sources with compact VLBI structures show intensity variations with amplitudes of at least $\sim 2\%$ whenever observed.

In order to find the typical time scales of the variations, structure functions $D(\tau) = \langle (R(t+\tau) - R(t))^2 \rangle$ have been used to describe the light curves. R here denotes the normalized residual flux density $R(t) = 100 \cdot (S(t)/\langle S \rangle - 1)$. Figure 2 shows two examples of structure functions with different shapes: Whereas the structure function of 2007+776 in Fig. 2(a) is monotonically increasing, indicating that the light curve is dominated by a linear trend over the total length of the observing run (i.e. ~ 5 days), the structure function of 0716+714 (Fig. 2(b)) shows a pronounced maximum, indicative of intraday variations. We call monotonic structure functions "type I", and those with a maximum "type II". Type II variability is found almost exclusively in sources with compact VLBI structure (with the only exception of the BL Lac object 0735+178), but it is quite common among these objects: 10 out of 12 such sources in our two complete samples have shown intraday variations in at least one observing run.

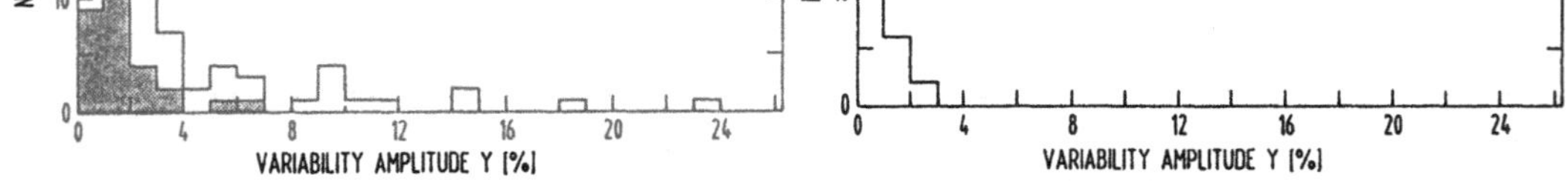

Fig. 1. Distributions of the variability amplitudes Y of (a) flat spectrum and (b) steep spectrum objects at 6 cm wavelength. The sources with a prominent VLBI jet or double stucture on the VLBI scale are hatched in (a).

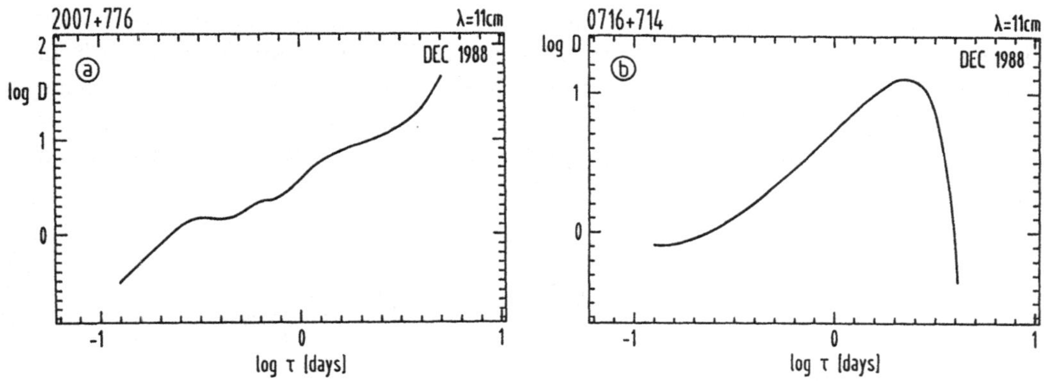

Fig. 2. Sample structure functions of (a) typical "type I" and (b) typical "type II" sources

4. The quasar 0917+624

The two quasars 0804+499 ($z = 1.43$) and 0917+624 ($z = 1.44$) showed the strongest variations seen so far; in 0804+499 a peak-to-peak amplitude of 31% and a steepest gradient of $\Delta R = 17\%$ within 2 hours were observed at 6 cm wavelength. In 0917+624, variations with amplitudes in excess of 20% have been found at several epochs (Quirrenbach et al. 1989a,b). Simultaneous light curves of this source at five frequencies between 2 and 20 cm obtained in May 1989 are presented in Fig. 3. It is obvious that there are correlated variations through the whole wavelength range covered, and that there is no significant time lag between corresponding peaks at shorter and longer wavelengths. The amplitude is highest at 11 cm with a continuous decrease to shorter wavelengths, but more fine structure is seen at the higher frequencies. At 20 cm, the amplitude is also lower than at 11 cm, and the variations appear "quenched".

In 0917+624, the variations of the total flux density S are accompanied by even more dramatic simultaneous changes of the polarized flux density P; the variations of P reach an amplitude of a factor ~ 3 (Quirrenbach et al. 1989b). All our polarization observations of this source have consistently shown an anti-correlation between S and P, an example is given in Fig. 4(a) and (b). For most of the time, the polarization position angle showed only small variations, indicating that the source component dominating the polarized flux density has a magnetic field parallel to the VLBI jet (Quirrenbach 1990). Around *J.D.* 2 447 256, however, the polarization vector made a larger excursion tracing a path that encircled the origin of the *QU*-plane; this "event" gives rise to the 180° swing of χ visible in Fig. 4(c).

From long-term monitoring of 0917+624 with the Green Bank interferometer it is known that this source did not show variations in excess of 25% between December 1987 and June 1988 (K. Johnston, private communication). This means that the full amplitude of the variations is present at a time scale of ~ 1 day; longer time scales do not contribute significantly. The structure function at 6 cm wavelength corresponding to a light curve obtained in the 4 week VLA monitoring campaign displayed in Fig. 5 confirms this finding: The structure function saturates at ~ 0.4 days, indicating that the spectral power of the variations is confined mainly to frequencies larger than 1 day^{-1}.

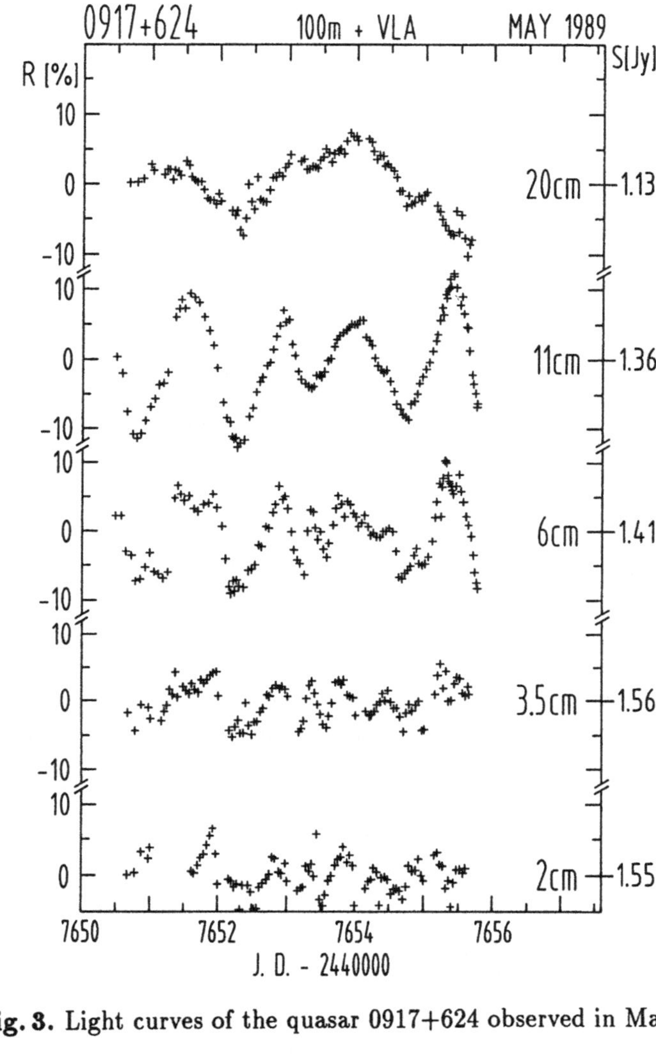

Fig. 3. Light curves of the quasar 0917+624 observed in May 1989

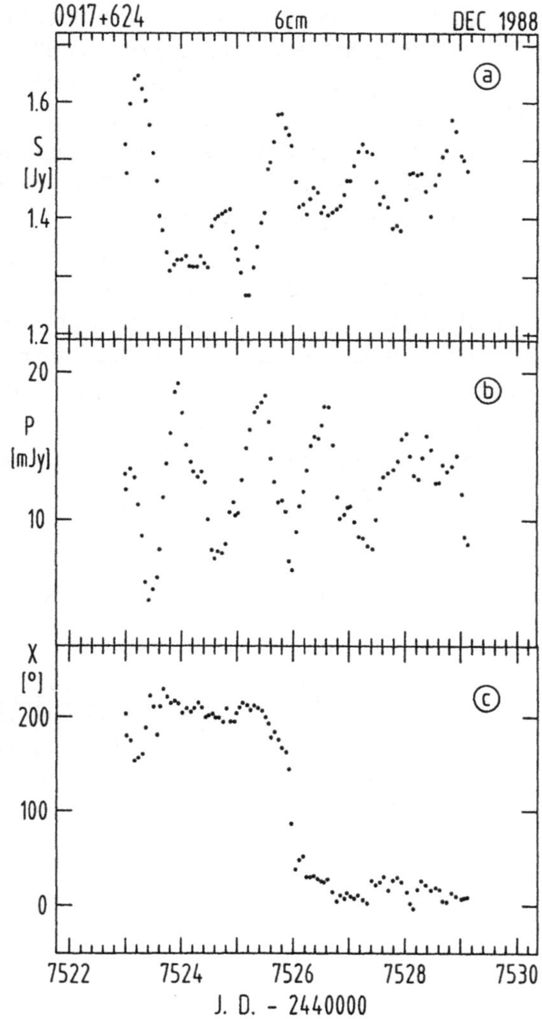

Fig. 4. (a) Total intensity S, (b) polarized flux density P, and (c) polarization position angle χ of 0917+624 at 6 cm in December 1988

5. Correlated radio and optical variability in the BL Lac object 0716+714

First parallel radio and optical observations of 6 sources selected from our high declination sample were carried out in May 1989 (Wagner et al. 1990). It was found that three of these sources showed intraday variations in both wavelength bands, whereas the other three objects were "quiet" in both regimes. In February 1990, the BL Lac object 0716+714 (z unknown) was observed for almost four weeks by the VLA, and by two optical telescopes in Heidelberg and on Calar Alto. In the radio, we obtained a regular sampling in 2 hour intervals, except for some gaps due to maintenance or other projects incompatible with our continuum observations. In the optical, a mean sampling of $\sim$ 30 minutes during nighttime could be achieved. In Fig. 6, the light curves obtained at 6500 Å and at 6 cm are displayed. From these figures it is apparent that the overall behaviour of the source was quite similar in both wavelength regimes: Whereas during the first part of our observing run (up to *J.D.* 2 447 930) 0716+714 was in a state characterized by high brightness and fast variations with a typical time scale of $\sim$ 1 day, it underwent a transition to a somewhat fainter state with slower variations (time scale $\sim$ 7 days) around *J.D.* 2 447 930. The fact that there is a change of the typical time scale consistently at radio and optical wavelengths is confirmed by a structure function analysis (Quirrenbach et al. 1990). From this near-simultaneous transition it is apparent that the time delay between optical and radio variations is of order a few days only; it is not possible to determine the delay without ambiguity, however. The cross-correlation functions of the two complete light curves as well as separate cross-correlations of the first and second parts all show several peaks at different lags due to the quasi-regular appearance of the variations.

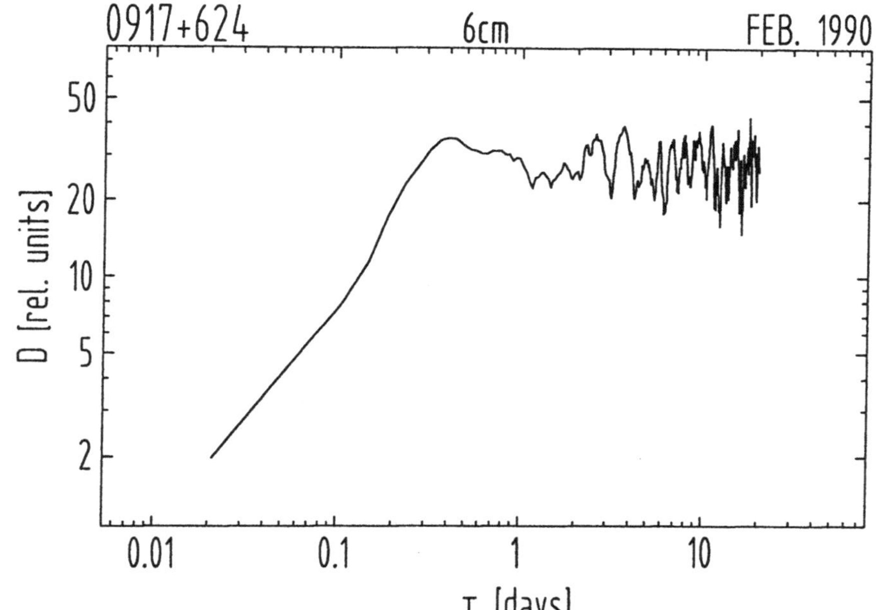

Fig. 5. Structure function of the quasar 0917+624 at 6 cm in February 1990

6. Discussion

The origin of intraday radio variations, specifically the question whether they are intrinsic to the sources or due to propagation effects, has been discussed in several papers (e.g. Heeschen et al. 1987, Quirrenbach et al. 1989a, Wambsganss et al. 1989). Several pieces of evidence point towards an intrinsic origin at least

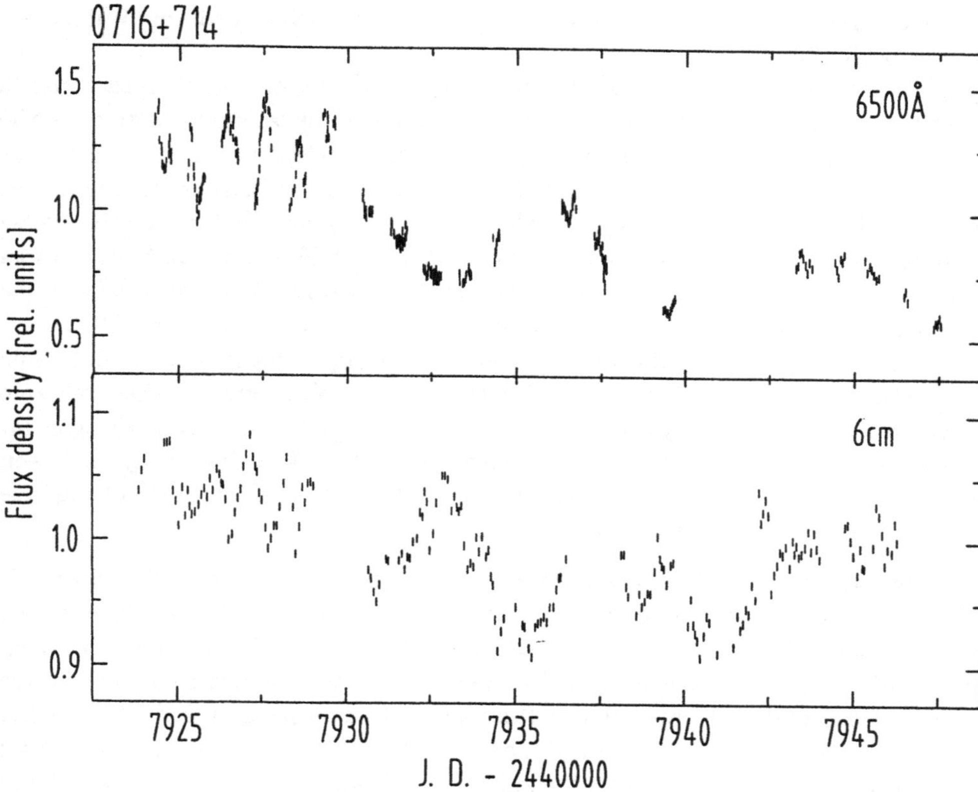

Fig. 6. Light curve of the BL Lac object 0716+714 at 6500 Å(R-band) and at 6 cm in February 1990

of the strong variations observed e.g. in 0716+714 and 0917+624; the strongest argument is given by the radio-optical correlation in 0716+714. In models based on RISS, it is also difficult (but probably not impossible) to reproduce the frequency dependence and the polarization variations observed in 0917+624. The statistical results on the two complete samples, especially the absence of non-variable sources with compact VLBI structures can readily be understood in terms of m.a.s.-size source components scintillating in the "standard" interstellar medium (Witzel and Quirrenbach 1990).

The frequency dependence of the variations and the polarization characteristics observed in 0917+624 can most easily be explained by a superposition of the "quiet" emission of a pc-scale jet and of a shock propagating in the innermost part of this jet (Qian et al. 1990, Witzel 1990). The very high brightness temperatures of $\gtrsim 10^{19}$ K implied by the rapid variations can be reconciled with the inverse Compton limit of $\sim 10^{12}$ K by assuming either very high bulk Lorentz factors $\gamma \approx 100$, or special non-spherical geometry of the emitting region, or non-isotropic pitch angle distribution of the relativistic electrons (Quirrenbach et al. 1990).

Further coordinated radio and optical monitoring of blazars with high time resolution, preferably including polarimetry, will certainly give further valuable information about the physical conditions in the regions producing the continuum emission of these sources. Proper measurements of the time delays between radio and optical variations are urgently needed to constrain models of particle acceleration and source geometries.

Acknowledgements: It is a pleasure to thank Drs. A. Alberdi, K. Anton, U. Erkens, M. Haehnelt, D. Heeschen, C. Hummel, K. Johnston, T. Krichbaum, M. Ott, S.J. Qian, M. Rioja, F. Sanchez-Pons, C. Schalinski, P. Schneider, S. Wagner, R. Wegner, J. Wambsganss, and A. Zensus for their contributions to various aspects of the material presented in this paper. I am especially indebted to Dr. A. Witzel, who suggested "intraday radio variability" to me when I was looking for a thesis work, and who patiently guided me through all the difficulties of the observations, data reduction, and interpretation.

References

Heeschen, D.S. 1984, *Astron. J.* **89**, 1111.

Heeschen, D.S., Krichbaum, T., Schalinski, C.J., and Witzel, A. 1987, *Astron. J.* **94**, 1493.

Heeschen, D.S., and Rickett, B.J. 1987, *Astron. J.* **93**, 589.

Hughes, P.A., Aller, H.D., and Aller, M.F. 1990, in *Parsec scale radio jets*, eds. Zensus, A., and Pearson, T., Cambridge University Press, p. 250.

Qian, S.J., Quirrenbach, A., Witzel, A., Krichbaum, T.P., Hummel, C.A., and Zensus, J.A. 1990, *Astron. Astrophys.*, in press

Quirrenbach, A. 1990, *Ph. D. thesis*, University of Bonn.

Quirrenbach, A., Witzel, A., Krichbaum, T., Hummel, C.A., Alberdi, A., and Schalinski, C. 1989a, *Nature* **337**, 442.

Quirrenbach, A., Witzel, A., Qian, S.J., Krichbaum, T., Hummel, C.A., and Alberdi, A. 1989b, *Astron. Astrophys.* **226**, L1.

Quirrenbach, A., Witzel, A., Wagner, S., Sanchez-Pons, F., Anton, K., Haehnelt, M., Johnston, K.J., Krichbaum, T.P., Wegner, R., Zensus, J.A. 1990, *in preparation.*

Pauliny-Toth, I.I.K., and Kellermann, K.I. 1966, *Astrophys. J.* **146**, 634.

Shapirovskaya, N. Ya. 1978, *Astron. Zh.* **55**, 953; *Sov. Astron.* **22**, 544.

Sholomitskii, G.B. 1965, *Astron. Zh.* **42**, 673; *Sov. Astron.-AJ* **9**, 516.

Wagner, S., Sanchez-Pons, F., Quirrenbach, A., and Witzel, A. 1990, *Astron. Astrophys.* **235**, L1.

Wambsganss, J., Schneider, P., Quirrenbach, A., and Witzel, A. 1989, *Astron. Astrophys.* **224**, L9.

Witzel, A. 1990, in *Parsec scale radio jets*, eds. Zensus, A., and Pearson, T., Cambridge University Press, p. 206.

Witzel, A., Heeschen, D. S., Schalinski, C., and Krichbaum, Th. 1986, *Mitt. Astron. Ges.* **65**, 239.

Witzel, A., and Quirrenbach, A. 1990, in *Propagation Effects in Space VLBI*, eds. Kardashev, and Gurvits, in press.

The High Frequency Properties of Synchrotron Radiation

Keith Ballard

Department of Astronomy, Blackford Hill, Edinburgh, UK. EH9 3HJ.

1. Introduction

The aim of this paper is to examine the relevance of models involving shocks to the high frequency near infrared and optical emission in BL–Lacs. A short study of the effect that losses can have upon non–thermal electron distributions is presented. This is combined with previous work relating to specific magnetic field geometries behind shock fronts to yield model spectra which can be compared to those actually observed. Agreement between the theory and observations is found only if the electrons are allowed to remain isotropic. Some of the limitations and implications for this model of the emission region are considered.

2. BL–Lac objects

Before proceeding to consider the details of a model for the optical and infrared emission of BL–Lacs (by which we refer to those objects where there is no suspicion that the optical light has been diluted by unpolarized thermal emission) it is worth stating briefly the observational properties of the high frequency emission, which a model is required to reproduce. For these we summarize the conclusions of Mead *et al.*, (1990) and Ballard *et al.*, (1990). These papers present the results of a program of monitoring 48 blazars (BL–Lacs & HPQ's) carried out on 14 nights between 1986 and 1988 at UKIRT. The observations were made simultaneously in the UBVRIJHK wavebands.

1.) In many cases the spectrum was a power law of constant slope, often steeper than $\alpha = 2.0$. There were, however, a significant number of observations ($\sim 30\%$) where the spectrum steepened with increasing frequency. This fact, coupled with the fact that the spectrum in the optical is steeper than in the radio (where $\alpha \simeq 0.2$, Valtaoja, this volume) is strongly suggestive of losses having modified the spectrum.

2.) Frequency dependence of polarization was a common occurrence, and, with only a couple of exceptions, in the case of the BL–Lacs it always increased towards the blue end of the spectrum. The level of polarization (up to 43%) was indicative of a partially ordered magnetic field when averaged over the entire emission region — although it is possible that it could be almost uniform in smaller regions. Frequency dependence of position angle was less common. Most of the frequency dependence of polarization is associated with the intrinsic curvature of the flux via the simple parameterization of Björnsson & Blumenthal (1982) *i.e.* as the spectral index increases so does the polarization according to the relationship:

$$\mathrm{pol}(\nu) = \Pi \frac{\alpha(\nu) + 1}{\alpha(\nu) + 5/3} \times 100\%$$

where $\alpha(\nu)$ is the spectral index and Π is a constant representing the degree of ordering in the magnetic field.

3.) Variability was usual over timescales of a single day. No obvious characteristic patterns to the variability were apparent.

These are the principal characteristics of BL–Lacs found from a large number of observations; any model must be able to explain these in order to be taken seriously. Having achieved that, a good model should also allow for the more occasional extreme events such as those seen by Holmes *et al.*, (1984) or Moore *et al.*, (1982).

3. The model

The remainder of this paper is devoted to the properties of shock models of BL–Lacs when losses are included. Such a study, although it does not reach any startlingly new conclusions, is instructive because it tests the limits of the applicability of models which purport to explain the blazar phenomenon.

There are two main motivations for a study of this kind. The first is simply an attempt to plug a gap in the knowledge of the properties of a simple synchrotron source. The original work in the field was published by Kardaschev (1962) who presented results for the evolution of a non–stationary synchrotron source, as the electrons lost or gained energy to various mechanisms, such as synchrotron emission, adiabatic expansion or the inverse Compton effect. Implicit in this work was the assumption that as the electrons lose energy they are tied to an element of the magnetic field and maintain their initial pitch angle with respect to the magnetic field. The only specific source fields considered were isotropic and consequently the overall electron distribution remained isotropic with time, and as there was no preferred field direction there was no polarization. The aim of this work is to evaluate any possible anisotropy which might develop in a more realistic field geometry and to investigate the effect that this has on the polarization.

The second motivation arises from observational studies of blazar emission, the interpretations placed upon the results and the necessity to adapt these to future observing programmes. Over the past decade the possibility that a shock passing along a jet represents the site of blazar emission has received much attention. Groups such as those situated in Preston and Michigan have obtained a large amount of data at millimetre and centimetre wavelengths and have shown convincingly that the day to day behaviour of the flux and polarization (Hughes, Aller & Aller, 1985) or the large flaring events in the flux around the self–absorption turnover which are periodically observed (Marscher & Gear, 1985) can be modelled by a fairly simple shock in a jet model. There is clearly a need to extend these works to higher frequencies in order to test their applicability in these regimes and to try therefore to understand the high frequency radiation. This will test the supposition that the optical emission originates in the same region as the radio emission. Furthermore, as submillimetre observations of the polarization become available it is important to be aware of the implications that they could have for our understanding of the emission region and mechanism, in particular the role played by losses in determining their properties.

In attempting to model blazars at high frequencies there are two elements which must be determined, the magnetic field and the electron distribution. Regarding the magnetic field, we have chosen in this case to extend naturally a model which has had success in the past *i.e.* that of Hughes, Aller & Aller (1985). In this model it is assumed that there is a shock passing along a jet close to the line of sight. The jet contains a magnetic field, taken to be isotropic with a uniform distribution of magnitudes upstream of the shock. Further, assume that the magnetic field is dynamically unimportant so that it plays no part in determining the jump conditions. As material passes through the shock, it is compressed and the magnetic field perpendicular to the shock normal is enhanced by a factor of r, the compression ratio of the shock. The material then emits synchrotron radiation with the Stokes parameters:

$$I(\nu,t) = L\int_E\int_\theta\int_\phi B\sin\alpha F\left(\frac{\nu}{\nu_c}\right)N(E,\alpha,t)d\phi\sin\theta d\theta dE$$

$$\begin{matrix} Q(\nu,t) \\ U(\nu,t) \end{matrix} = L\int_E\int_\theta\int_\phi B\sin\alpha G\left(\frac{\nu}{\nu_c}\right)\begin{matrix} \cos 2\chi \\ \sin 2\chi \end{matrix} N(E,\alpha,t)d\phi\sin\theta d\theta dE$$

$$V(\nu,t) = 0$$

e.g. Pacholczyk, (1970). The various terms in these equations are as follows: L is the line of sight depth of the source; note that throughout this work it has been assumed that the source is optically thin as we are primarily interested in the high frequency emission where this should be a reasonable assumption. $F(x)$ and $G(x)$ are standard synchrotron functions expressing the spectrum of radiation from a single electron, E is the electron energy. The angle χ is the position angle on the sky of the radiation measured relative to some fixed direction. $B\sin\alpha$ is the magnetic field projected onto the plane of the sky as a function of α, the angle between the line of sight and the element of the magnetic field. In order to evaluate α it is necessary to take the components of the magnetic field downstream of the shock (which lies in the xy plane) in terms of the polar co-ordinates θ, ϕ where θ is the angle between the element of field and the shock normal and ϕ is the azimuthal angle, and rotate it through an angle, ϵ to the x–axis so that the electron's velocity vector lies along the new x'–axis. From this an expression for $B \sin \alpha$ can be derived:

$$B_2 \sin_2 \alpha = B_2(r_2 \sin_2 \theta \sin_2 \phi + r_2 \sin_2 \theta \cos_2 \phi \sin_2 \epsilon \\ + \cos_2 \theta cos_2 \epsilon - 2r \sin \theta \cos \theta \cos \phi \sin \epsilon \cos \epsilon)$$

And then by integrating over θ, ϕ one is effectively integrating the emission from an electron over the entire magnetic field distribution. This introduces two extra parameters into the model, the shock compression ratio r and the angle between the line of sight and the shock face ϵ. Finally, one has to supply $N(E, \alpha, t)$, the distribution of electrons with energy, pitch angle and time. There are several possibilities for this which will be considered in turn in the next section.

4. Results

4.1. No scattering, decaying source

We assume an initially isotropic distribution of electrons, with a power law energy distribution, with a cutoff at a low energy, E_b, and high energy, E_c. Also in this case assume that the electrons have a pitch angle which is constant with time. Then the decay of a simple power law source with synchrotron losses is governed by:

$$N(E, \alpha, t) = kE_{-\gamma}(1 - \beta t E)_{\gamma-2} \Theta \left(\frac{E_c}{1 + \beta t E_c} - E \right)$$

(Kardaschev, 1962) where the step function has been included to follow the evolution of the high energy cutoff. $\beta = (6\pi\epsilon_0 m_{e4} c_5 / e_4)_{-1} B_2 \sin_2 \alpha = bB_2 \sin_2 \alpha$. The results of a sample integration showing flux, spectral index (the negative slope of the log(Flux)–log(ν) plot), the polarization and the position angle are shown in Fig 1.

This case, where the electron distribution becomes anisotropic can be regarded as occurring whenever the electron pitch angle remains constant for a longer time than the time it takes the lowest energy electron to lose energy, perhaps because the field is essentially uniform over large distances. The parameters chosen are $r = 4.0$ and $\epsilon = 0_\circ$ (*i.e.* looking along the shock face). This choice maximizes the degree of polarization in order to emphasize the properties of the source. The most striking feature is the decrease in polarization and position angle rotation of $90_\circ$ which develops as the spectrum evolves. There is a simple explanation for this feature. At low frequencies and early times the enhanced magnetic field perpendicular to the shock normal dominates the emission giving a polarization position angle aligned with the shock normal. As the spectrum evolves the electrons which are tied to these field lines lose more energy than those electrons lying on field lines along the shock normal. At high frequencies it is these electrons, moving around field lines

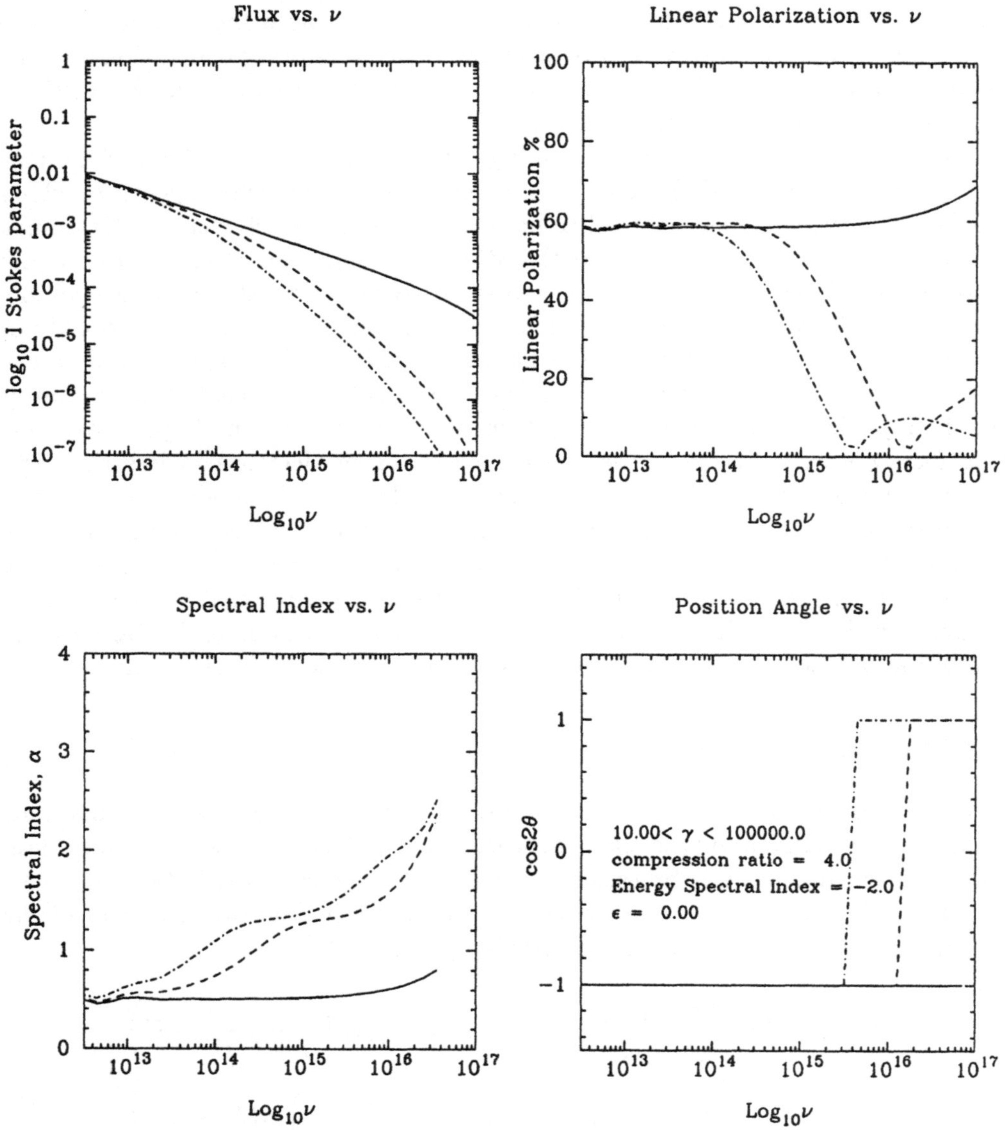

Fig. 1. This shows the decay of a source with synchrotron losses & no electron scattering. The three lines are for $t = 0$ (solid) and two equally spaced later times (dashed then dot–dashed). The electrons maintain their original pitch angle and the distribution rapidly becomes anisotropic. The decrease in polarization is contrary to what is generally observed at optical wavelengths, and there is an abrupt position angle flip of $90_{\circ}$.

along the shock normal which are left and which dominate the emission. At the frequency where the polarized fluxes are equal, there is zero polarization and the position angle rotates. The emission at the highest frequencies is characteristic of synchrotron emission around a cutoff.

However, the spectra presented in Fig 1 are dissimilar to the form of the majority of observations of blazars. There is only a single published spectrum which looks qualitatively similar, that of Holmes *et al.*, (1984) (which incidentally cannot be satisfactorily fitted by the model). Although it might be unwise to reject this model out of hand for all observations, it is clearly not relevant for the vast majority. It should also be mentioned at this point that the model so far is probably of little relevance in understanding BL–Lacs because in the first instance only a decaying source has been considered whereas one should also include the possibility of the injection of fresh energetic electrons if one is to have any hope of explaining flaring events. Secondly there is no scattering of electrons in less than the synchrotron loss timescale of the electrons considered. This necessarily precludes the possibility of the electrons being accelerated through the Fermi mechanism, or indeed any process which alters the pitch angle of the particle. It is therefore not too surprising that the results in Fig 1 bear little resemblance to the spectra which are observed. Although the results discussed above are useful in understanding the evolution of a source, an attempt to drop these restrictions should be made.

4.2. Electron scattering

We will now consider briefly the alternative limit where the electrons maintain isotropy; this corresponds to the case where the scattering timescale is much less than the synchrotron loss timescale. In deriving an electron distribution one must integrate the energy lost to an element of the magnetic field over the whole field distribution equally. One obtains a distribution of the form:

$$N(E,\alpha,t) = \frac{kE_{-\gamma+1}}{\beta(\gamma-1)}\left([1-(1-\beta tE)_{\gamma-1}]\,\Theta\left(\frac{E_c}{1+\beta tE_c}-E\right)\right.$$
$$\left.+\left[1-\left(\frac{E}{E_c}\right)_{\gamma-1}\right]\Theta\left(E-\frac{E_c}{1+\beta tE_c}\right)\Theta(E_c-E)\right)$$

where $\beta = -(4\pi/9)(5r_2+1)bB_2$ and a constant injection of electrons with an isotropic power–law distribution has been included. The electron distribution is independent of pitch angle. The results of a sample integration are shown in Fig 2.

The flux is straightforward to explain; at low frequencies it rises continuously as fresh electrons are injected, at high frequencies above the break the injection is balanced by the energy loss mechanism. There is no rotation of the position angle as no anisotropies develop in the electron distribution (see Björnsson, 1985). The interesting behaviour here is the form of the polarization which increases gradually. This behaviour is quite similar qualitatively to what one actually observes for BL–Lacs, which is encouraging. The final plot included in Fig 2 is the polarization

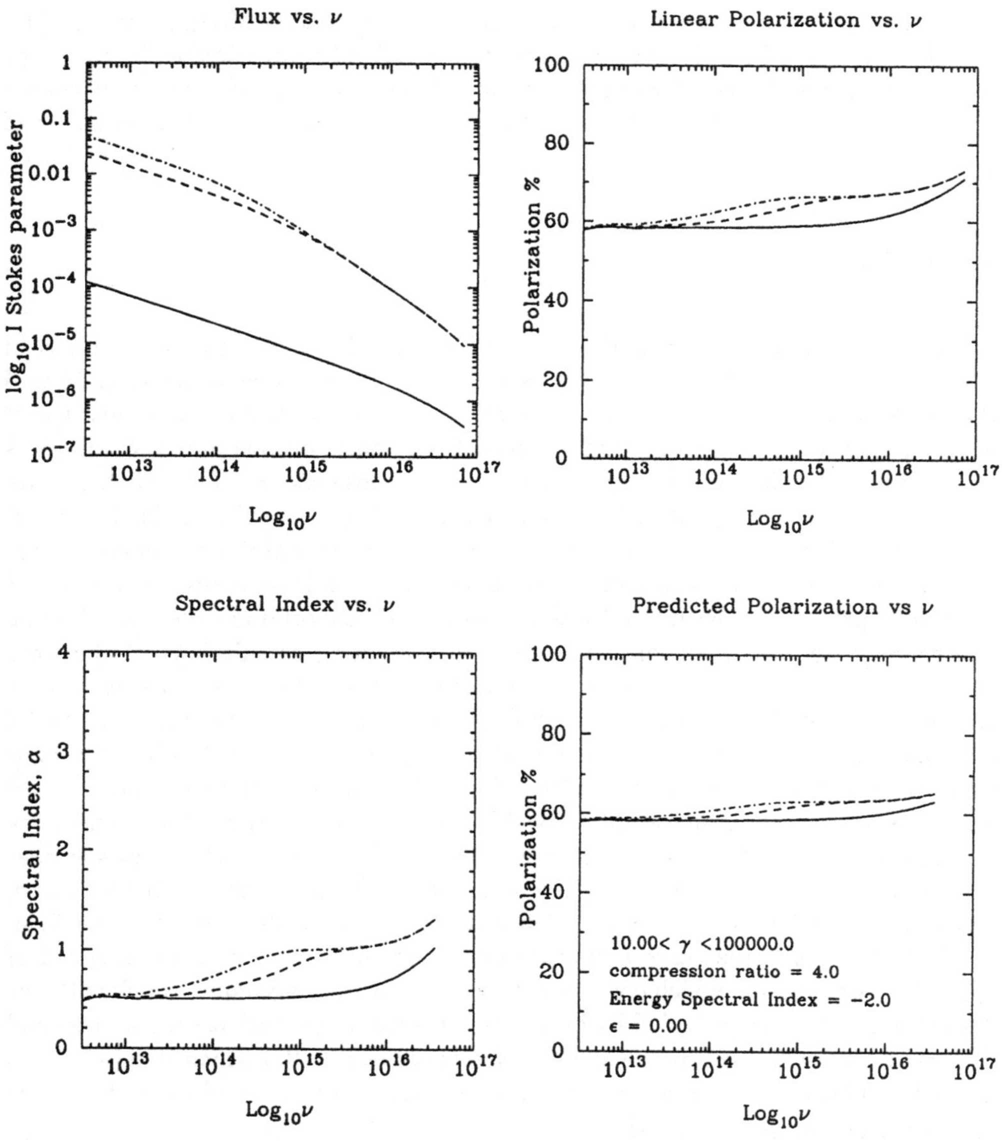

Fig. 2. This shows the evolution of a source with synchrotron losses & electron scattering included, together with the steady injection of 'fresh' electrons with a power law form. The electrons now scatter in much less time than they lose energy and the distribution remains isotropic. The final plot is the polarization predicted from a parameterization of the form of equation 1.

predicted by the parameterization of synchrotron radiation of Björnsson & Blumenthal (1982). The polarization changes are derived from changes in the slope of the power–law distribution of the electrons. As mentioned earlier the observed polarization could be fitted well by this parameterization in a large number of cases.

5. Discussion

We have considered the properties of the emission from two possible models of the source structure of BL–Lacs and have shown that if models which explain well the emission at radio wavelengths are correct, and are applicable to optical wavelengths, then qualitative agreement between the observations and the theory is found only if the electrons remain isotropic as they lose energy. If the electrons are scattered in the vicinity of a shock front, then the first order Fermi shock acceleration mechanism will work. Thus there is further circumstantial evidence at least that this mechanism is occurring in BL–Lacs. The Fermi mechanism is attractive because it provides us with a natural explanation of the origin of the non–thermal electrons and a prediction of their slope. In the main, calculations of the form of the particle distribution assume a uniform magnetic field, this assumption is another that should be dropped. In the case of a completely disordered magnetic field one can use numerical simulations of the orbits of electrons to obtain a value for the spectral index (Achterberg, 1988; Ballard *et al.*, in preparation); these turn out to be steeper than one would expect for an ordinary oblique shock where the electrons get trapped upstream (Kirk & Heavens, 1989). In fact the results appear to be similar to those for the simplest parallel shock case. The spectrum of electrons accelerated at a shock front with a disordered magnetic field are shown in Fig 3. In this particular case a compression ratio of $r = 4.0$ and shock speed of $u_{sh} = 0.4c$ were taken, which corresponds to a strong hydrodynamic shock with the downstream pressure provided by ions only. This example produces a non–thermal spectral index which is not the same as the value of $\alpha \simeq 0.2$ which is observed in the radio. This could be more accurately reproduced with a simulation at a shock with a higher compression ratio.

There are still considerable problems with this model of the optical emission region, for example it cannot explain the relatively common occurrence of frequency dependent position angles. One could presumably reproduce this by employing either anisotropic distributions at some level (Björnsson, 1985) or multiple component models, however unappealing that may be. However that is to miss the point of this work which is to start to explain the global properties of BL–Lacs as outlined at the beginning using a simple physical model, rather than to rely upon more empirical pictures which are less informative about the actual physics. Phenomenological models where a number of randomly oriented components are taken and passed through a shock can explain the observed correlations between the various properties of the emission and by rather artificial adjustments to the parameters can reproduce the incidence rate of frequency dependent polarization

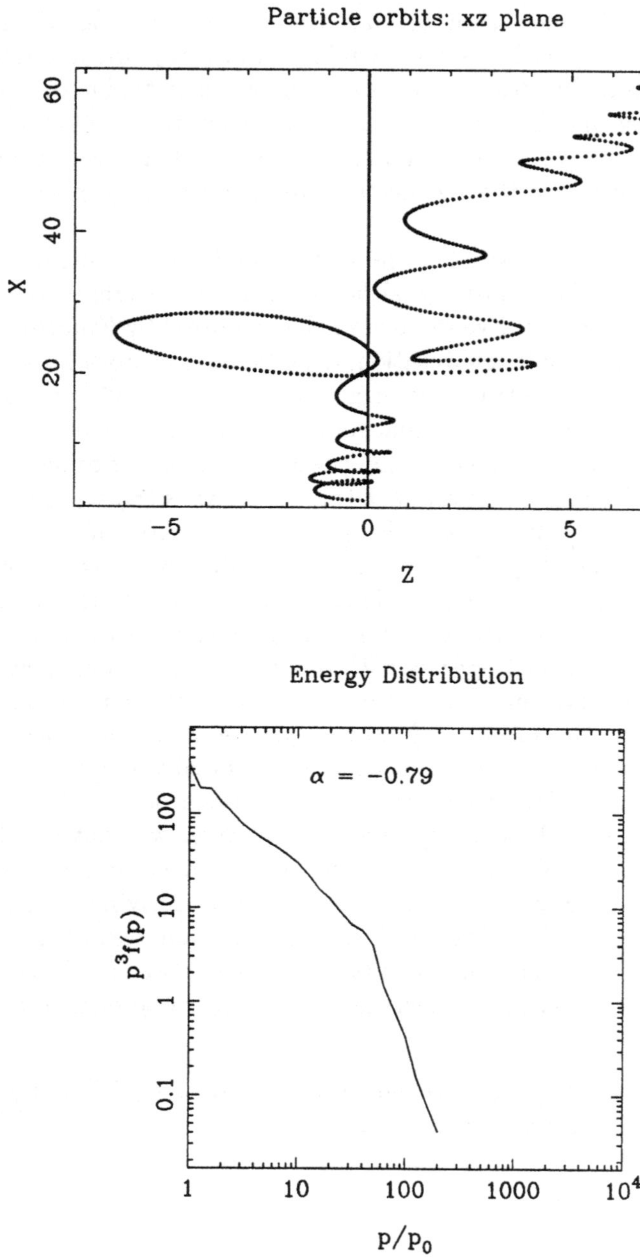

Fig. 3. The top figure shows an example of an electron accelerated by the Fermi mechanism in a randomly oriented magnetic field. The electron starts from the shock front and is injected upstream (z negative). It crosses the shock repeatedly until lost from the shock vicinity. The bottom figure shows the resultant spectrum. The synchrotron spectral index is $\alpha = 0.79$ (see Ballard *et al.*, in preparation for details).

and position angle. In such a picture one would effectively be summing over a finite number of discrete cells, each containing a randomly oriented uniform magnetic field, rather than integrating over a continuous distribution of magnetic field elements. This work may be considered the limit when the number of elements is infinite. It may well be that a model with an infinite number of cells is only directly relevant to the radio where the synchrotron lifetimes are very much greater than at optical wavelengths.

Finally, the connection between the optical and radio emitting regions has not yet been established conclusively, although there are several pieces of evidence which favour this being true. Events such as that reported by Kikuchi *et al.* (1988) for OJ287 where a position angle rotation, simultaneous at both radio and optical wavelengths was observed are highly suggestive. Position angle rotations such as this may be a result either of an actual rotation of the magnetic field within the source, or of aberration effects, where the speed of the jet itself changes (Blandford & Königl, 1979). However the mere fact that it occurs suggests that the optical light is originating in the same region in space as the radio emission. In addition, the work of Landau *et al.* (1986) who find the radio to optical spectra change smoothly supports this. If the emission in the two wavebands did arise in separate regions and varied independently then the spectra would not be smooth. The fact that Landau *et al.* fitted their spectra with parabolae is not in conflict with sharp breaks in the energy distribution because of the nature of the polynomials F(x) & G(x) which represent the spectrum from a single electron and which are smooth relative to the energy distribution; hence any sharp breaks tend to be washed out, giving a much smoother flux distribution as indeed can be seen from Figs 1 and 2. Finally there is the work of Impey (1987) who has shown that the distribution of the difference between VLBI structure axis and preferred optical position angle tends to be peaked around $90_{\circ}$, again suggestive of underlying source structure. In this respect measurements of the polarization at submm wavelengths, would be helpful, in addition to testing the validity of the above work. These are not yet available to observers although an instrument is undergoing final tests at present at the JCMT.

Acknowlegements: KRB acknowledges the University of Edinburgh for travel assistance and the SERC for a studentship.

References

Achterberg, A., 1988. *Mon. Not. R. astr. Soc.*, **232**, 323.

Ballard, K.R., Mead, A.R.G., Brand, P.W.J.L., Hough, J.H., 1989. *Mon. Not. R. astr. Soc.*, **243**, 640.

Björnsson, C.-I., 1982. *Mon. Not. R. astr. Soc.*, **216**, 241.

Björnsson, C.-I. & Blumenthal, G.R., 1982. *Astrophys. J.*, **259**, 805.

Blandford, R.D. & Königl, A., 1979. *Astrophys. J.*, **232**, 34.

Holmes, P.A., Brand, P.W.J.L., Impey, C.D., Williams, P.M., Smith, P., Elston, R., Balonek, T., Zeilik, M., Burns, J., Heckert, P., Barvainis, R., Kenny, J., Schmidt, G. & Puschell, J., 1984. *Mon. Not. R. astr. Soc.*, **211**, 497.

Hughes P.A., Aller H.D. & Aller, M.F., 1985. *Astrophys. J.*,**298**, 301.

Impey, C.D., 1987. In : *Superluminal Radio Sources*, p. 233, eds. Zensus, J.A. & Pearson, T.J., Cambridge University Press.

Kardashev, N.S., 1962. *Sov. Astron. - A.J.*, **6**, 317.

Kikuchi, S., Inoue, M., Mikami, Y., Tabara, H. & Kato, T., 1988, *Astr. Astrophys.*, **190**, L8.

Kirk, J.G. & Heavens, A.F., 1989. *Mon. Not. R. astr. Soc., in press*

Landau, R., Grolsch, B., Jones, T.J., Jones, T.W., Pedelty, J., Rudnick, L., Sitko, M.L., Kenney, J., Roellig, T., Salonen, E., Urpo, S., Schmidt, G., Neugebauer, G., Matthews, K., Elias, J.H., Impey, C.D., Clegg, P. & Harris, S., 1986. *Astrophys. J.*, **308**, 78.

Marscher, A.P. & Gear, W.K., 1985. *Astrophys. J.*, **298**, 114.

Mead, A.R.G., Ballard, K.R., Brand, P.W.J.L., Hough, J.H., Bailey, J.A. & Brindle, C., 1989. *Astr. Astrophys.*, **83**, 183.

Moore, R.L., McGraw, J.T., Angel, J.R.P., Duerr, R., Lebofsky, M.J., Rieke, G.H., Wiśniewski, W.Z., Axon, D.J., Bailey, J.A., Hough, J.H., Thompson, I., Breger, M., Schulz, H., Clayton, G.C., Martin, P.G., Miller, J.S., Schmidt, G.D., Africano, J. & Miller, H.R., 1982. *Astrophys. J.*, **260**, 415.

Pacholczyk, A.G., 1970, Radio Astrophysics, Freeman: San Francisco.

Synchrotron Source Models and the Infrared–Optical Variability of Blazars

Klaus Meisenheimer

Max-Planck-Institut für Astronomie, Königstuhl 17, 6900 Heidelberg 1, Germany

Abstract: I explore whether the source models developed for extended radio sources can be scaled down to explain the polarized infrared–optical continua of blazars and their dramatic variations. I discuss in some detail the overall polarization and flux from a jet with a helical field containing a shock. It turns out that even very simple models of this class are able to explain the exotic behaviour observed in OJ 287 in January 1983.

1. Introduction

Observations with linked radio interferometers (e.g. VLA, MERLIN, and the Australian Telescope) show the structure of extended radio sources in great detail (Perley *et al.* 1984, Carilli *et al.* 1989, Owen *et al.* 1989). In some cases the low frequency measurements are supplemented by observations in the high frequency regime (reviewed in Röser 1989, Meisenheimer 1989), thus we have some information about the magnetic field structure and field strengths, the energy distribution of the relativistic electrons responsible for the synchrotron radiation, and the places where those electrons are accelerated. In this contribution I would like to line out how the source models which have been developed for these extended sources may help to understand the strange behaviour observed in the compact and highly variable synchrotron sources – the blazars.

Given that it is generally accepted that the flat *radio* continua of blazars are produced by a superposition of several self-absorped components with different frequencies ν_1 below which the synchrotron radiation becomes optically thick ($\nu_1 = 10^9...10^{12}$ Hz), it is clear that the synchrotron source models which apply to (optically thin) extended sources can only be applied to the high frequency domain of a compact flat spectrum radio source. Thus I will focus on the *infrared–to–optical* regime of the blazar continuum only.

A similar approach has led to the "shocked jet" models proposed in the past by several authors (see Blandford & Königl 1979, Königl & Choudhuri 1985*a*, *b*,

Hughes, Aller & Aller 1985, 1989, Marscher & Gear 1985). However, these models try to describe the cm to millimetre variations. Thus optical depth effects which are very hard to quantity may dominate the observed spectra and polarizations. Models which explicitly try to reproduce the high frequency observations by *one* selfconsistent source have been sparse (e.g. Björnsson 1985).

2. The synchrotron source model

My source model is cooked–up from the following ingredients:

(a) a cylindrical jet (coordinates: r, φ, z) of which only one shell $[r, \Delta r]$ with small gradients of the magnetic field $\partial B/\partial r \cong 0$ and the relativistic particle density $\partial n_0/\partial r \cong 0$ is considered.

(b) A particle acceleration process which produces a shock acceleration like spectrum $n(E) \sim E^{-q}(1.5 \leq q \lesssim 2.5)$ with a high energy cutoff around $E = E_c$ where synchrotron losses balance the acceleration gains (Webb *et al.* 1984, Heavens & Meisenheimer 1987, Kirk & Schneider 1987). The electrons may either be accelerated everywhere in the source or in a thin layer at $z = 0$. The pitch angle distribution of the electron momentum is *isotropic*.

(c) The dominant loss process of the relativistic particles is synchrotron emission resulting in a temporal or spatial evolution of the freshly accelerated electron distribution. Thus radiated power and energy loss of the particle population are equal.*

(d) In order to keep the model simple and illustrative a purely helical field $\mathbf{B} = (0, B_\varphi, B_z)$ with vanishing radial or random components is assumed. Both the toroidal component B_φ and the poloidal field B_z may change along the z–axis.

In Figure 1 I try to visualize the field structure of a jet with a helical field $B_\varphi = B_z$ (45°–helix). Only one bunch of field lines is shown but one should imagine that the whole jet mantle is filled with the field. At some place z_s I have introduced an axial compression by a factor $r = 4$ which corresponds to a strong planar shock ("knot"). The toroidal field is enhanced by 4 while the poloidal field remains unchanged: $B_\varphi = 4B_z$. Thus the helix looks more tightly bound. As you can see by inspecting the figure, in the **jet** ($z < z_s$) the projected field is dominated by a longitudinal component while in the **knot** ($z > z_s$) most of the projected field lines are orientated perpendicular to the jet axis. This provides a natural explanation for the polarisation angle flips observed in knots or hotspots of radio jets. Note that the mean projected field in the jet is orientated along the axis although the toroidal component is as strong as the poloidal one. Highly resolved radio maps of the jet in M 87 (Owen *et al.* 1989) indicate that a helical field structure is indeed realized in nature.

---

* Inverse Compton losses are not considered. They would increase the loss rate but would not contribute significantly to the infrared–to–optical radiation.

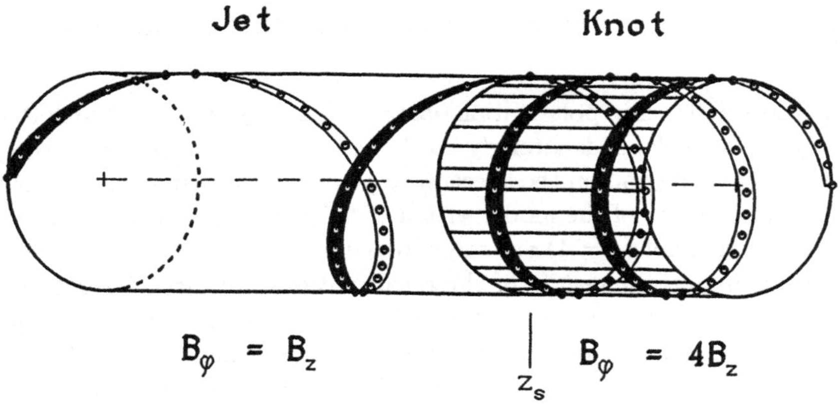

Fig. 1. Jet with helical magnetic field (see text).

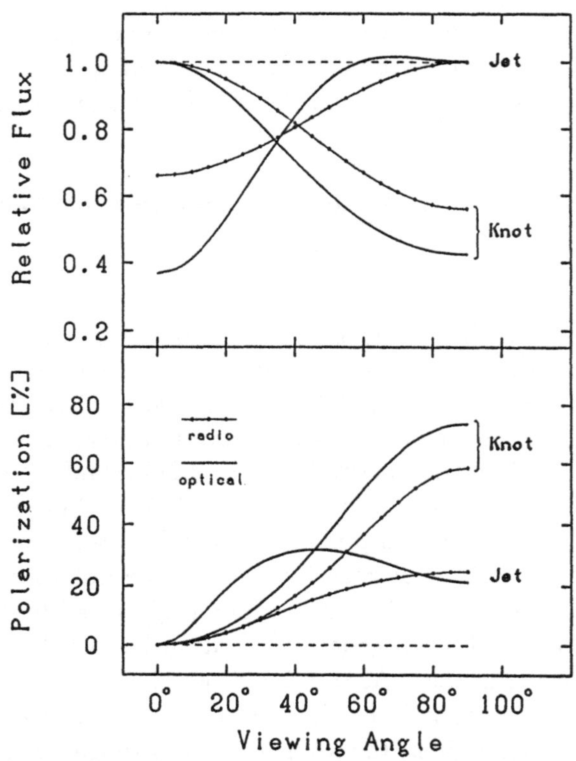

Fig. 2. Relative flux and polarization from a jet with helical magnetic field. "Jet" refers to a cylindrical mantle with $B_\phi = B_z$ while "knot" represents a region of enhanced toroidal field $B_\phi = 4\ B_z$. Two different spectral regimes are considered: "radio" refers to a region where the spectrum follows a straight power law $S_\nu \sim \nu^{-1}$, "optical" refers to a regime in the spectrum near a high frequency cutoff as observed in many hotspots and blazars ($\alpha \cong -3$).

3. Synchrotron radiation from a jet with helical field

The synchrotron flux and polarization observed from a jet and knot with the helical field given in Fig. 1 depends strongly on the viewing angle Θ (between the jet axis and the line of sight) and on the spectral slope of the spectrum near the observing frequency. The results of the model with a 45°–helix for the flux and polarization of the jet and its knot ($r = 4$) are summarized in Fig. 2.

The most important results of these calculations are: first, the fractional polarization P is always low ($P < 33\% \ll P_{max}$) although the field is perfectly ordered. The dependence of P on the viewing angle changes with the observing frequency. Most striking is the flat maximum at $P \cong 30\%$ at $\Theta > 30°$ observed in the optical regime. This behaviour can produce a very wide variety of frequency–dependent polarization if one combines the light from an unresolved jet and knot with perpendicular polarization.

Second, the total flux also depends on the viewing angle without assuming any anisotropy in the pitch angle distribution or relativistic bulk motion. This comes purely from the $S_\nu \sim B^{1-\alpha}$ dependence of optically thin synchrotron emission. In the optical regime one may change the flux ratio jet : knot by a factor of 10 when varying the aspect angle from 0° (head–on) to 90° (jet axis in the plane of sky).

4. Modelling the frequency–dependent polarization of blazars

As an example I apply the synchrotron source model with helical field to the most spectacular case of rapid variations of frequency dependent polarization – the polarization flare of OJ 287 reported by Holmes *et al.* (1984). They observed OJ 287 on four consecutive nights in January 1983 with several telescopes simultaneously to measure the flux and polarization between the K (2.2μm) and U band (0.35μm). Fig. 3 displays the strange polarization behaviour of OJ 287. On the first night (7-Jan) a polarization of about 9% with position angle (E-vector) at about 80° was observed. On the second night a dramatic change in the polarization occurred: The mean polarization dropped to $P \lesssim 2\%$, apparently caused by the balancing effect of two components with orthogonal polarization but similar spectra. This also explains the 80° swing in the P.A. at $\lambda \cong 600$ nm (5×10^{14} Hz). On the next day (9-01) the amount of polarization has increased again and the position angle is near to its original value throughout the whole frequency range observed. Another day later (10-01; not shown) the observed polarization has changed very little.

The smooth lines through the data points demonstrate that a simple model assuming a shock in a helical field is able to produce the observed polarization. The model assumes a steady injection of relativistic particles with a power–law spectrum $n(E) \sim E^{-1.7}$ with a sharp cutoff at $E_c = 17$ GeV at $z = 0$ and the variable field configurations as displayed in Figure 4. The region of the jet which is affected by the observed polarization flare is small (< 0.02 pc) compared with

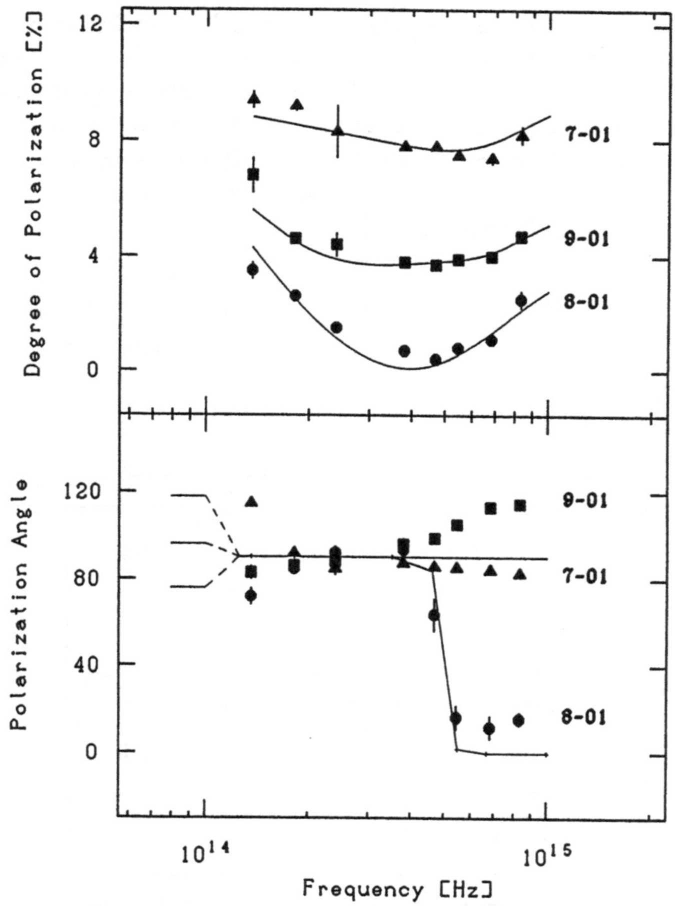

Fig. 3. The infrared–to–optical polarization of OJ 287 as observed on three consecutive nights in January 1983 (Holmes *et al.* 1984). The smooth lines through the data points show the result of my model calculations. To allow the direct comparison with the strictly axisymmetric model I have removed a steady position angle swing from $P.A. = 75°$ (7–01) to $P.A. = 116°$ (9–01) from the observed polarization angles.

the entire length of the emission region. From the wider set of field configurations which could reproduce the observations, I choose the one requiring the least changes between the observed epochs. This model implies that the perturbation which is responsible for the observed event had already occurred before January 7th. On the first night (7-01) the poloidal field immediately downstream of the particle injection (at $z = 0$) is enhanced above its mean value (Fig. 4). At $z = 0.003\,\mathrm{pc}$ downstream a shock occurs which enhances the toroidal component by a factor of $r = 4$ and causes a rapid decay of the poloidal component. On the 8th of January, the poloidal field at $z < 0.003\,\mathrm{pc}$ has weakened while the toroidal component has not changed. The further development is characterized by a broadening and weakening of the shocked region, progressing further downstream. The model reproduces not only the observed polarization rather nicely but also agrees with the observed flux variations: the model predicts that the R band flux on the three subsequent nights is 0.85:1:1. Holmes *et al.* observed 0.82:1:0.9. Note that the entire photo–polarimetric variability is generated in this model by magnetic field changes alone. The rotation of the polarization angle from $P.A. = 75°$ to $P.A. = 116°$ observed at $\lambda \geq 1\mu m$ cannot be reproduced by a strictly symmetric model but points to a helical structure of the emission regions as observed for example in the jet of M 87. (See also Königl & Choudhuri 1985).

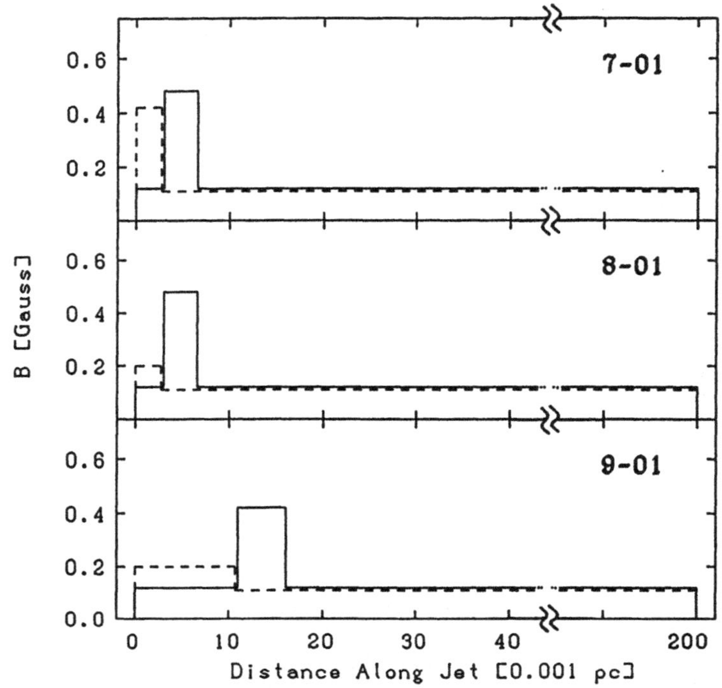

Fig. 4. Magnetic field variation during the polarization flare of OJ 287. The toroidal field B_ϕ (——) and poloidal field B_z (- - -) for the three nights are shown. The relativistic electrons are injected at $z = 0$. The model allows one to determine the absolute field strength (see text). In calculating the linear scale I assumed a jet velocity near the speed of light.

5. Field strength and linear dimensions

You may have noticed that Fig. 4 gives absolute values of the magnetic field strength and the linear dimensions of the synchrotron emitting regions. These numerical values follow from the model in the following way. I find that the model requires a cutoff frequency of $\nu_c = 6 \cdot 10^{15}$ Hz. In the model, the relativistic electrons lose half of their energy within about the same length scale on which day–to–day variations of the field structure take place. So the synchrotron loss time is $\tau_{1/2}(obs) \cong 1$ day (in the observers frame). If we allow for some relativistic bulk motion with $\gamma \cong 5$ towards us* we get a Doppler blue shift of $D \cong 9$. In this case the observed loss time is

$$\tau_{1/2}(obs) = 1.2\ days \left(\frac{D \times \nu_c}{10^{14}\mathrm{Hz}}\right)^{1/2} \left(\frac{B_\perp}{Gauss}\right)^{3/2}$$

resulting in a mean transversel field

$$\langle B_\perp \rangle = 0.15\ Gauss.$$

The maximum injection energy at $z = 0$ is $\gamma_c = 33000$, and the thickness of the shock is 3×10^{-3} pc or less. Note that the viewing angle $\Theta = 30°$ assumed in the model would correspond to $\Theta_{obs} = arcsin\left[\frac{sin\Theta}{D}\right] = 3°$ between the jet axis and the line of sight.

It should be noted, finally, that the model is somewhat idealized because it describes the variability of OJ 287 in terms of a series of steady–state situations. A fully realistic model would have to account for the fact that the energy distribution at $z \gg 0$ depends on the historical development of the magnetic field structure rather than on the particle energy losses in the actual magnetic field. However, it is not the purpose of this contribution to give a detailed model of OJ 287 but only to demonstrate that a helical field geometry describes the qualitative behaviour very well.

6. Conclusion

In this contribution I have shown that helical magnetic fields which seem to be present in several extended extragalactic radio jets provide a framework in which the variable frequency–dependent infrared–to–optical polarization observed in many blazars can be explained. The most important result of this study is that maximum polarizations of $P \cong 25\%$ are naturally predicted for a wide range of viewing angles and that the occurance of knots (i.e. planar shocks or pinch instabilities) may decrease this value. This would explain why very often the polarization of a blazar is low ($P = 5...10\%$) during the most active phases. Application of the model to a spectacular polarization flare of OJ 287 shows that the synchrotron loss time $\tau_{1/2}$ is the relevant time scale for flux variations. Thus ultra–rapid radio variability (Quirrenbach *et al.* , this proceedings) would require either very high fields ($B \gg 100\ Gauss$) or extreme Lorentz factors for the bulk relativistic motion ($\gamma_{jet} \gg 100$). Intrinsic explanations of the very fast radio variability, therefore, seem to require either well focussed relativistic particle beams (as in pulsars) or some coherent emission process.

---

* Roberts & Wardle (1987) report superluminal motion in OJ 287 with $v_{app} = 6.6c$ (for $H_0 = 50\ km\ s^{-1} Mpc^{-1}$).

References

Blandford, R. D. and Königl, A., 1979, *Astrophys. J.* **232**, 34

Björnsson, C.I., 1985, *Mon. Not. R. astr. Soc.* **216**, 241

Carilli, C.L., Dreher, J.W. and Perley, R.A., 1989, in K. Meisenheimer and H.-J. Röser (eds.): *Hot Spots in Extragalactic Radio Sources.* Lecture Notes in Physics Vol. **327**, Springer Heidelberg, p. 51

Heavens, A.F. and Meisenheimer, K., 1987, *Mon. Not. R. astr. Soc.* **225**,, 335

Holmes, P.A. *et al.* 1984, *Mon. Not. R. astr. Soc.* **211**, 497

Hughes, P.A., Aller, H.D. and Aller, M.F., 1985, *Astrophys. J.* **298**, 301

Hughes, P.A., Aller, H.D. and Aller, M.F., 1989, *Astrophys. J.* **341**, 68

Kirk, J.G. and Schneider, P., 1987, *Astrophys. J.* **322**, 256

Königl, A. and Choudhuri, A.R., 1985a, *Astrophys. J.* **289**, 173

Königl, A. and Choudhuri, A.R., 1985b, *Astrophys. J.* **289**, 188

Marscher, A.P. and Gear, W.K., 1985, *Astrophys. J.* **298**, 114

Meisenheimer, K., 1989, *Rev. Mod. Astronomy* **2**, 129

Owen, F.N., Hardee, P.E. and Cornwell, T.J., 1989, *Astrophys. J.* **340**, 698

Perley, R.A., Bridle, A.H. and Willis, A.G., 1984, *Astrophys. J. Suppl.* **54**, 291

Roberts, D.H. and Wardle, J.F.C., 1987, in J.A. Zensus and T.J. Pearson (eds.): *Superluminal Radio Sources*, Cambridge University Press, p. 193

Röser, H.-J., 1989,in K. Meisenheimer and H.-J. Röser (eds.): *Hot Spots in Extragalactic Radio Sources.* Lecture Notes in Physics Vol. **327**, Springer Heidelberg, p. 91

Webb, G.M., Drury, L.O'C. and Biermann, P., 1984, *Astron. Astrophys.* **137**, 185

Rapid Variability of BL Lac Objects in the Optical Regime

Stefan J. Wagner

Landessternwarte Königstuhl, 6900 Heidelberg, Germany

Abstract: Some of the results of an extensive four-week monitoring campaign of the variability of several BL Lac objects on short time-scales, simultaneously at radio and optical frequencies, are described. In some objects the variability provides strong evidence that the optical emission is beamed as has been demonstrated for the continuum emission at radio frequencies. In the BL Lac object 0716+714 the character and the time-scales of short period variability changed significantly within less than 48 hours during the four week campaign. The similarity of radio and optical lightcurves during both states and the simultanity of the change indicates a close correlation of the variability over more than five orders of magnitude in frequency. This correlation and the short time-scales rule out any variability introduced by propagation effects and requires an intrinsic origin. Some implications are briefly discussed.

1. Introduction

BL Lac objects are defined (Strittmatter et al., 1972) as a class of active galactic nuclei that exhibit large amplitude variability of their continuum emission on short time-scales. Although being used to define this class of AGN, the variability is still poorly understood. During the seventies many studies aimed at a better understanding of this variability in the then known $\sim$ 3 dozend BL Lac objects but remained without fundamental impact on physical models of these AGN. During the eighties multifrequency studies revealed that variability is common in BL Lacs throughout the entire electromagnetic spectrum with -in most cases - decreasing time-scales towards higher frequencies. In several cases correlations between the variability on time-scales of weeks to months, observed at different frequencies were discovered (see e.g. Bregman, 1986, Brown et al., 1989). In most of these cases the radio emission lags behind the optical emission by several months (e.g. Belov et al., 1990). In order to study the spectral behaviour of variability on time-scales shorter than one week, we performed simultaneous optical-radio observations of several flat-spectrum radio sources previously identified as to be variable on comparable time-scales in both waveband regimes.

2. Observations

Following earlier studies which identified some BL Lac objects to be variable at similar time-scales in the optical and radio regime during the same epoch (Wagner et al., 1990) we carried out an intense multi-frequency campaign in February 1990. Radio observations at four different frequencies were performed with the VLA and are described by A. Quirrenbach elsewhere in this volume. During nighttime (in Europe) throughout the four weeks starting with J.D. 2447920 simultaneous observations were carried out in the optical at the observatories on Calar Alto, Spain, and Königstuhl, Heidelberg, Germany.

During J.D. 2447922 - 2447936 we used the 2.2 m telescope on Calar Alto. A CCD camera equipped with an RCA chip was mounted to the Cassegrain focus to perform relative photometry. The brighter sources of the sample were also monitored during J.D. 2447920-2447952 with the 0.71 m telescope of the Landessternwarte Königstuhl in Heidelberg. Again a CCD camera (equipped with a GEC chip) was used for relative photometry. Further details on the observational procedures will be given elsewhere.

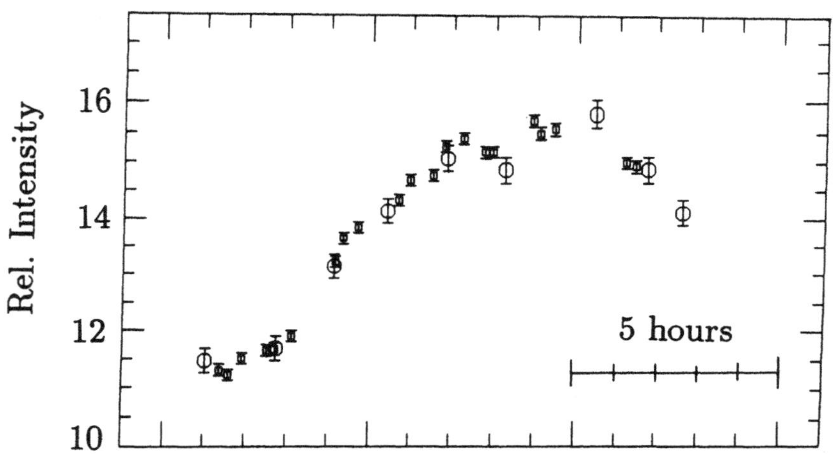

Fig. 1. Comparison of the measurements of 0716+714 performed at the two telescopes. Circles represent measurements obtained with the 0.7m telescope at Heidelberg - Königstuhl (mean exposure-time 15 min), boxes indicate measurements with the 2.2 m telescope at Calar Alto (exposure-time about 100 sec). Both data sets were normalized to the same reference star such that the mean ratio during the whole campaign equals unity. During this particular night of J.D. 2447927, the 40 % change in flux is well covered by both telescopes. The lightcurves agree with another within the errors, confirming the reality of even low-amplitude modulations on top of the all-over behaviour.

With both data-sets aperture photometry was simulated on the reduced CCD frames. Normalized flux ratios (F_{QSO} / $F_{Star\ i}$) / <F_{QSO}/ $F_{Star\ i}$ > were computed for several comparison stars. As an estimate of the errors involved, different combinations of reference stars were compared with each other. A few variable comparison stars were identified and eliminated. For all other stars the normalized lightcurves are identical with each other within the errors and were averaged for further analysis.

An important step towards higher accuracy was an effective reduction of bias-, dark- and flatfield-noise. The large number of science frames (10^3) taken with an identical instrumental set-up allowed the construction of high S/N flatfields.

The data of the two telescopes agree with each other within the errors. In 6 nights at least 5 frames were taken at both telescopes. An example of the perfect match is shown in figure 1.

3. Variability of 0954 + 65

This BL Lac object has not yet been studied in detail at optical frequencies. Our test-campaign revealed variability of 100 % in less then 24 h (Wagner et al., 1990). A redshift of $z = 0.368$ has been reported by Lawrence et al., 1986 and was recently confirmed by Stickel et al., 1991. The lightcurve of 0954+65 during the four week campaign is presented in figure 2.

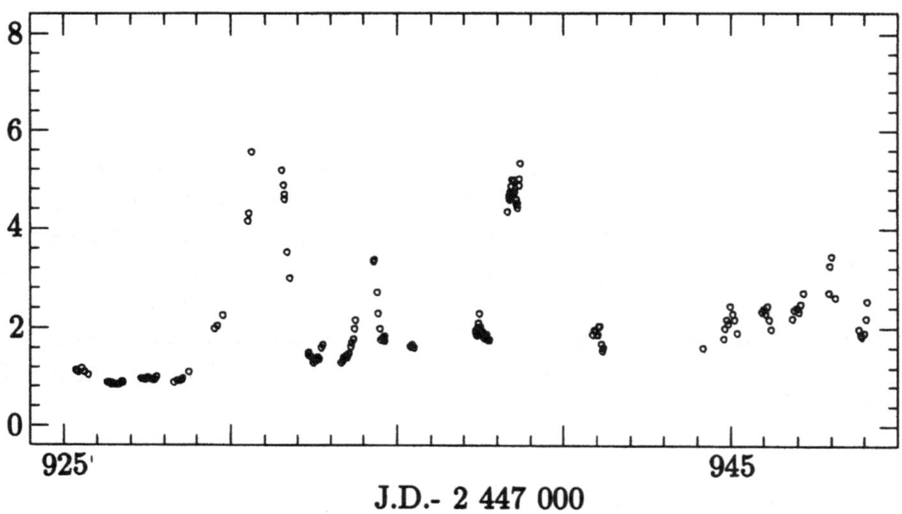

Fig. 2. Lightcurve of 0954+65 during February 1990, measured in the R-band (650 nm). The data of the two telescopes used in the study are combined without individual coding. Errors of all measured fluxes are smaller that the size of the symbols. Exposure-times were smaller than the widths of the symbols in all cases. Rapid fluctuations and multiple reversals within single nights are resolved and real.

The character of the variability is dominated by rapid outbursts, exceeding the luminosity of the quiescent state by more than a factor of six. Apart from the major, well documented flare at J.D. 2447930, additional outbursts of lower amplitude were observed as well. Unfortunately, neither the peak of the major outburst nor any of the lesser flares were sampled continuously due to the restriction of observing at nighttime. During the well-sampled part the doubling-time rose from 24 h during the initial stages of the outbursts to 6 hours during the fastest parts. This, however, is only a lower limit. The decay of the flares was comparatively fast. The shape of the flares is well described by an exponential increase and decay over the whole observed range in flux. All sufficiently well sampled flares have a self-similar shape. During the night of J.D. 2447938 two flares occured with a time-difference much smaller than the decay time such that a minimum and a maximum occured within one night. During this night the observed flux increased, decreased and rose again by more than 50% of the quiescent flux within less than 3 hours each. Towards the (less densely sampled) end of the campaign the lightcurve seems to be dominated by lower amplitude flickering.

With a cosmological correction of $\Delta\ t_{intrinsic} = \Delta\ t_{observed} \cdot \frac{1}{1.368}$, the intrinsic e-folding times are reduced to 4 h. With a luminosity of 10^{46} erg/s at quiescent level the appearent luminosity fluctuations amount to 10^{48} erg/s^2 during the steepest increase (assuming isotropic emission and correcting for the cosmological time contraction of 0.731).

This is comparable to the flux increase observed in 3C273 by Courvoisier et al., 1988 and implies brightness temperatures which would violate the Compton limit for isotropic incoherent synchrotron radiation.

4. Variability of 0716 + 714

The source was also found to be variable at optical and radio frequencies on similar time-scales during a preliminary run in May 1989 (Wagner et al., 1990). 0716+714 had been known to be variable at radio frequencies on short time-scales for several years (Heeschen et al., 1987) but received less attention by optical observers. It is an extreme BL Lac object which does not show any spectral features even in very deep spectrographic studies (Stickel, private communication). Images taken at high resolution during our campaign do not show indications of a host galaxy which implies a distance of $z > 0.2$.

4.1. Optical observations

The complete lightcurve derived in the R-band ($\lambda_{obs} = 650$ mm) is shown in figure 3. Two features are immediately evident:

a) During the four week interval of observations the character of the variability changed from a very rapid mode (time-scales of about 1 day) to a slower one around J.D. 2447930.

b) Before and after this change the variability exhibits a quasiperiodic character. While the amplitudes of both modes are compatible, the characteristic time-scales differ by a factor of ~ 7.

The difference in character and the preferred time-scales are also evident from fourier- and structure-function analyses. Figure 4 displays the optical power spectra during the two parts of the campaign. The radio data show very similar changes.

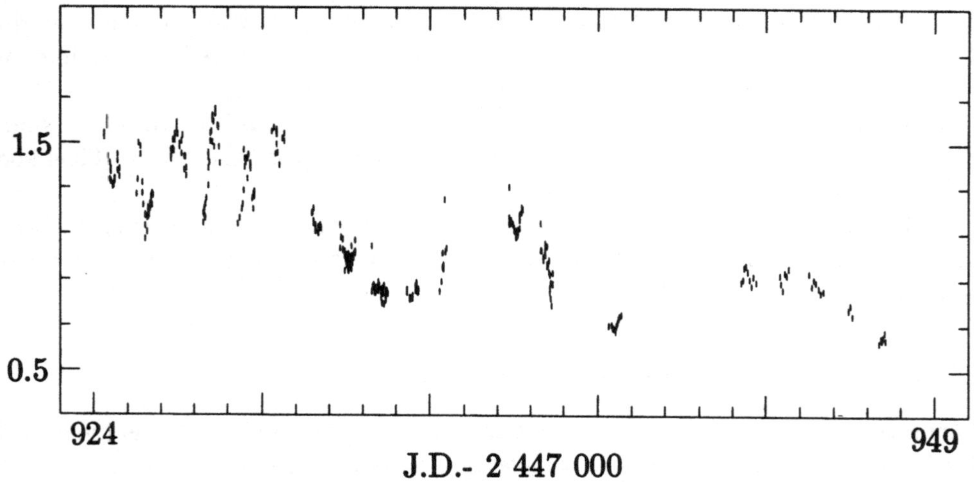

Fig. 3. Normalized lightcurve of 0716+714 during February 1990, measured in the R-band (650 nm). Again the data of the two telescopes are combined. Photometric errors are 0.8 and 1.5 % for the data sets of the two telescopes. The character of the variability clearly changes on J.D. 2447930 from a rapid quasiperiodic state with a typical time-scale of 1 day into a somewhat fainter state of significantly longer time-scale (7 days).

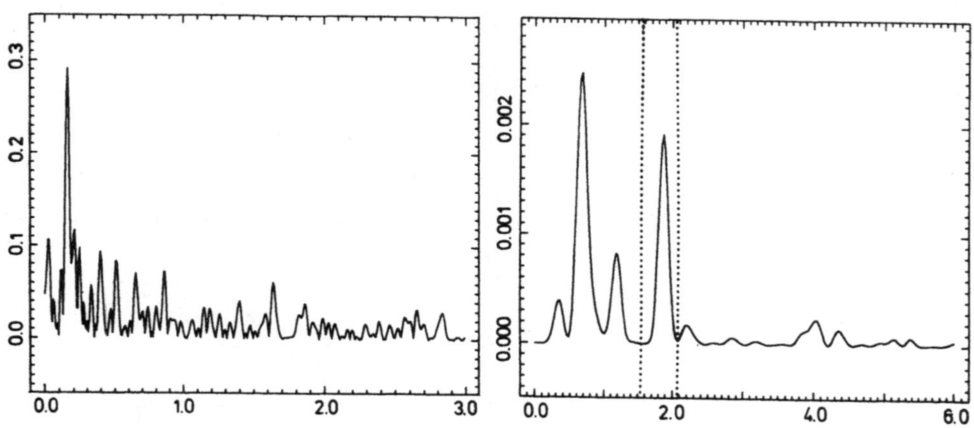

Fig. 4. Power spectra for the two parts of the campaign. For the second part of the campaign (left) only the maximum at 7 days is significant. The power spectrum of the first 8 days (right), during which the source underwent rapid variation is modified by the window-function. Daytime enforces gaps lasting about 13 hours, during which no data are taken. These interruptions produce an artificial peak at the corresponding frequency. This window in the frequency-domain is indicated by the dashed limits.

The coverage does not allow to distinguish whether the drop in mean brightness during the observations is linked to the change in modes or whether there is a steadily declining component (time-scales $\geq$ 2 months) on which equally bright states of variability are superposed. The amplitude of the variability is similar for both parts of the lightcurve (40%).

Apart from this all-over decline departures from the mean brightness are symmetric with respect to this average value, i.e. different from the case of 0954+65 the means and median values are similar.

During the first week the lightcurve remains identical under time-reversal. During the second part only two cycles are observed which does not allow any statement about time-reversability.

4.2. Comparison with radio observations

During the same interval as for the optical observations, 0716+714 was measured at 4 frequencies with the VLA at an average temporal resolution of 2 h.

At 6 cm the lightcurve is very similar to the optical measurements (Quirrenbach, this volume). Again variability on time-scales of $\sim$ 1.2 d is seen during the first week which is followed by fluctuations on a time-scale of $\sim$ 7 days during the remaining 3 weeks (figure 5). As in the optical regime, the average luminosity drops between the two parts by an amount comparable to the amplitudes in either of the two epochs, which again - as in the optical - are comparable to each other.

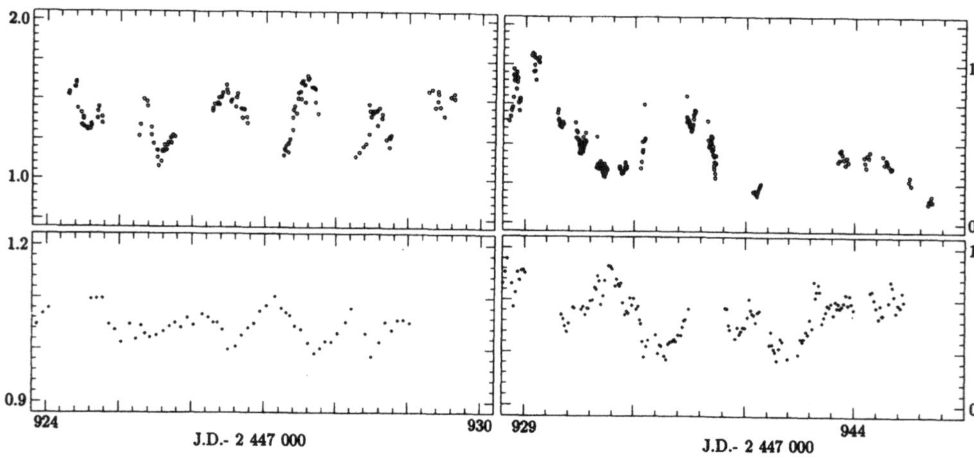

Fig. 5. Comparison of the optical and radio lightcurves for the two different parts of the whole campaign. Note that the time axis is compressed in the right part with respect to the first week of the campaign. The scales in intensity are different for the radio (bottom) and the optical (top) but identical in the left and right parts respectively.

The fact that the character of the variability clearly changed rather abruptly but simultaneously in both frequency regimes already indicates some correlation between the two frequencies. This is even more supported by the similarity in timescales both before and after this change in both frequency bands. Furthermore, both in the radio and in the optical the ratio of the amplitudes before and after

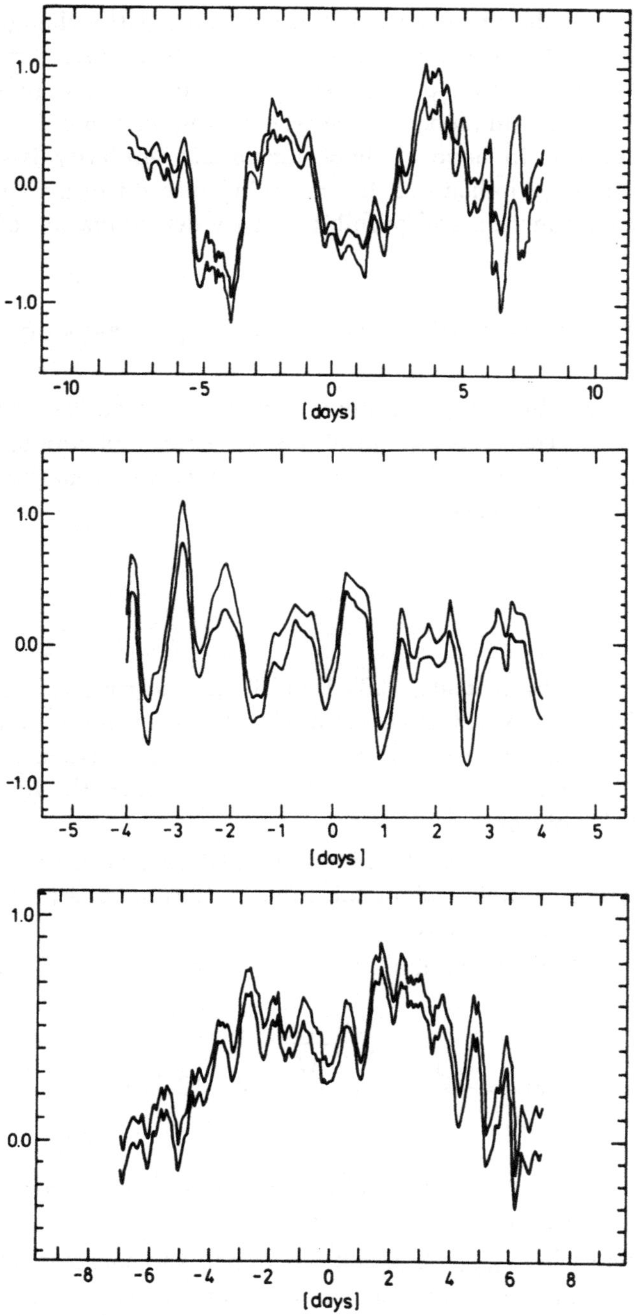

Fig. 6. The cross-correlation of the first week (top) of rapid variability, of the slower mode (center) and the whole compaign (bottom). In all cases the two lines indicate the envelopes of the error bars. While multiple maxima, interspaced at the characteristic quasi-period, can be clearly identified in the two upper, individual ccf's, no clear lag can be identified. The complete ccf at the bottom is a superposition of the two individual modes.

the change as well as the drop of the mean level during the change are identical to within the errors. All of these similarities indicate a close correlation of the variability in the two regimes, over 5 decades in frequency. Figure 6 shows the formal cross-correlation functions for the whole campaign and for the rapid and slow period separately. In none of these curves the similarity is reflected by a single peak. The quasiperiodicity of the variability introduces multiple maxima in the cross-correlation function and prohibit an easy determination of the lag.

4.3. Constraints on variability introduced by propagation effects

One of the basic motivations of the simultaneous radio and optical observations was the hope to constrain the observed variations to intrinsic or propagation-induced effects. While the models of intrinsic origin are numerous, scenarios for the latter case can be devided into 2 classes:

Microlensing

It has been suggested by Nottale, 1986 that BL Lac objects are (micro-) lensed AGN at large distances. Apart from amplifying the continuum emission the microlensing effect introduces variability of the observed continuum emission due to variable amplification of stellar caustics passing through the line of sight (see Borgeest et al., this volume).

Although an attractive explanation for several BL Lac features (see Stickel, this volume) microlensing cannot explain the rapid variations observed in 0716+714 and 0945+65 (Wagner et al., 1991).

The time-scales Δt of the variability are given as

$$\Delta t = \frac{1}{v_{trans}} \cdot \sqrt{\frac{M}{M_{\odot}}} \cdot 5 \cdot 10^{14}\, m$$

Even for low-mass lenses ($M = 0.1\, M_{\odot}$) the observed time-scales require $v_{trans} \simeq c$, inconsistent with random velocities of stars.

Transversal velocities $v_{trans} \simeq c$ could occur in the source plane with knots moving on tracks that allow apparent superluminal motion. This, however, requires the multiple ejection of knots with relativistic speeds, a behaviour which makes intrinsic variability on the observed time-scales a necessary requirement for amplification by microlensing. The latter would only be an additional, amplifying, effect. Since there is no direct evidence for microlensing, it is considered to be unlikely for these sources.

Interstellar scintillation:

Especially low frequency variability of radio sources is often explained by interstellar scintillation. While both 0716+714 (Heeschen et al., 1987) and 0954+65 (Fiedler et al., 1987) were found to show RISS-induced variability, the spectral and time-behaviour of the intraday variability of these sources requires rather unusual characteristics of the interstellar medium (Quirrenbach et al., 1989 and Wambsganss et al., 1989). At $10^{14.5}$ Hz interstellar scintillation has no effect. Therefore, the close relationship between the optical and radio lightcurves in 0716+714 is direct evidence that RISS-induced variability is of minor importance also at 6 cm.

We therefore rule out that the variability described in chapters 3 and 4 is dominated by propagation effects. The variability has to be explained by intrinsic mechanisms which have to work rather coherently over the 5 orders of magnitude spanned by our observations at 0.6 μm and 6 cm.

5. Implications for BL Lac models

The similarity in character and the simultaneous change in modes observed in the optical and radio waveband regimes in 0716+714 cannot be explained within the classical scheme of expanding sources which predict much longer time-scales for the lower frequencies and significantly larger time-lags between individual events observed in the two wavelength ranges. Any model needs to account for the similar sizes and identical sites of the regions emitting the variable part of the optical and radio radiation.

It is tempting to interpret the quasiperiodicity as indication of the orbital frequency of an emitting region rotating about the nucleus. In this case the variable component would be emitted from (a subvolume within) the circumnuclear disk. This geometry, however, does not allow beaming from relativistic bulk motion. The short time-scales would be intrinsic, and require extremely efficient acceleration. Again, classical mechanisms such as the diffusive shock acceleration (Drury, 1988) are too slow. Furthermore the radiation losses set severe limits at the attainable energies. Outbursts of coherent emission processes provide an attractive means to explain the short time-scales and the violation of the Compton limit. The flares are likely to erupt in the corona of the disk. This would allow quasi-simultaneous variability over a large range in frequency because optical depth effects are less severe. Since the disk has to be magnetized (Camenzind, 1990) its corona will be connected to the disk via flux-tubes which could store large amounts of energy. Microscopic processes, as described e.g. by Lesch, this volume, could release this energy in local thunderstorms. It is still not clear, however, whether such flares are efficient enough.

The most attractive alternative situates the variable component in a jet pointing towards the observer. This allows relativistic contractions of the apparent time-scales and an increase in line-of-sight luminosities due to relativistic beaming. The bends indicated by VLBI observations (e.g. Hummel, this volume) introduce changes of the orientation and are expected to cause secular changes in the variability pattern. Within such a scheme, the change in modes observed in 0716+714 might be identified with a changing misdirection.

Königl and Choudhuri (1985) suggested that shocks propagating down helical jets could subsequently illuminate successive cross-sections and thus cause even quasi-periodic variations. In their paper this model was applied to variability on much longer time-scales. Variability on short time-scales can be explained in the same way if the jets are bent and turbulent (Marscher, 1991). Qian et al., 1990 modelled the rapid variations of 0917+65 successfully within this framework. Apart from the less severe limits on the e-folding times introduced by beaming, the observed time-scales do not correspond to transverse light-travel times. The special geometry doesn't require causal connection of the emitting area (Quirrenbach, this volume). Such a geometric interpretation of the short time-scales involved is supported by the detailed shape of the lightcurve of 0716+714 during the rapid mode in the first part of the campaign. The nearly "sawtooth-shaped" lightcurve is indicative of a geometric illumination effect. It is not yet clear, however, whether propagating shocks have the required structure of optical depths that can explain the short time-lags between the optical and radio lightcurves.

Acknowledgements. I would like to thank K. Anton, S. Appl, A. Barzewski, M. Camenzind, U. Erkens, M. Haehnelt, R. Khanna, A. Quirrenbach, F. Sanchez-Pons and A. Witzel for help during the observations, reduction and interpretation.This work was supported by the Deutsche Forschungsgemeinschaft (SFB 328).

References

Belov V., Hagen-Torn, V. and Marchenko, S.: 1989, *Astrophysics* **30**, 1

Borgeest, U., Kayser, R., Refsdal, S., Schramm, J. and Schramm, T.: 1990, this volume

Bregman, J.: 1986, in *Continuum Emission in Active Galactic Nuclei*, M. Sitko (Ed.), NOAO, p. 102

Brown, L., Robson, E., Gear, W. and Smith, M.: 1989, *Astrophys. J.* **340**, 150

Camenzind, M.: 1990 in *Reviews in Modern Astronomy* **3**

Courvoisier, T., Robson, E., Blecha, A., Bouchet, P., Hughes, D., Krisciunas, K. and Schwarz, H.: 1988, *Nature* **335**, 330

Drury, L.: 1988, in *Hot Spots in Extragalactic Radio Sources*, K. Meisenheimer, H.-J. Röser, (Eds.), Springer, p. 279

Fiedler, R., Dennison, B., Johnston, K. and Hewish, A.: 1987, *Nature* **326**, 675

Heeschen, D., Krichbaum, T., Schalinski, C. and Witzel, A.:1987, *Astron. J.* **94**, 1493

Hummel, C.: 1990, this volume

Königl, A. and Choudhuri, A.: 1985, *Astrophys. J.* **289**, 188

Lawrence, C., Pearson, T., Readhead, A. and Unwin, S.: 1986, *Astron. J.* **91**, 494

Lesch, H.: 1990, this volume

Marscher, A.: 1991, in *Variability of Active Galactic Nuclei*, R.H. Miller and P. Wiita (Eds.), Cambridge Univ. Press.

Qian, S., Quirrenbach, A., Witzel, A., Krichbaum, T., Hummel, C. and Zensus, A.: 1990, *Astron. Astrophys.* in press

Quirrenbach, A.: 1990, this volume
Quirrenbach, A., Witzel, A., Qian, S., Krichbaum, T., Hummel, C. and Alberdi, A.: 1989, *Astron. Astrophys.* **226**, L1
Nottale, L.: 1986, *Astron. Astrophys.* **157**, 383
Stickel, M.: 1990, this volume
Stickel, M. et al.: 1991, in preparation
Strittmatter, P., Serkowski, K., Carswell, R., Stein, W., Merrill, K. and Burbidge, E.: 1972, *Astrophys. J.* **175**, L7
Wagner, S., Sanchez-Pons, F., Anton, K., Quirrenbach, A. and Witzel, A.: 1991, in *Variability of Active Galactic Nuclei*, R.H. Miller and P. Wiita (Eds.), Cambridge Univ. Press.
Wagner, S., Sanchez-Pons, F., Quirrenbach, A. and Witzel, A.: 1990, *Astron. Astrophys.* **235**, L1
Wambsganss, J., Schneider, P., Quirrenbach, A. and Witzel, A.: 1989, *Astron. Astrophys.* **224**, L9

UV Variability of Blazars

Aldo Treves and E. Girardi

Dipartimento di Fisica, Università di Milano, Italy

1. Introduction

Variability of Blazers is a powerful tool for constraining theoretical models, because it may enlighten on the structure of the inner part of the source where direct imaging is impossible. In particular correlated variability at different frequencies can demonstrate the existence of various components and their possible interrelation. However, variability studies are very time consuming, and multifrequency observations are difficult to organize, so that the field is still essentially in its infancy.

An important result was obtained by Impey and Neugebauer (1988) who considered IRAS observations of Blazars combined with observations in radio, optical, UV bands as given by the literature. They took the ratio of the maximum to the minimum state at the various frequency averaged on the entire population and showed that such ratio increases with frequency, indicating therefore that the variability of Blazars increases with increasing energy. Treves et al. (1989) considered a set of 9 simultaneous observations in the optical, UV, low energy (0.2 keV) and medium energy (3 keV) X-rays of PKS 2155-304 and showed that the variability-energy dependence in this case extends to the keV region. Giommi et al. (1990) examined the EXOSAT archives finding that in the 36 Blazars under examination the medium energy X-ray variability is systematically larger than the low energy one. Maraschi et al. (1989) and Celotti et al. (1990) interpreted the variability-energy relation extending the inhomogeneous jet model for BL Lac described by Ghisellini et al. (1985), where higher frequencies are produced in an inner region, while lower ones in regions of increasing dimensions.

In the frame of this kind of discussions we present here the first results of a systematic study of variability of Blazars as from the data contained in the IUE archives.

2. UV data

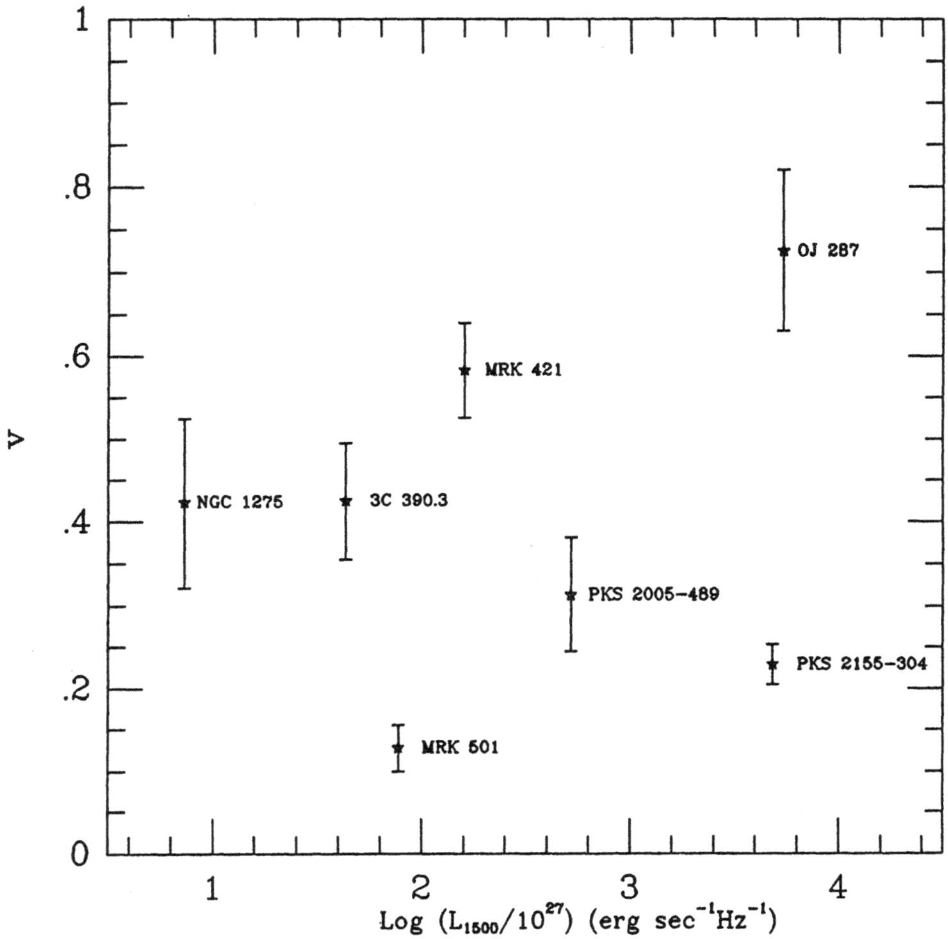

Fig. 1. Variability index vs. luminosity for the 7 blazars with known redshift.

IUE has observed thus far 40 Blazars in the band 1200-3000 Å. For a discussion of data up to 1984 see Ghisellini et al. (1986). Altogether the IUE Blazar spectra are 430. A systematic study of this data set in course in Milan (for parallel studies see e. g. Kinney et al. (1990)). Here we concentrate on 8 objects which have been observed more than 5 times in the short wave length band (1200-1900 Å) and in the long wavelength band (1900-3000 Å). The observations are very unevenly spaced within the 12 y lifetime of the satellite. The spectra have been acquired through the ULDA facility and examined visually for the elimination of obvious flaws. For each spectrum the flux was measured in a band of 50 Å around 1500 Å and 2500 Å. The photometric error is estimated of about 10%. Variability has been first searched for calculating the χ^2 with respect to the mean flux. As apparent in Table 1 variability is highly significant in all sources but Mrk 501.

As variability parameters we have chosen the rms fractional variability defined as $v=\sigma/<F>$ and the ratio F_{max}/F_{min}. The uncertainty on v was estimated through a bootstrap simulation (Simpson and Meyer–Hasselwander (1986)), which yields the 68% uncertainty. The results are reported in Table 1.

3. Discussion

The following points derive from an examination of Table 1.

1) There is no obvious difference in the variability properties at 1500 Å and 2500 Å, this is indicated by the mean and median value of the distributions of the v parameter and by the KS test. However, if one considers the uncertainties in each v value as it is done in the weight average one finds significantly larger variability at 1500 Å than at 2500 Å ($<v>_{1500}=0.23 \pm 0.01$; $<v>_{2500}=0.18 \pm 0.01$). Note tha t the weighted average is dominated by PKS 2155-304.

2) If one combines the results on variability in UV with those of variability in X-rays as from in general non–simultaneous observations, no clear evidence is present of a variability energy correlation. This contrasts with the cases of simultaneous observations (see above and Celotti at al. (1990)).

3) There is no clear dependence of the variability on the luminosity of the source. This is illustrated by Fig. 1. It is noticeable that the variability behaviour of Blazars appears somewhat different from that of Seyferts. In Seyferts one finds in general a clear hardening of the spectrum with increasing intensity (e. g. Ulrich at this meeting), to be compared with the confuse situation described in point 1. Contrary to the case of BL Lacs, Seyferts exhibit a variability luminosity correlation as shown for instance by Edelson et al. (1990). Finally, in Seyferts the variability-energy correlation found in Blazars for simultaneous observations over a broad spectral band is in general absent (see e. g. Treves et al. (1990)).

References

Celotti, A., Maraschi L., Treves A.: 1990, submitted to Ap. J.
Edelson, R.A., Krolik, J.H., Pike, G.F.: 1990 *Ap.J.* **359**, 86
Ghisellini, G., Maraschi, L., Treves, A.: 1985 *A.A.* **146**, 204
Ghisellini, G., et al.: 1986 *Ap.J.* **310**, 317
Giommi, P., et al.: 1990 *Ap.J.* **356**, 432
Impey, C.D. and Neugebauer, G.: 1988, *A.J.* **95**, 307
Kinney, A. L., et al.: 1990, *Ap.J.S.S.* (in press)
Maraschi, L., Celotti, A., Treves, A.: 1989 *ESA-SP* **296**, p. 825
Simpson, G., Mayer–Hasselwander, H.: 1986 *A.A.* **162**, 340
Treves, A., et al.: 1989 *Ap.J.* **341**, 733
Treves, A., et al.: 1990 it Ap.J. **359**, 98

Table 1.

Source	λ	χ^2	$P^{(1)}$	v	σ_V	$\frac{F_{max}}{F_{min}}$	$N^{(2)}$
NGC1275	2500	0.310E+04	0.000E+00	0.55	+/–0.21	10.87	6
	1500	0.211E+04	0.000E+00	0.42	+/–0.10	7.76	13
0735+178	2500	0.335E+04	0.000E+00	0.53	+/–0.29	10.71	5
	1500	0.367E+03	0.000E+00	0.65	+/–0.18	5.99	7
OJ287	2500	0.101E+05	0.000E+00	0.57	+/–0.06	18.85	29
	1500	0.102E+05	0.000E+00	0.73	+/–0.10	24.36	23
MRK421	2500	0.721E+03	0.000E+00	0.47	+/–0.06	4.52	36
	1500	0.977E+03	0.000E+00	0.58	+/–0.06	6.22	32
MRK501	2500	0.187E+01	0.867E+00	0.06	+/–0.02	1.17	6
	1500	0.159E+02	0.101E+00	0.13	+/–0.02	1.49	11
3C390.3	2500	0.108E+03	0.117E–18	0.37	+/–0.07	2.55	9
	1500	0.133E+04	0.000E+00	0.43	+/–0.07	7.26	18
2005–489	2500	0.944E+02	0.151E–16	0.33	+/–0.06	2.58	8
	1500	0.121E+03	0.287E–20	0.31	+/–0.05	2.78	11
2155–304	2500	0.128E+03	0.190E–13	0.19	+/–0.02	2.48	30
	1500	0.240E+03	0.907E–33	0.23	+/–0.02	2.59	33

$^{(1)}P$ is the chance probability for constant flux.

$^{(2)}N$ is the number of IUE spectra.

The Connection Between Frequency Dependent Polarization (FDP) and Growing Radio Shocks in Blazars

Leena Valtaoja

Turku University Observatory, Tuorla, SF-21500 Piikkiö, Finland

Abstract: Because of the rapid variability of polarization in blazars, it is generally assumed that the polarization is due to synchrotron radiation, which in itself produces frequency independent polarization. Dilution by the host galaxy can explain part of the FDP (decrease to red), but cannot explain all kinds of observed FDP and frequency dependence of the position angle. To interpret these observations, models consisting of two separate synchrotron sources have been used with succes. Even if the two components have not been identified physically, the numerical model fittings generally show that the more variable synchrotron component has both flatter spectrum and higher intrinsic polarization. This fits together with the scenario where a shock moves in the jet, with the compressed shock being the more polarized component and the jet the less polarized and less variable steeper spectrum component.

The shocked jet models have had success in explaining the detailed variations associated with radio outbursts. We suggest that the early stages of the radio shocks could cause the observed variable frequency dependence in the optical polarization.

In principle, the radio shock model allows different degrees of FDP and flexible time scales for the observed variability. It can also explain frequency dependence both decreasing and increasing towards red wavelengths as well as the existence of the frequency dependence in position angle. But unlike the previous two-component models ,which introduce a new set of initial parameters for every new observation, the jet - shock model, having a physical background, includes evolution with time and thus gives us a firm basis to define a single set of parameters. In the paper we describe the overall scenario for the evolution of such a radio shock and give predictions for the wavelength dependence of the optical polarization during different stages of the radio flare.

To compare our predictions with the observations we investigated the four sources (BL Lac, 3C 279, DA 237 and 0109+224) for which optical and radio data covering the same radio outburst were available. We tried to follow the development of FDP during radio outbursts using the 22 and 37 GHz radio data from the Metsähovi Radio Research Station quasar monitoring program and optical data from Crimean Astrophysical Observatory and from Mead et al. (1990). Altogether we looked at 33 individual observations, comparing them with four different radio flares. The detailed testing of the model would demand much longer series of optical polarization observations in a wide range of wavelengths. But at least we can conclude from the comparisons, that the existing data are not against the idea that the observed variations in the FDP could be caused by the same growing shocks which are responsible for the radio outbursts.

References

Mead, A.R.G. et al. 1990, Astr. and Astrophys.Suppl. 83, 183
Valtaoja, L. et al. 1990, Astron. J. (in press)

Optical Observations of Rapid Variability in the QSO PG 0117+213

R. Ekins, S. Plunkett and Brian McBreen

Physics Department, University College, Dublin 4, Ireland

Abstract: CCD imaging photometry data has been obtained on a sample of sources to search for rapid variability. Light curves and photometric data are presented for the QSO PG 0117+213 and the BL Lacertae objects 0414+009 and 0836+182. The B and V magnitudes of the QSO PG 0117+213 decreased by 0.86 and 0.30 in a time interval of 46 hours. In the rest frame of the QSO (z=1.493) these variations occurred in only 18.5 hours. Furthermore this QSO varied significantly over a period of only 25 minutes. No short timescale variability was detected in the two BL Lacertae objects.

1. Introduction

Observations of rapid optical variability in quasars and BL Lacertae objects may provide crucial constraints on the emission mechanism and the physical size of the emitting region. Rapid optical variability is a well-known identifying characteristic for BL Lacertae objects (blazars). Large amplitude variations on timescales ranging from days to decades have been well documented (Webb et al. 1988, Smith et al. 1987) and recently optical and X-ray variations on timescales significantly shorter than a day have been reported (Carini, Miller and Goodrich 1990; Xie et al. 1988; Wagner et al. 1990; Giommi et al. 1990). However optical variability in quasars on timescales shorter than a day appears to be rare (Grauer 1984). In one unusual case Courvoisier et al. (1988) observed a factor of two change in the near infrared emission from the quasar 3C273 on a timescale of one day. In order to constrain models for the rapid variability in quasars and BL Lacertae objects, we have investigated the optical behaviour of some of these sources.

2. Observations

The observations were made with the 1.0 metre JK telescope at Observatorio del Roque de los Muchachos equipped with a GEC CCD and an autoguider. The field of view of the CCD was 3.5 by 2.5 arcminutes. The observations were made through B-band and V-band filters. Repeated exposures of duration between 3 minutes and 10 minutes were obtained of the starfield containing the source. The stars in the frame provide comparisons for use in the data reduction process. The magnitudes of the sources and reference stars were obtained from frames of standard stars in M67.

The frames were reduced using the method of Howell and Jacoby (1986). Each object in the frame is processed through an aperture-photometry routine. Differential magnitudes can be calculated for any pair of objects observed in the frame. The simultaneous observations of the object of interest and several comparison stars permit the removal of variations arising from fluctuations in atmospheric transmission and extinction. The 1950 coordinates of the three reference stars used in the analysis of PG 0117+213 are R1(01^h 17^m 35^s, 21° $17'$ $08''$), R2(01^h 17^m 27^s, 21° $17'$ $12''$) and R3(01^h 17^m 30^s, 21° $18'$ $03''$).

3. Results

The QSO PG 0117+213 (Schmidt and Green 1983, Neugebauer et al. 1987) was monitored during two separate observing runs in January 1989 and January 1990. The QSO was observed on four consecutive nights between 13 and 16 January 1989. No internight variability was detected at a level of 0.04 magnitudes. The B and V magnitudes of the QSO were 16.08 and 15.92 respectively in agreement with the photometry of Schmidt and Green (1983). The QSO was observed one year later on 1 January 1990 and 3 January 1990. On 1 January 1990 the QSO was found to have increased in brightness above the 1989 level and the B and V magnitudes were 15.23 and 15.53 respectively. Three sets of observations were obtained in both the B and V filter over a time of two hours and no variations in brightness of the QSO were observed.

Two nights later on 3 January 1990 the QSO had faded dramatically. The B and V magnitudes were 16.09 and 15.86 respectively. In a period of 46 hours the QSO had decreased by 0.86 magnitudes in B and 0.30 magnitudes in V. In the QSO rest frame (z=1.493) the intensity changes occurred during a time interval of only 18.5 hours. The variations could have occurred on a faster timescale because the light curve of the QSO was undersampled.

The QSO was monitored on 3 January 1990 for two hours and eight exposures were obtained in both the B and V filters. The results obtained are shown in Figures 1 and 2. When the QSO is compared with the reference stars R1 and R3 in the B-band (Figure 1), it is apparent that the QSO brightened by about 0.3 magnitudes in a time interval of only 25 minutes. During this time there was no significant variation in the magnitude difference between the reference stars i.e. R3

- R1. In the V-band (Figure 2) the magnitude difference between the reference stars (R1 - R2) is constant within experimental error. However the QSO has declined in brightness by about 0.1 magnitudes when compared with R1 and R2. This variation in the V-band occurred in the same time interval as the increase in the B magnitude.

The BL Lacertae object 0414+009 (Burbidge and Hewitt 1987; Ulmer et al. 1983) was observed repeatedly for two hours on 3 January 1990 immediately following the observations of PG 0117+213. The results of the V-band photometry are displayed in Figure 3. No significant variability was observed over the period of the observations. The BL Lacertae object 0836+182 (Burbidge and Hewitt, 1987; Wills and Wills 1979) was observed repeatedly on 4 January 1990. The V-band photometry is displayed in Figure 4 and no significant variability was observed.

4. Discussion

If it is assumed that the variability detected in the QSO is generated in the vicinity of a supermassive black hole, the size of the emitting region can be estimated as R = ct assuming no relativistic beaming (Miller, Carini and Goodrich 1989). We shall consider initially the timescale of the dramatic variation that occurred between 1 January and 3 January 1990. The B magnitude of PG 0117+213 decreased by a factor of two in a maximum time of 46 hours. The maximum duration of this event is 18.5 hours in the rest frame of the QSO. Assuming the radiation is generated in the vicinity of a black hole at $R \sim 3R_s$ where $R_s = 2GM/c^2$ is the Schwarzschild radius, then the mass of the supermassive black hole is given by $M \leq c^3t/6G$. This argument leads to an upper limit of 2×10^9 $M_\odot$ for the black hole mass. The Eddington limit for a black hole of mass M is given by (Witta 1985) $L_E = 1.3 \times 10^{38}$ $M/M_\odot$ erg s^{-1} which in this case leads to $L_E \simeq 2.6 \times 10^{47}$ erg s^{-1}. For a brightness of 15.2 magnitudes, the luminosity of the QSO is of order 2×10^{47} erg s^{-1}. Hence the luminosity of the QSO is comparable to the Eddington luminosity. However the light curve of the QSO between 1 January and 3 January 1990 was undersampled and the variations could have occurred on a much faster timescale. From the above arguments it is clear that the 25 minute variation observed on 3 January 1990 implies relativistic beaming for its explanation. A similar conclusion has been reached for some BL Lacertae objects (Carrasco et al. 1985; Xie et al. 1989).

The magnitude changes observed in PG0117+213 are sparsely sampled and further observations are required to fully delineate the light curve of the flares. Observations at radio frequencies would also be useful to search for delayed radio emission from the optical flares observed in the radioquiet QSO (Kellermann et al 1989).

Acknowledgements: The Jacobus Kapteyn telescope on the island of La Palma is operated by the Royal Greenwich Observatory at the Spanish Observatorio del Roque de los Muchachos of the Instituto de Astrofisica de Canarias.

References

Burbidge, G. and Hewitt, A. 1987, Astron. J. **92**, 1.
Carini, M.T., Miller, H.R. and Goodrich, B.D. 1990, Astron. J. **100**, 347.
Carrasco, L. Dultzia-Hacyun, D. and Cruz-Gonzalez, I. 1985, Nature **314**, 146.
Courvoisier, T.J.L. et al. 1988, Nature **335**, 330.
Giommi, P. et al. 1990, Astrophys. J. **356**, 432.
Grauer, A.D. 1984, Astrophys. J. **277**, 77.
Howell , S.B. and Jacoby, G.J. 1986, Publ. Astr. Soc. Pacif. **98**, 802.
Miller, H.R., Carini, M.T. and Goodrich, B.D. 1989, Nature **337**, 627.
Kellermann, K.I. et al. 1989, Astron. J., **98**, 1195.
Neugebauer, G. et al. 1987, Astrophys. J. Suppl. **63**, 615.
Schmidt, M. and Green, R.F., 1983, Astrophys. J., **269**, 352.
Smith, P.S. et al. 1987, Astrophys. J. Suppl., **64**, 459.
Ulmer, M.L. et al. 1983, Astrophys. J., **270**, L1.
Wagner, S. et al. 1990, Astron. Astrophys., **235**, L1.
Webb, J.R. et al. 1988, Astron. J., **95**, 374.
Wills, B. and Wills, D. 1979, Astrophys. J. Suppl., **41**, 689.
Witta, P.J. 1985, Phys. Rep. **123**, 117.
Xie, G.Z. et al. 1988, Astron. J., **96**, 24.
Xie, G.Z. et al. 1989, Astron. Astrophys., **220**, 89.

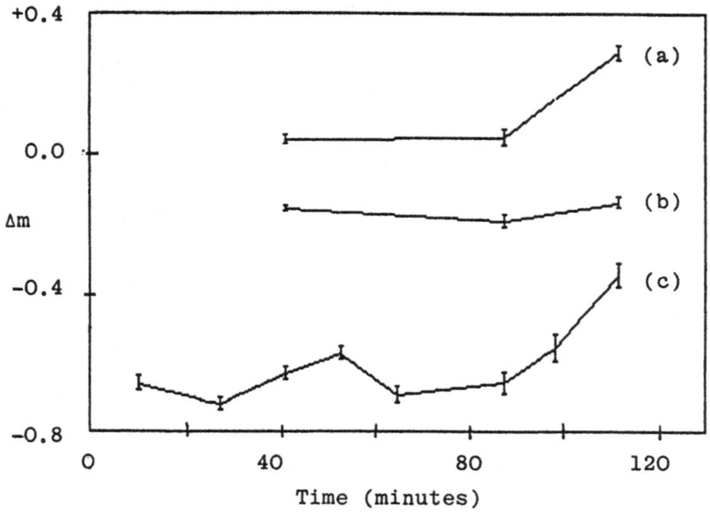

Fig. 1. The difference in B magnitude (Δm) between (a) the reference star R3 and the QSO PG 0117+213 (R3-Q) (b) the reference stars R3 and R1 (R3 - R1) and (c) the reference star R1 and the QSO (R1-Q). For display purposes 4.0 and 3.5 magnitudes have been added to R3-Q and R3 - R1 respectively. The origin of the time axis is 19^h27^m UT on 3 January 1990.

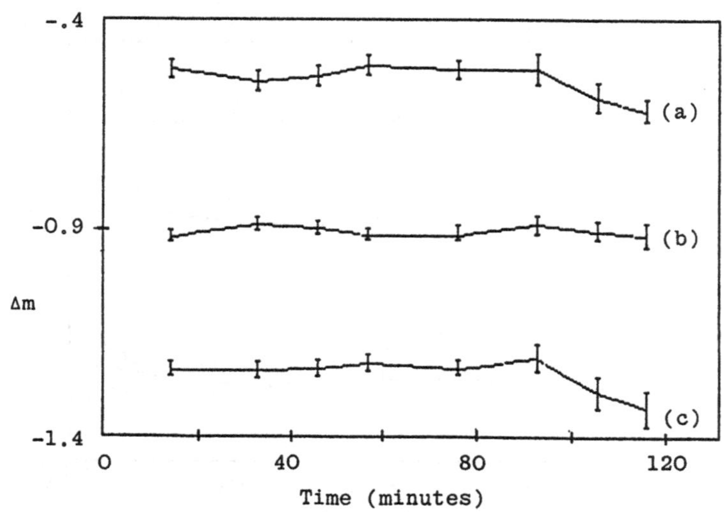

Fig. 2. The difference in V magnitude (Δm) between (a) the reference star R2 and the QSO PG 0117+213 (R2-Q), (b) the reference stars R1 and R2 (R1 - R2) and (c) the reference star R1 and the QSO (R1-Q). The origin of the time axis is the same as in Figure 1. For display purposes 0.2 magnitudes has been subtracted from R1 - R2.

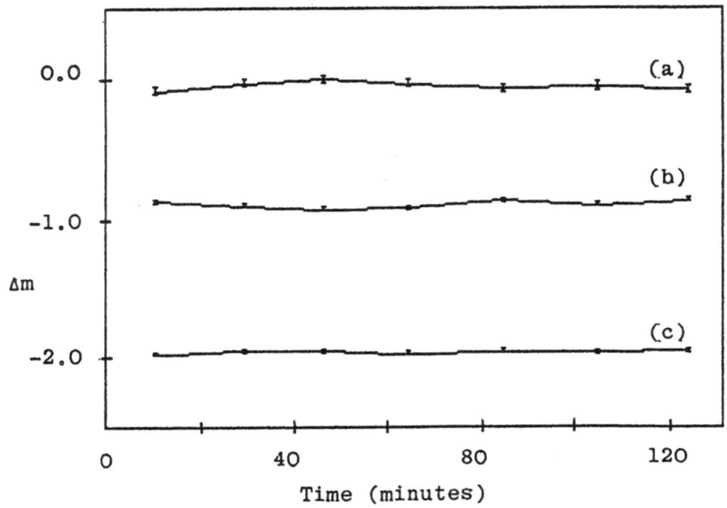

Fig. 3. The difference in V magnitude (Δm) between (a) the reference star R2 and the BL Lacertae object 0414+009, (b) the reference stars R1 and R2, and (c) the reference star R1 and the BL Lacertae object. The origin of the time axis 21^h28^mUT on 3 January 1990. For display purposes 2.0 and 1.0 magnitudes have been added to curves (b) and (c) respectively.

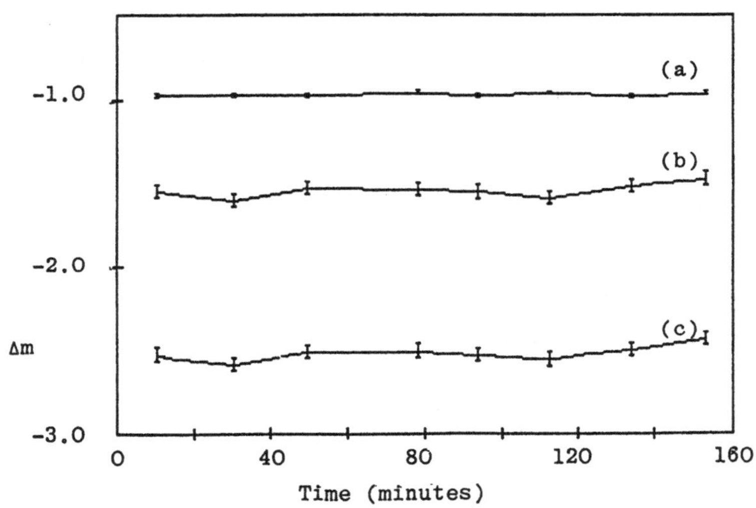

Fig. 4. The difference in V magnitude between (a) the reference stars R2 and R1, (b) the reference star R2 and the BL Lacertae object 0836+182 and (c) the reference star R1 and the BL Lacertae object. The origin of the time axis is 02^h14^mUT on 4 January 1990.

The Connection between Blazars and Compact Radio Sources

Claes-Ingvar Björnsson

Stockholm Observatory, S-133 36 Saltsjöbaden, Sweden

Abstract: It is argued that the unresolved VLBI-cores in flat spectrum, compact radio sources have properties very similar to the associated blazar emission region: Large scale magnetic field, anisotropic pitch-angle distribution and the importance of relativistic streaming motion. Problems afflicting models invoking Compton-cooling behind a relativistic shock in a jet to explain the initial phase of out-bursts in the mm/far-infrared regime are discussed. As an alternative, a qualitative discussion is given of a model where relativistic streaming and pitch-angle scattering are the main ingredients.

1. Blazars vs. compact radio sources

The close connection between blazars and compact radio sources (CRS) has only become evident in the last few years. Although it has been known for a long time that blazars always are associated with CRS, it was the work of Fugmann (1988) and Impey & Tapia (1988) which first gave strong indication that also the reverse is true; in complete samples of CRS it was found that at least 2/3 of all sources are blazars. The direct link between blazars and CRS is emphasized by the observations of Landau et al. (1986) and Impey & Neugebauer (1988), which show that these sources have smooth spectra from radio to optical; this is in marked contrast to non-blazar/CRS, which reveal evidence for spectral breaks or several components. In fact, there are indications that in some objects or at some occasions, the radio and optical emission region coincide; Kikuchi et al. (1988) found that a large swing in position angle (PA) occurred simultaneously in the radio and in the optical. Gabuzda, Wardle & Roberts (1989) detected intraday variations in the unresolved VLBI-core of 0735-178. Using the 100m Effelsberg telescope, Witzel & Quirrenbach (1990) found that intraday variations are quite common in CRS. This variability timescale is similar to the one seen in blazars.

Circular polarization (CP) in CRS has a clear signature: For a given source, CP show a preferred sign both as a function of time (Komesaroff et al. 1984) and frequency (Weiler & de Pater 1983). CP most likely comes from the optically thick core. This is corroborated by the fact that CP exhibits more rapid and larger relative variations than do either the degree of polarization (P) or the flux (Komesaroff et al. 1984). Björnsson (1990) has used these properties of CP to argue that although the magnitude of CP is probably due to radiative transfer effects, the sign is almost certainly due to the magnetic field geometry, i.e. the magnetic field in the unresolved VLBI-core is not turbulent but has large scale coherence.

Analysing the frequency dependence of P and PA in blazars Björnsson (1985) concluded that it can not (except for small changes in P) be due to spectral curvature but is best explained as resulting from an anisotropic pitch-angle distribution of the electrons with a sharp high energy cut-off. In particular, a two component model is unlikely since (i) the spectral index $\alpha (= -d \ln F / d \ln \nu)$ increases monotonically with frequency (Impey & Neugebauer (1988)), (ii) the expected correlation between a minimum in P and a maximum in $dPA/d\nu$ is not seen. Together with a large scale magnetic field, indicated by the large values of P, an anisotropic pitch-angle distribution suggests that relativistic streaming can be an important feature of blazars/CRS.

2. Streaming and pitch-angle scattering

Although the later stages of outbursts in the radio regime can be explained as due to an adiabatically expanding, optically thick, inhomogeneous source (e. g. van der Laan 1966, Blandford and Königl (1979)), the initial phases in the mm/far-infrared can not. In order to fit the observed variations with time of the flux and the frequency where the flux peaks, rather contrived injection of electrons and/or magnetic field needs to be invoked. An alternative is to explain the initial increase of flux as due to the the decreasing importance of Compton-cooling behind a shock which propagates outwards in a jet (Marscher and Gear 1985), i.e. the relative efficiency of synchrotron cooling (radio emission) increases with time. Since the efficiency of first order Fermi-acceleration at relativistic shocks probably is low unless the shock is parallel, the main increase of electron and magnetic field energy density behind the shock is likely to be due to the compression. Hence, the shape of the electron distribution does not change. Therefore, due to the cooling behind the shock, the spectral index a of the optically thin radiation in the outburst is steeper by 0.5 than in the quiescent jet. This is indeed the case for the 1983 outburst in 3C 273 (Clegg et al. 1983, Robson et al. 1983), where α increased from 0.7 to 1.2. However, this is not generally true; Valtaoja et al. (1988) studied 17 outbursts in 15 different sources and found that in the cooling phase all outbursts were consistent with an optically thin spectral index of $\alpha = 0.2$, indicating no spectral steepening since $0.2 < \alpha < 0.5$ in the quiescent phase. Furthermore, the relative importance of cooling due to Compton-scattering and synchrotron radiation is given by U_{ph}/U_B, where U_{ph} and U_B are the energy densities of photons and magnetic field, respectively. Since the peak flux is assumed to occur where the source becomes optically thick, the increase of U_{ph} due to the shock can not be much larger than the observed increase in flux. Thus, even for a parallel shock (i.e. no increase in B), the relative increase of Compton-cooling in the outburst should be no more than a factor of a few. Accordingly, it is expected that if Compton-cooling is unimportant in the quiescent jet, it should remain so in an outburst due to a relativistic shock.

Due to this potential problems in ascribing the properties of the initial phases of outbursts to relativistic shocks, I will now discuss some of the characteristics of a model based on the conclusions arrived at in the previous section, namely: Let mono-energetic (for simplicity) electrons with small pitch-angles be injected instantaneously into a jet containing a large scale magnetic field. Assume further that the energy density of the injected electrons is small compared to that of the magnetic field. The electrons will then stream along the magnetic field lines until pitch-angle scattering makes synchrotron cooling efficient. The important timescales are: (i) Synchrotron cooling, $t_s = 3 \times 10^8/\gamma B^2 \sin^2\theta$, where γ is the Lorentz factor of the injected electrons and θ their pitch-angle. (ii) "Source crossing" time, r/c, where r is the scale height of the jet. (iii) Time scale for pitch-angle scattering (e.g. Spangler 1979),

$$t_\theta \simeq \frac{10^3}{w}\left(\frac{\gamma_3\, r_{pc}}{B}\right)^{1/2}, \tag{1}$$

where w is the turbulent energy density divided by the total magnetic energy density, $\gamma_3 = \gamma/10^3$ and $r_{pc} = r/3 \times 10^{18}\,\mathrm{cm}$. In deriving equation (1) it was assumed that the energy driving the turbulence is put in on a scale equal to r and that the turbulent spectrum is of the Kraichnan type, which is appropriate for weak turbulence ($w \ll 1$). As the electrons stream along the field lines, the importance of pitch-angle scattering is determined by

$$S = \frac{r}{c\, t_\theta} = w_{-5}\left(\frac{B\, r_{pc}}{\gamma_3}\right)^{1/2}, \tag{2}$$

where $w_{-5} = w/10^{-5}$ in equation (2) includes turbulence with non-relativistic phase velocities only, since waves with relativistic phase velocities tend to make the streaming electrons loose energy rather than scatter.

It will be assumed that w increases faster with r than $(B\, r_{pc})^{-1/2}$, so that S increases with time. The properties of the outburst depends critically on the role of cooling.

No cooling ($t_s > t_\theta$): The emitted flux is proportional to the number of electrons scattered to large pitch-angles. Thus, the outburst has three phases.

(i) $S < 1$: flux increases with time as pitch-angle scattering becomes more efficient.

(ii) $S \simeq 1$: the flux reaches its maximum.

(iii) $S > 1$: the electrons are isotropized and the later stage of the outburst is similar to an adiabatic, van der Laan type expansion.

Cooling efficient ($t_s < t_\theta$): The electrons cool before reaching large pitch-angles. Hence, streaming is important and the last adiabatic phase does not occur. Most of the energy is radiated away at $S \simeq 1$. Using the condition $t_s(\sin\theta = 1) = t_\theta$ at $S = 1$, a lower limit to the frequency where streaming is important can be obtained

$$\nu_9^{\min} = \frac{10^{-2}}{B^3\, r_{pc}^2}\,, \tag{3}$$

where $\nu_9^{\min}$ is the minimum frequency in units of 10^9 Hz. Equating t_s with t_θ gives the typical frequency of the radiation

$$\nu_9 = 5\,\frac{B^{1/4}\,\gamma_3^{5/4}}{r_{pc}^{1/4}}\,. \tag{4}$$

In an electron-proton plasma streaming enhances CP,

$$CP(\%) \simeq \sqrt{\frac{B}{\nu_9\ \sin\theta}}\,. \tag{5}$$

If the rapid variations with an amplitude $\simeq 10\%$ observed by Witzel and Quirrenbach (1990) is due to streaming, $CP \leq 1\%$ is needed since typical values for the total flux is $CP \simeq 0.1\%$ (Komesaroff et al. 1984). From

$$\nu_9\ \sin\theta = 3\left(\frac{\gamma_3}{B\, r_{pc}}\right)^{1/2} = 3\, w_{-5}(S=1) \tag{6}$$

it is seen that $w_{-5}(S=1) \geq B$ is required.

With streaming large Doppler boosting factors ($D \simeq (\sin\theta)^{-1}$) can be obtained, the upper limit is set by the Lorentz factor of the emitting electrons. The need for large values of D has been emphasized by Witzel and Quirrenbach (1990); in order to explain the observations, using standard synchrotron theory, they deduced $D \simeq 200$. It is important to note that for streaming, the Doppler factor deduced from super-luminal motion (Γ) can be much smaller than D; e.g. assume that the magnetic field has a helical structure with pitch-angle β, then $\Gamma = (\sin\beta)^{-1}$ if $\Gamma < D$.

References

Blandford, R.D., Königl, A.: 1979, *Ap.J.* **232**, 34.
Björnsson, C-I.: 1985, *M.N.R.A.S.* **216**, 241.
Björnsson, C-I.: 1990, *M.N.R.A.S.* **242**, 158.
Clegg, P.E., et al.: 1983, *Ap.J.* **273**, 58.
Fugmann, W., 1988: *Astr.Ap.* **205**, 86.
Gabuzda, D.C., Wardle, J.F.C., Roberts, D.H.: 1989, *Ap.J.* **338**, 743.
Impey, C.D., Neugebaur, G.: 1988, *Astr.J.* **95**, 307.
Impey, C.D., Tapia, S.: 1988, *Ap.J.* **333**, 666.
Kikuchi; S., Inoue, M., Mikami, Y., Tabara, H., Kato, T.: 1988, *Astr.Ap.* **190**, L8.
Komesaroff, M.M., Roberts, J.A., Milne, D.K., Rayner, P.T., Cooke, D.J.: 1984, *M.N.R.A.S.* **208**, 409.
Landau, R., et al.: 1986, *Ap.J.* **308**, 78.
Marscher, A.P., Gear, W.K.: 1985, *Ap.J.* **298**, 114.
Robson, E.I., et al.: 1983, *Nature* **305**, 194.

Spangler, R.S.: 1979, *Ap.J.* **232**, L7.
Valtaoja, E., et al.: 1988, *Astr.Ap.* **203**, 1.
van der Laan, H.: 1966, *Nature* **211**, 1131.
Weiler, K.W., de Pater, I.: 1983, *Ap. J. Suppl.* **52**, 293.
Witzel, A., Quirrenbach, A.: 1990, preprint (to be published in the proceedings of "Propagation effects in space VLBI")

VLBI Knots and Superluminal Motion

Christian A. Hummel

Max-Planck-Institut für Radioastronomie, Auf dem Hügel 69,
5300 Bonn 1, Germany

Abstract: We briefly describe high resolution radio imaging work carried out on "active galactic nuclei" (AGN). The relation between radio flux density variability and structural variability, and jet dynamics are investigated. New observations of structural characteristics of the jets close to the "central engine" are presented.

1. Introduction

Early catalogues of extragalactic radio sources, selected at a comparatively low radio frequency ν (e.g. the 3CR-catalogue at ν=178 MHz), contain a population of sources which mostly have steep power law radio spectra, with spectral indices $\alpha_{11\mathrm{cm}}^{3\mathrm{cm}}$ (total flux density $S \propto \nu^{\alpha}$) distributed closely around -0.8. These sources do not generally show any variability and turned out to be extended "doubles" in most cases. By contrast, the spectral index distribution of sources selected at comparatively high radio frequencies ($\nu \geq 5$ GHz) is bimodal, with a sizeable population of sources showing statistically flat radio spectra. Flat spectrum sources were soon found to show variations in flux density on time scales of months to years. Historically, this was the first evidence that they must contain small scale structures (see Kellermann and Pauliny-Toth 1981, and references therein). The high brightness temperatures implied suggested that the origin of the radiation was the incoherent synchrotron process of relativistic electrons, gyrating in a magnetic field. Subsequently, the Very Long Baseline Interferometry technique (VLBI) was developed, and has led to the detection of the expected, extremely compact source components (knots, i.e. emission features of less than a few milliarcseconds extent, see Figure 2), whose self absorbed synchrotron spectra peake at different frequencies and thus produce the flat overall spectra. The monitoring of structural variations, with angular resolutions still unsurpassed by other methods, revealed apparent superluminal motions along elongated, jet-like structures, away from a stationary (e.g. Bartel et al. 1986) core component with an inverted spectrum. Causal connection of the emergence of superluminal components from the cores to flux density outbursts has been demonstrated by several authors (recently also at mm-wavelengths for 3C 273: Krichbaum 1990). The VLBI jet, emerging from the central engine and penetrating the BLR and NLR, turns into the kpc scale jet which powers hot spots and lobes. A model based on the relativistic motion of plasma at a small angle to the line of sight (LOS) to the observer, which explains various observational facts such as the apparent one-sidedness of many jets, apparent superluminal motion, excess of expected over measured inverse-Compton X-ray flux, and the short variability timescales, has become, as the "relativistic beaming model", a widely accepted framework for jet models (e.g. Blandford and Königl 1979). By using shocks, rather than individual expanding plasmons propagating down an underlying relativistic jet, Hughes, Aller and Aller (1989) were able to model the total and polarized flux density variability of BL Lac and to relate the shock structure to observed positions of knots in VLBI maps. Shocks are a useful concept for particle reacceleration, and the compression of the perpendicular magnetic field component in the knots has recently been demonstrated by Gabuzda et al. (1989) using polarization VLBI measurements (see also Roberts et al. 1990).

2. Statistical Investigations

In the early days of observing individual strong and prominent sources, a program was initiated to investigate the occurrence of superluminal motion in a statistically complete, flux density limited sample of flat spectrum radio sources (e.g. Witzel 1987). The sample was selected from the S5 survey (Kühr et al. 1981), which contains 6 BL Lac objects and 7 quasars north of 70 degrees declination, with a flux density greater than 1 Jy at 6 cm at the time of the survey, and a spectral index $\alpha_{11\mathrm{cm}}^{6\mathrm{cm}}$ flatter than -0.5. The sources have been observed over a wide range of frequencies, from radio through X-rays (see Eckart et al. 1986, 1987, and references therein). Radio maps have been obtained at frequencies from 327 MHz to 43 GHz, with resolutions from a few arcseconds to 0.1 mas, using the VLA, MERLIN and VLBI networks. Recent status reports on the VLBI observations are given by Witzel (1987) and Schalinski et al. (1988). Source structures at 5 GHz range from very compact (2 sources), slightly extended or double (7 sources) to one-sided jet morphology (3 sources). One source has a complex double structure. The morphological types found are representative of flat spectrum sources, as larger samples show (see Pearson and Readhead 1988, who studied a flux density limited sample without the restriction of flat spectral index). This investigation has led to the conclusion that a high fraction of these sources display effects due to bulk relativistic motion (Witzel et al. 1988). Within the relativistic beaming model, this means that sources preferentially aligned with the LOS have been selected. Relativistic beaming, therefore, can be a reason for a strong bias in samples of sources because of its ability to alter the apparent properties of the sources drastically.

In order to study the occurrence of bulk relativistic motion in randomly oriented sources, Zensus and Porcas (1987), as well as Hough and Readhead (1989), defined flux density limited samples of quasars on the basis of the extended source structure, which is not expected to be relativistically beamed. As these sources contain only weak cores (Doppler enhancement less prominent!) observations need even higher technical effort and take many years to finish. Superluminal motion has been revealed in the cores of three sources so far; the velocities found are lower than the average velocity found in flat spectrum sources.

The relativistic beaming model has been used to explain the variety of morphologies and properties of radio sources as being due to one population of similar sources pointing at different angles to the LOS. Relative numbers of flat and steep spectrum radio sources have been predicted by Orr and Browne (1982). Barthel suggested in 1989 that classical symmetric double radio galaxies of type Fanaroff-Riley 2 form the parent population of quasars. The latter then are radiogalaxies pointing towards the observer at an angle less than 45° to the LOS.

3. Individual sources

Modern, high quality VLBI maps show a great complexity in the pc-scale structures of the jets, and now allow us to address questions concerning the details of jet dynamics and emission mechanisms, thereby testing and refining the relativistic beaming model in more detail: Krichbaum et al. (1990) and Hummel (1990) find stationary and moving patterns (components) in the jet of the very luminous quasar 0836+71, which can be explained most easily as stationary and moving shocks, embedded in a preexisting underlying relativistic fluid, whose velocity determines the boosting factor of the radiation. Similar coexistence of moving and stationary knots has also been found in other sources (e.g. 4C39.25 [e.g. Marcaide et al. 1989], 3C 395 [Simon et al. 1987], 3C 454.3 [Pauliny-Toth et al. 1987]). The transverse width of the jet of 0836+71, though nearly unresolved, can be measured and correlated with the appearance of the features in accordance with a model by Daly and Marscher (1988) of a relativistic jet expanding into a surrounding medium of lower pressure (the BLR and NLR). The pressure ratio can be determined from the observations.

The outermost component of 0836+71 (at the given dynamic range of the VLBI-maps) is at a deprojected distance of 220 pc from the core and exhibits superluminal motion. Thus the jet remains relativistic to at least this distance, and seems to stay so even to a few kpc because of its continuity and one-sidedness observed at larger distances. The investigation of the jets to such large distances at high resolution has become possible by the combination of data from different arrays observing simultaneously, e.g. VLBI and MERLIN (e.g. 0836+71: Hummel et al. in prep., M87: Muxlow et al. 1988).

Lateral displacements, wiggles and bends in the jets are quite common and serve as further probes of the surrounding medium. Different physical processes, even in the same source on different scales, might be at work. In the quasars 3C 273 and 3C 345 (Cohen et al. 1987, Krichbaum et al. 1990, Biretta et al. 1986), components, which define the curved jet structure, seem to move along the bent ridge line. First results from VLBI at high frequencies (43 GHz) show that strong jet curvature near the cores might be more common than expected. Several sources of the S5 sample have been detected at 7 mm. Owing to sensitivity limitations, only one source yielded enough data so that a map of its structure could be made. The BL Lac object 1803+78 displays a sinusoidally bent jet, possibly containing superluminally moving components, between two stationary components. Since the resolution at 7 mm is of the order of 100 μarcsec or less, one is able to look at structures closely related to their origin in the central engine (linear resolutions ranging from 30 light days [3C 84, $z = 0.018$] to 1.3 light years [1803+78, $z = 0.68$]).

Monitoring observations of the quasar 1928+73 ($z = 0.30$) at 13 mm since 1985 trace motion in what might be the purest sinusoidally bent jet found so far. In Figure 1 we show a display of maps of 1928+73 covering all scales from kpc to sub-pc. We are almost able to trace the jets along their whole extent, from the pc-scale using VLBI — giving additionally kinematic information — to kpc-scale using MERLIN and VLA. Obviously, bends and wiggles occur on all scales and might be due to different physical processes. In view of the linearity of the jet observed to a distance of about 10 to 20 kpc from the core (on both counterjet and jet side), one might think of the wiggly pc-scale jet as being boosted filaments embedded in a broader jet.

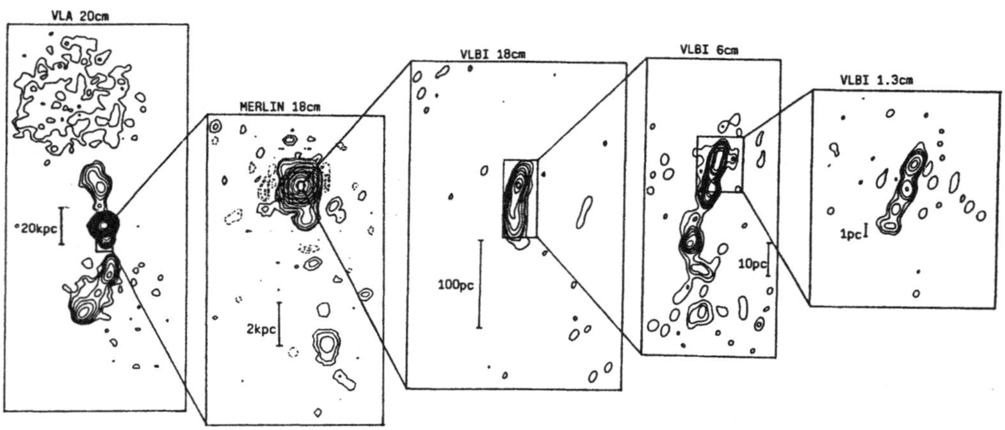

Fig. 1. Panorama of maps of the S5-quasar 1928+738. From left to right: map at 20 cm obtained with the VLA (Johnston et al. 1987), map at 18 cm obtained with MERLIN (T.W.B. Muxlow, priv. comm., Hummel et al., in prep.), VLBI maps at 18 cm (R. Wegner, priv. comm.), 6 cm (Eckart et al. 1987), 1.3 cm (Hummel et al., in prep.). Boxes indicate relative magnifications; note the linear scales specified!

At the highest resolution obtained so far for 1928+73, one can trace superluminal motion of the components in the sinusoidally bent jet at 13 mm in the display of Figure 2. The components appear to move along parallel ballistic trajectories at a common velocity of 4c (significantly lower than observed at larger distances

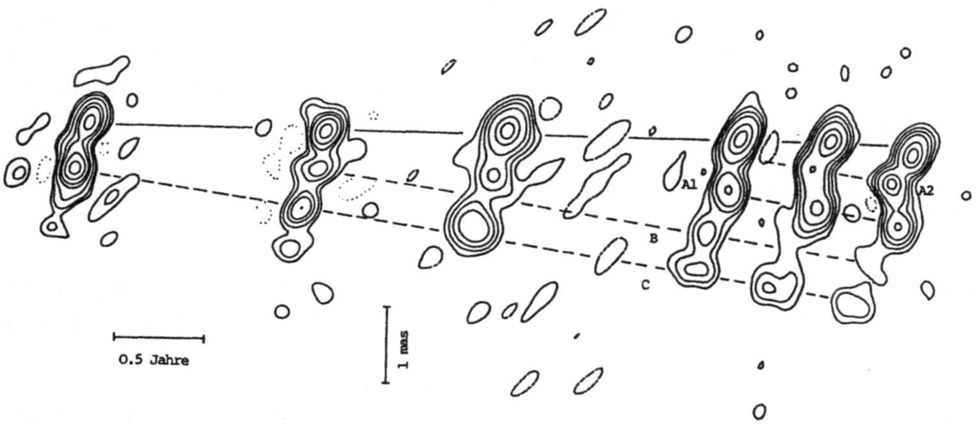

Fig. 2. Panorama of VLBI-maps of the S5-quasar 1928+738 at 1.3 cm. Epochs of observation are from left to right: 1985.1, 1986.4, 1987.4, 1988.7, 1989.2, 1989.7. The maps are aligned at the northernmost component, which is the inverted spectrum core (the last map had to be shifted by a small amount southward because of a shift of the core brightness centroid due to an emerging new component). Dashed lines identify knots on the basis of a common velocity scheme. Note the sinusoidal wiggle of the jet ridge line, most obvious in the last three epochs!

from the core [Witzel et al. 1988]), thereby preserving the sinusoidal structure. This behaviour is therefore different from the case of 3C 345 or 3C 273. The regular sinusoidal (helical?) pattern with a lateral amplitude of 0.1 mas suggests a causal connection either to orbital motion of the central engine or rotation of the accretion disk, the latter producing a helical twisting of the magnetic field lines, to which the observed VLBI components might be confined (see Camenzind 1989).

4. Conclusions

High quality radio imaging of AGN using VLBI today not only allows us to study jet dynamics in more detail, but also achieves at the higher observing frequencies the extremely high resolution required to connect the jet phenomena to processes going on in the very center of AGN. Interaction of the jets with the BLR and NLR can also be studied, giving constraints on physical parameters of jets and the ambient medium. A considerable amount of data concerning the structure and kinematics has been obtained, not only for individual sources, but also for complete samples, which give statistically meaningful results. Information from other wavelength bands has been and should be further added in order to reach a comprehensive understanding of AGN.

Acknowledgements: I would like to thank Drs. T.P. Krichbaum, A. Quirrenbach, C.J. Schalinski, and A. Witzel for their contributions and help. Critical reading of the manuscript by Drs. I.I.K. Pauliny-Toth and R.W. Porcas is gratefully acknowledged.

References

Bartel, N., Herring, T.A., Ratner, M.I., Shapiro, I.I., and Corey, B.E. (1986): "VLBI limits on the proper motion of the 'core' of the superluminal quasar 3C 345", *Nature* **319**, 733

Barthel, P.D. (1989): "Is every quasar beamed?", *Astrophys. J.* **336**, 606

Biretta, J.A., Moore, R.L., and Cohen, M.H. (1986): "The evolution of the compact radio source in 3C 345. I. VLBI observations", *Astrophys. J.* **308**, 93

Blandford, R.D., and Königl, A. (1979): "Relativistic jets as compact radio sources", *Astrophys. J.* **232**, 34

Camenzind, M. (1989): "Hydromagnetic flows from rapidly rotating compact objects and the formation of relativistic jets", in *Accretion Disks and Magnetic Fields in Astrophysics*, ed. by G. Belvedere (Dordrecht: Kluwer), p.129

Cohen, M.H., Zensus, J.A., Biretta, J.A., Comoretto, G., Kaufmann, P., and Abraham, Z. (1987): "Evolution of 3C 273 at 10.7 GHz", *Astrophys. J.* **315**, L89

Daly, R.A., and Marscher, A.P. (1988): "The gasdynamics of compact relativistic jets", *Astrophys. J.* **334**, 539

Eckart, A., Witzel, A., Biermann, P., Johnston, K.J., Simon, R., Schalinski, C., and Kühr, H. (1986): "Investigation of a complete sample of flat spectrum radio sources from the S5 Survey I. Analysis", *Astron. Astrophys.* **168**, 17

Eckart, A., Witzel, A., Biermann, P., Johnston, K.J., Simon, R., Schalinski, C., and Kühr, H. (1987): "Investigation of a complete sample of flat spectrum radio sources from the S5 Survey II. Results", *Astron. Astrophys. Suppl. Ser.* **67**, 121

Gabuzda, D.C., Cawthorne, T.V., Roberts, D.H., and Wardle, J.F.C. (1989): "The milliarcsecond polarization structure of six BL Lacertae Objects", *Astrophys. J.* **347**, 701

Hough, D.H., and Readhead, A.C.S. (1989): "A complete sample of double-lobed radio quasars for VLBI tests of source models: Definition and statistics", *Astron. J.* **98**, 1208

Hummel, C.A., and Krichbaum, T.P. (1990): "The Jet in Quasar 0836+71", in *Parsec-scale radio jets*, ed. by J.A. Zensus and T.J. Pearson (Cambridge: Cambridge University Press), p. 38

Hughes, P.A., Aller, H.D., and Aller, M.F. (1989): "Synchrotron emission from shocked relativistic jets. II. A model for the centimeter wave band quiescent and burst emission from BL Lacertae", *Astrophys. J.* **341**, 68

Johnston, K.J., Simon, R.S., Eckart, A., Biermann, P., Schalinski, C., Witzel, A., and Strom, R.G. (1987): "1928+738: A superluminal source with large-scale structure", *Astrophys. J.* **313**, L85

Kellermann, K.I., and Pauliny-Toth, I.I.K. (1981): "Compact Radio Sources", *Ann. Rev. Astron. Astrophys.* **19**, 373

Krichbaum, T.P. (1990): "Images of Parsec-Scale Jets from 7 mm VLBI", in *Parsec-scale radio jets*, ed. by J.A. Zensus and T.J. Pearson (Cambridge: Cambridge University Press), p. 83

Krichbaum, T.P., Hummel, C.A., Quirrenbach, A., Schalinski, C.J., Witzel, A., Johnston, K.J., Muxlow, T.W.B., and Qian, S.J. (1990): "The complex jet associated with the quasar 0836+71", *Astron. Astrophys.* **230**, 271

Krichbaum, T.P., Booth, R.S., Kus, A.J., Rönnäng, B.O., Witzel, A., Graham, D., Pauliny-Toth, I.I.K., Quirrenbach, A., Hummel, C., Alberdi, A., Zensus, J.A., Johnston, K., Spencer, J., Rogers, A.E.E., Lawrence, C., Readhead, A., Hirabayashi, H., Inoue, M., Dhawan, V., Bartel, N., Shapiro, I., Burke, B., and Marcaide, J. (1990): "43 GHz-VLBI observations of 3C 273 after a flux density outburst in 1988", *Astron. Astrophys.*, in press

Kühr, H., Witzel, A., Pauliny-Toth, I.I.K., and Nauber, U. (1981): "A catalog of extragalactic radio sources having flux densities greater than 1 Jy at 5 GHz", *Astron. Astrophys. Suppl. Ser.* **45**, 367

Marcaide, J.M., Alberdi, A., Elósegui, P., Schalinski, C.J., Jackson, N., and Witzel, A. (1989), "The peculiar superluminal radio source 4C 39.25: observations and model", *Astron. Astrophys.* **211**, L23

Muxlow, T.W.B., Junor, W., Spencer, R.E., Simon, R., Benson, J., and Reid, M. (1988): "High dynamic range mapping of a complex radio source - M87", in *IAU Symposium 129, The Impact of VLBI on Astrophysics and Geophysics*, ed. by M.J. Reid and J.M. Moran (Dordrecht: Kluwer), p. 131

Orr, M.J.L., and Browne, I.W.A. (1982): "Relativistic beaming and quasar statistics", *Mon. Not. R. astr. Soc.* **200**, 1067

Pauliny-Toth, I.I.K., Porcas, R.W., Zensus, J.A., Kellermann, K.I., Wu, S.Y., Nicolson, G.D., and Mantovani, F. (1987): "Peculiar variations in the structure of the quasar 3C 454.3", *Nature* **328**, 778

Pearson, T.J., and Readhead, A.C.S. (1988): "The milliarcsecond structure of a complete sample of radio sources. II. First-epoch maps at 5 GHz", *Astrophys. J.* **328**, 114

Roberts, D.H., Kollgaard, R.I., Brown, L.F., Gabuzda, D.C., and Wardle, J.F. (1990): "Milliarcsecond polarization structure of the superluminal quasar 3C 273", *Astrophys. J.* **360**, 408

Schalinski, C.J., Witzel, A., Krichbaum, Th.P., Hummel, C.A., Biermann, P.L., Johnston, K.J., and Simon, R.S. (1988): "Bulk relativistic motion in a complete sample of flat spectrum radio sources", in *IAU Symposium 129, The Impact of VLBI on Astrophysics and Geophysics*, ed. by M.J. Reid and J.M. Moran (Dordrecht: Kluwer), p. 71

Simon, R.S., Johnston, K.J., Hall, J., Spencer, J.H., and Waak, J.A. (1987): "Superluminal Motion Towards a Stationary Component in Quasar 3C 395", in *Superluminal Radio Sources*, ed. by J.A. Zensus and T.J. Pearson (Cambridge: Cambridge University Press), p. 72

Witzel, A. (1987): "A complete Sample from the S5 Survey", in *Superluminal Radio Sources*, ed. by J.A. Zensus and T.J. Pearson (Cambridge: Cambridge University Press), p.83

Witzel, A., Schalinski, C.J, Johnston, K.J., Biermann, P., Krichbaum, T.P., Hummel, C.A., and Eckart, A. (1988): "The occurence of bulk relativistic motion in compact radio sources", *Astron. Astrophys.*, **206**, 245

Zensus, J.A., and Porcas, R.W. (1987): "Superluminal motion in a randomly oriented quasar sample", in *Superluminal Radio Sources*, ed. by J.A. Zensus and T.J. Pearson (Cambridge: Cambridge University Press), p.126

Relativistic Jet Models and Variable Knot Emission

Max Camenzind

Landessternwarte Königstuhl, 6900 Heidelberg, Germany

Abstract: Variability in the radio waveband and moving features detected by VLBI observations are commonly explained in terms of shocks moving in relativistic collimated flows. We discuss the origin of these jets with their relation to the magnetic structure of accretion disks. Dynamo action in the disk creates a rotating magnetosphere which is filled with disk–plasma. Along open flux tubes, the plasma is accelerated to relativistic speeds and collimated outside the light cylinder of the magnetosphere. The relaxation of these newly formed jets towards a minimal energy equilibrium is the origin for strong X-ray emission, ejection of knots along the jets and rapid variability in X-rays and the infrared–optical region.

1. Introduction

The most widely accepted explanation for superluminal motion is that *emission knots* which appear to separate from a *stationary core* at transverse speeds exceeding the speed of light, are relativistically moving features in a collimated flow of synchrotron–emitting plasma. Such a model does not only explain superluminal motion, but it also accounts for the high brightness temperature (via Doppler boosting of the radiation) and the short time–scale variations (via time–delay associated with an emitter propagating nearly along the line of sight). Since Blandford and Königl (1979) discussed the possibility that these propagating components on VLBI maps and the variability in the radio waveband might be explained by shocks in a relativistic collimated flow, a number of models have been proposed based on this hypothesis. Marscher and Gear (1985; see also Marscher, 1987) found that the total flux variation shown by the superluminal source 3C 273 in 1983 could be understood in terms of the shock model. The evolution of the spectrum in the millimeter to infrared region required that the dominant energy loss evolves from inverse Compton to synchrotron and adiabatic cooling as a shock propagates along a diverging channel. Hughes, Aller and Aller (1989a, b) found that the total and polarized flux outbursts in BL Lac and 3C 279 could be explained as arising from a series of closely spaced shocks in a relativistic flow.

The existence of these shocks is derived from numerical simulations of hydrodynamic jets which often show the formation of internal shocks. It is then supposed that VLBI knots are the hot downstream regions of such flow structures. The rise in polarized emission is associated with the compression of a previously tangled magnetic field in the jet. This is derived from the fact that many sources display in the quiescent phase a polarized flux that is only 1% – 2 % of the total flux, with a position angle perpendicular to the VLBI structure. This fact is usually explained as due to a source magnetic field which is predominantly turbulent and has only a weak mean component parallel to the source axis. A shock perpendicular to the flow axis compresses the field component in the plane. Thus an observer viewing radiation emitted in that plane in the source frame would see an effectively ordered magnetic field which gives rise to the highly polarized emission if the compression were large enough.

Relativistic jets are however formed under much more complicated conditions than originally thought (Blandford and Payne, 1982; Lovelace et al., 1987). In particular, relativistic speeds cannot be attained in pure hydrodynamic acceleration, magnetic fields are a necessary element of jet formation. The most likely origin of these jets is due to rapidly rotating magnetospheres built up by accretion disks around supermassive objects (Camenzind, 1986, 1989, 1990). In the following we first discuss the origin of the magnetic structure of accretion disks and in particular the exterior force–free region of the magnetosphere of the disk. The structure of the outflowing disk–plasma explains the formation of jets, X-ray emission, ejection of knots and various variable phenomena.

2. The magnetic structure of AGN accretion disks

The canonical model for the explanation of the emission properties of AGN consists of a supermassive rotating black hole with mass M_H and angular momentum J_H surrounded by an axisymmetric accretion disk specified by its accretion rate $\dot{M}_{\rm acc}$. Direct evidence for the existence of accretion disks in AGN follows from the presence of a strong optical–UVX excess in the continuum of Seyfert galaxies and quasars (Sun and Malkan, 1989; Ulrich, 1989; Laor and Netzer, 1989; Courvoisier et al., 1990). In this respect, each source is specified in terms of 4 parameters: the mass M_H of the central black hole, the Kerr parameter $a_H = J_H/J_{\rm max} < 1$ for the black hole angular momentum ($J_{\rm max} = GM_H/c$), the accretion rate $\dot{M}_{\rm acc}$ for the disk and the angle θ for the disk inclination with respect to the observer's line of sight. Masses and accretion rates can essentially be determined from model fittings of the big blue bump, but there is no indication whether black holes are fast rotating ($a_H \simeq 0.98$) or only moderately rotating ($a_H \simeq 0.8$) objects. The rotational energy of extremely fast rotating holes is potentially an additional source of energy, $E_{\rm rot} \leq 0.29\, M_H c^2 \simeq 4 \times 10^{62}\,{\rm erg}\, M_{H,9}$.

A Kerr hole surrounded by an accretion disk disposes of two different energy sources. The first one is due to the accretion process with a total accretion power

$$L_{\rm acc} = \epsilon_H(a_H)\, \dot{M}_{\rm acc} c^2 = 5 \times 10^{45}\,{\rm erg\,s^{-1}}\, \epsilon_{H,-1}\, \dot{M}_{\rm acc}/M_\odot\,{\rm yr}^{-1}\,, \qquad (1)$$

which is proportional to the accretion rate and the relativistic disk–efficiency $\epsilon_H < 0.42$. The latent spin energy of a rotating hole can be tapped by externally applied magnetic fields with a maximum power extraction given by (Thorne et al., 1986)

$$L_H \simeq \frac{c}{128} B_H^2 r_H^2 a_H^2 \simeq 10^{45} \, \mathrm{erg\,s^{-1}} \, B_{H,4}^2 M_{H,8}^2 a_H^2 \,. \tag{2}$$

B_H is the magnetic field strength wich covers the black hole and r_H is the radius of the horizon. This magnetic energy is, however, only disposable if the hole is rotating faster than the inner accretion disk ($a_H > 0.8$). This could only occur for newly born supermassive holes. This fact would be a natural explanation for the distinction between radio–loud and radio–weak objects among sources with a strong UVX–bump. Since the centers in spiral galaxies are accreting for quite a long time, these black holes should be in equilibrium with their surrounding disk. Black holes in newly formed centers of elliptical galaxies could still have a high angular momentum, when recently formed by merging.

The interstellar matter which accretes onto these supermassive objects in the center of these galaxies carries along magnetic fields. The evolution and influence of these fields on the structure of the disk are usually neglected in the theory of standard accretion disks. The formation of jets and the entire non–thermal activity of AGN can however only be understood in terms of the existence of strong magnetic fields in the innermost region of AGN (Blandford and Königl, 1979; Blandford and Payne, 1982; Lovelace et al., 1987; Camenzind, 1986, 1989, 1990; Courvoisier and Camenzind, 1989; Lesch et al., 1989).

Fully turbulent accretion disks are ideal set–ups for magnetic dynamos (Pudritz, 1981; Camenzind, 1990), quite in analogy to the dynamo action in the disks of spiral galaxies (Ruzmaikin et al., 1988). The essential ingredients for these dynamos are the strong shear of a Keplerian disk and the density gradient perpendicular to the shear. Any small seed field accreted from the molecular torus at the parsec scale will decay exponentially or be amplified by the dynamo until it reaches equipartition with the pressure $P_D(R)$ in the disk

$$P_D(R) \simeq \frac{\dot{M}_{\mathrm{acc}}}{4\pi\,\alpha\,R^2} \left(\frac{GM_H}{R}\right)^{1/2} \left(\frac{H}{R}\right)^{-1} . \tag{3}$$

In geometrically thin disks, the toroidal component B_ϕ is the dominant component of the magnetic field in the disk, the poloidal component B_p is regenerated from the toroidal one via the dynamo effect. A disk dynamo generates a large scale structure for the magnetic fields which is superposed by turbulent small scale structures (Fig. 1). This large scale field is essentially time–dependent, and its structure depends largely on the dynamo numbers (Pudritz, 1981; Camenzind, 1990). For typical parameters of geometrically thin accretion disks, $H/R \leq 0.5$, seed fields in the disk are rapidly amplified and form a quadrupolar structure near the inner edge of the accretion disk (Fig. 1). Dipolar fields are in general unstable against the inclusion of the dynamo effect. Many models have been put forward for the structure of disk fields in terms of dipolar fields (Blandford and Payne, 1982;

Königl, 1989; Wang et al., 1990). We find that dipolar fields should be unstable, if the accretion disk works as a true dynamo (Camenzind, 1990).

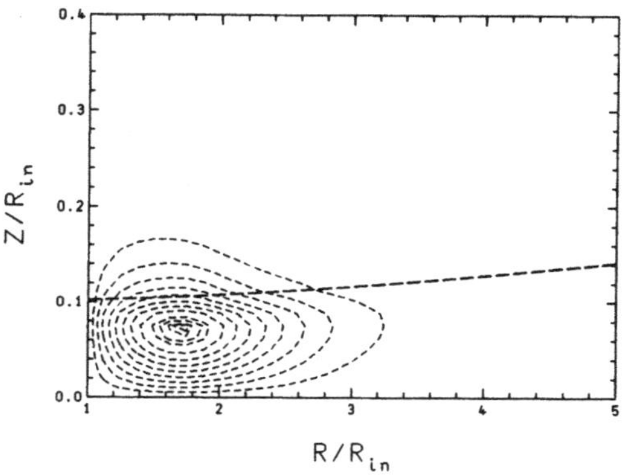

Fig. 1a. Quadrupolar structure of the poloidal disk fields in geometrically thin disks for a boundary layer rotation law. The dashed line marks the scale height $H(R)$ of the disk (Camenzind, 1990).

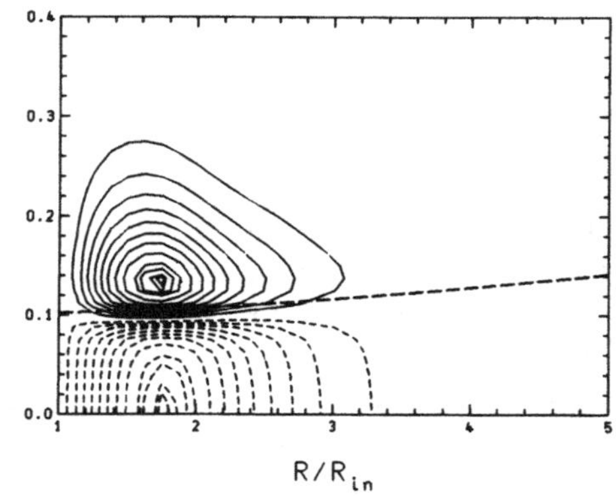

Fig. 1b. The distribution of the toroidal fields RB_ϕ in the disk corona system. Solid lines are for positive values, dashed lines for negative values, the corona is assumed to have the same rotation law as the disk.

The essential parameter for a possible classification of AGN is not the absolute accretion rate $\dot{M}_{\rm acc}$, but is value relative to the Eddington accretion rate $\dot{M}_{\rm Ed}$, $m = \dot{M}_{\rm acc}/\dot{M}_{\rm Ed} \propto \dot{M}_{\rm acc}/M_H$. In Fig. 2 we show the most likely positions of various types of AGN, when parametrized in terms of their masses M_H, disk magnetic fields B_D, as estimated from the pressure in the disk, Eq. (3), and the accretion rate m. The upper limit $B_{\rm Ed}$ for the strength of the magnetic field around black holes has an energy density comparable with that of radiation (Rees, 1984)

$$B_{\rm Ed} = \sqrt{L_{\rm Ed}/m_e^2 c} \simeq 4 \times 10^4 \,\text{Gauss}\, M_{H,8}^{-1/2} \,. \tag{4}$$

The effective maximal field strengths derived from standard accretion disk theory are in fact lower and can be suitably parametrized in terms of the mass M_H and the relative accretion rate m ($R_D \simeq 5R_S$)

$$B_D \simeq 6 \times 10^3 \,\text{Gauss}\, \sqrt{m\, M_{H,8}}\, \sqrt{1/\alpha\, \epsilon_{H,-1}}\, \sqrt{1/(H/R_D)} \,. \tag{5}$$

Sources characterized by the presence of a strong big blue bump must have high accretion rates (Sun and Malkan, 1989), since the inner disk which is the origin of the UVX–bump becomes optically thin for low accretion rates (Rees et al., 1982). Under the action of an accretion burst in the inner part of an elliptical galaxy, the central source is expected to move from the quasar and FR II radio galaxy region to BL Lac objects and FR I radio galaxies, when the accretion rate decreases. The appearance of a source depends largely on the value of the relative accretion rate m. If dynamo action occurs in the disk, the above field strengths correspond to maximal magnetic fluxes Ψ_D carried by the inner disk, $B_p \simeq B_D\,(H/R)^p$,

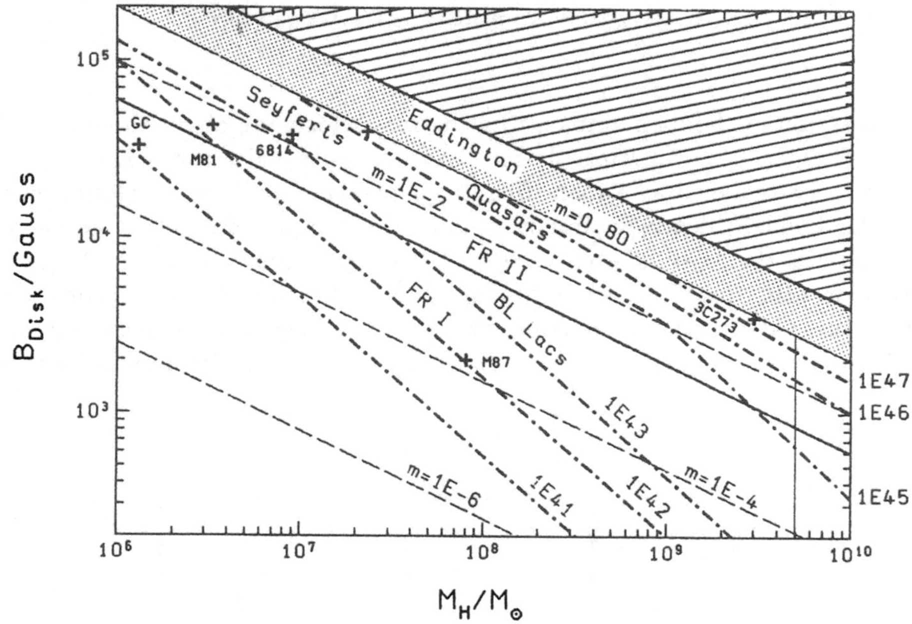

Fig. 2. The classification of AGN in terms of central masses M_H and the typical toroidal magnetic field B_D of the accretion disk. The accretion rate is parametrized in terms of the Eddington accretion rate, $m = \dot{M}_{\rm acc}/\dot{M}_{\rm Ed} < 1.0$. Seyfert galaxies, QSOs and quasars all have high accretion rates ($m \geq 0.01$), while FR I radio galaxies and BL Lac objects have rather low accretion rates ($m \simeq 0.001$). Also shown are lines of constant accretion power in units of erg/s (dot–dashed lines).

$$\Psi_D \simeq 10^{34}\,{\rm G\,cm^2}\,\sqrt{m\,M_{H,9}^3}\,\sqrt{1/\alpha\epsilon_{H,-1}}\,\sqrt{(H/R_D)^p}\,. \tag{6}$$

This magnetic flux scales strongly with the central mass so that the magnetic flux carried by the accretion disk of a weak Seyfert galaxy is at least 4 orders of magnitude lower than in the case of 3C 273. Similarly, we get from Eq. (5) the total poloidal current driven by the differential rotation of a Keplerian disk

$$I_D \simeq 9\times 10^{18}\,{\rm Amp}\,\sqrt{m\,M_{H,9}}\,\sqrt{1/\alpha\,\epsilon_{H,-1}}\,\sqrt{1/(H/R_D)}\,. \tag{7}$$

Magnetic flux Ψ_D and current I_D are the essential new elements of magnetized disks.

3. The structure of parsec–scale jets in AGN

Dynamo action in the disk forms a rotating magnetosphere above the disk which will be deformed by outflowing plasma. Exterior to the disk the magnetic structure has to be modeled as a force–free magnetosphere given as a solution of the relativistic Grad–Schlüter–Shafranov equation for the poloidal fields (see Camenzind, 1990). In the axisymmetric case, such a magnetosphere consists of a family of nested rotating magnetic surfaces (Fig. 3 and 4). Under the conditions found in AGN, these structures are roughly force–free meaning that the Alfvén point is near the light cylinder $R_L \simeq 10\, R_S$. It is the presence of this light cylinder which

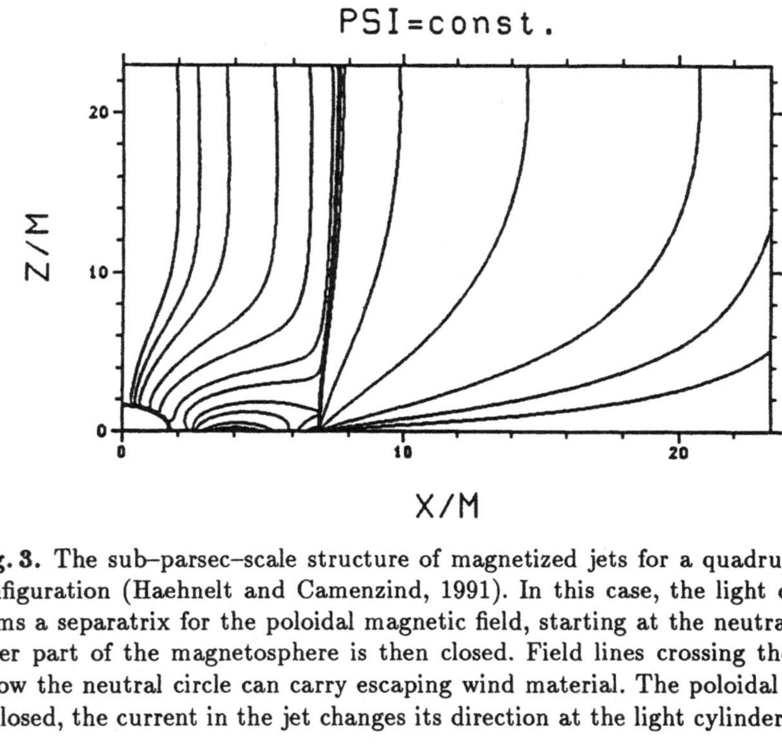

Fig. 3. The sub–parsec–scale structure of magnetized jets for a quadrupolar disk–field configuration (Haehnelt and Camenzind, 1991). In this case, the light cylinder surface forms a separatrix for the poloidal magnetic field, starting at the neutral circle N. The inner part of the magnetosphere is then closed. Field lines crossing the light cylinder below the neutral circle can carry escaping wind material. The poloidal current system is closed, the current in the jet changes its direction at the light cylinder.

strongly deforms the exterior part of the magnetosphere shown in Fig. 1. In particular, we find for quadrupolar fields a critical magnetic surface which closes to the light cylinder and separates the closed part of the magnetosphere from the open one. The outer part of the field lines leaving the disk must cross the light cylinder below this surface and enter the light cylinder at large distances from the central source and closing back to the black hole. In this quadrupolar topology, all field lines are closed. They can therefore also carry currents driven either by the differential rotation of the disk or the rapid rotation of the black hole itself. In either case, these currents leave the disk along the open field structure and return inside the light cylinder towards the hole. The black hole participates in this current system, quite in analogy to current systems driven by magnetized rotators (Camenzind, 1990).

The open field line structure carries a wind from the disk which is collimated by magnetic forces to semi–relativistic Lorentz factors given by the magnetization of the disk

$$\sigma_D = \frac{\Psi_D^2}{2\dot{M}_{\rm jet} c R_L^2} \simeq \frac{1}{\alpha\alpha_w}\frac{M_H}{R_D}\left(\frac{R_D}{R_F}\right)^2\left(\frac{H}{R_D}\right)^{2p-1} \simeq \frac{0.1}{\alpha\alpha_w}\left(\frac{H}{R_D}\right)^{2p-1} . \quad (8)$$

The mass flux in the jet is bounded by angular momentum conservation, $\dot{M}_{\rm jet} = \alpha_w \dot{M}_{\rm acc}(R_F/R_L)^2$, $\alpha_w \leq 1$. When $\sigma_D > 1$, we find $\gamma_{\rm jet} \simeq \sigma_D$ (Camenzind, 1989). This expression for the magnetization shows that relativistic flows can only occur, if $\sigma_D > 1$, i.e. if the disk is strongly magnetized, $H/R_D \simeq O(1), m \simeq O(1)$ and the angular momentum of the disk is not entirely extracted by the escaping wind, $\alpha_w \ll 1$, or $\alpha \ll 1$. Accretion disks in Seyfert galaxies, QSOs and FR I sources have $\sigma_D < 1$ $(m < 1)$ with only semi–relativistic outflow speeds, $v_j \leq 0.5$ c.

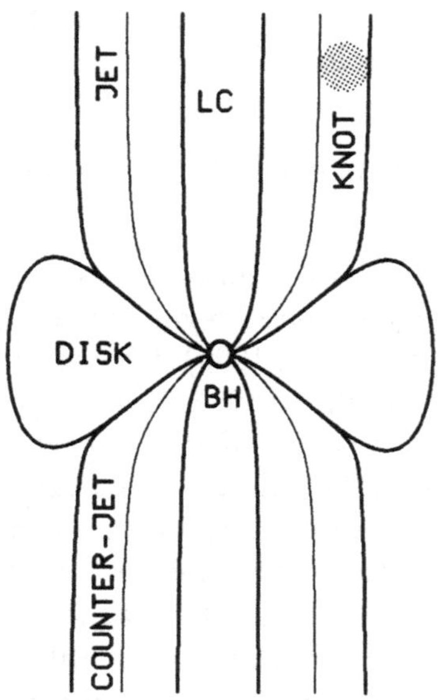

Fig. 4. The structure of magnetized jets driven by the rapidly rotating magnetosphere of Fig. 3. The magnetized disk wind is collimated outside the light cylinder ($\simeq 10^{15}$ cm for a 10^8 $M_\odot$ black hole) into bipolar jets on the parsec–scale. The magnetic surfaces inside the light cylinder close mostly to the black hole. The light cylinder itself is a separatrix for the poloidal magnetic structure.

Since the outflowing plasma carries away angular momentum, the plasma in the jet will also rotate with a typical period $P_{\rm jet} \simeq P_L (R/R_L)^2$, where P_L is the rotation period at the light cylinder R_L (R is the radius with respect to the jet axis)

$$P_L \simeq P_{\mathrm{K,Foot}} \simeq 10'000\,\mathrm{sec}\, M_{H,7} \left(\frac{R_{\mathrm{Foot}}}{5R_S}\right)^{3/2} . \tag{9}$$

This would be the rotation period of a non–axisymmetric perturbation near the light cylinder. Evidence for such a quasi–periodicity exists for the sources NGC 6814 ($P_L \simeq 12'000$ sec in X-rays), 0917+624 ($P_L \simeq 1.2$ days in the radio regime) and in optical flares of 3C 273 ($P_L \simeq 20$ days). These periods agree with the expected masses of these sources: NGC 6814 ($M_H \simeq 10^7\ \mathrm{M}_\odot$), 0917+624 ($M_H \simeq 10^8\ \mathrm{M}_\odot$) and 3C 273 ($M_H \simeq 2 \times 10^9\ \mathrm{M}_\odot$).

VLBI observations of the jet in 3C 273 can be explained by models of the above type where the central machine produces a relativistic flow guided by a magnetic channel with a mild curvature (Zensus et al., 1988). VLBI maps at 5 GHz show a narrow jet extending to a projected distance $\simeq 125h^{-1}$ parsecs from the core, corresponding to a true distance of $\simeq 1.0$ kpc. Superluminal motion has been detected out to at least half of this distance. The projected opening angle of the jet is less than 4°, i.e. the intrinsic angle is less than 1°. This means that this jet forms a straight cylinder with a radius of a few parsecs at a distance of a few hundred parsecs. The moving emission knots can be explained as features entrained by the underlying relativistic plasma. The base of the jet (Fig. 4) is somewhat narrower with a dimension of a fraction of a light year, corresponding to $\simeq 100R_L$ for a mass of several $10^9\ \mathrm{M}_\odot$.

4. X-rays, knot ejection and flux outbursts in 3C 273

The magnetized jets produced by the magnetosphere of the disk are roughly in a force–free equilibrium in the collimated regime. This equilibrium is, however, not in a relaxed state (Taylor, 1986; Appl, 1990). We expect therefore the occurrence of a resistive relaxation process in the immediate vicinity of the jet formation region. During this process, a great fraction of the magnetic energy can be dissipated e.g. by reconnection processes. The structure emerging from the light cylinder is then quite similar to the configuration of a reversed field pinch well known from plasma physics. Resistive relaxation will dissipate most of the magnetic energy in the vicinity of the light cylinder, since the magnetic structure is quite strongly stressed in that region. In the following, we identify this region with the X-ray source region in AGN. There are indeed many indications for the fact that the X-rays of AGN are not emitted in the immediate vicinity of the black hole, but in a region with a dimension of at least 10 Schwarzschild radii (Courvoisier and Camenzind, 1989; Kunieda et al., 1990; Pounds, 1990).

The initial magnetic energy carried away by the disk–flow can be estimated from the expression for the Poynting flux (Camenzind, 1990)

$$L_m \simeq \frac{c}{2}(-B_\phi)_D \frac{R_D}{R_L} \Psi_D \simeq I_D \Psi_D / R_L \simeq 10^{46}\,\mathrm{erg\,s^{-1}}\,(-B_\phi)_{D,3}\,\Psi_{D,34} . \tag{10}$$

Magnetized disks have two energy channels, on the one hand thermal emission from the surface of the disk and on the other hand emission of magnetic energy.

Part of this second channel can be dissipated near the light cylinder into particle acceleration with a maximum Lorentz factor $\gamma_{\max}$ given by magnetic confinement (for more details, see Lesch, 1990)

$$\gamma_{\max} \simeq \frac{B_L^2}{8\pi m_e c^2 n_r} \simeq \frac{1}{2}\frac{m_p}{m_e}\frac{v_A^2}{c^2} \leq 900\,. \tag{11}$$

Since the energy density in the UV–photons dominates over the magnetic energy, these particles cool over inverse Compton emission producing a power law spectrum in the X-ray region (Svensson, 1990) with a total luminosity

$$L_X \simeq \frac{4}{3}\,\sigma_T c\,\gamma_b^2\, u_{\rm ph}\, n_r\, V_L \simeq 10^{46}\,{\rm erg\,s^{-1}}\, R_{L,15}^3\, n_{r,6}\, u_{ph,4}\gamma_{b,2.5}^2\,. \tag{12}$$

The entire hard X-ray emission of Seyfert galaxies and quasars can be obtained in this process, provided that a large fraction of the magnetic luminosity of the disk is transformed into particle heating near the light cylinder. Due to the restriction of the acceleration process by Eq. (11), EUV photons from the disk are not much scattered beyond 1 MeV with little pair production (for details of the spectra, see Svensson, 1990). In Blazars, the synchrotron–self–Compton process is more important than scattering of UV–photons (Marscher, 1987).

The quasar 3C 273 underwent strong optical and infrared outbursts in the beginning of 1984 and of 1988, first observed in the infrared–optical bands (Courvoisier et al., 1988) and subsequently in the mm–regime. The time–delay between the outburst seen in the optical and that seen at 37 GHz was about 6 months (Courvoisier et al., 1990). VLBI maps taken at 43 GHz show quite clearly the correlation between the flux outbursts and the propagation of knots away from the central core (Krichbaum et al., 1990). From these maps there is therefore compelling evidence for a close connection between the birth of a new knot in the parsec–scale jet of 3C 273 (and of similar sources, 3C 120, 3C 345, 3C 279) and a preceding strong activity in the optical–infrared region. From the extensive dataset compiled by Courvoisier et al. (1990) we can also conclude that the actual birth of the knot is accompanied by strong X-ray flares. Such flares have been detected in the 2 – 10 keV flux in the beginning of 1984 (birth of knot C8) and in the beginning of 1988 (birth of knot C9). This activity also seems to precede the optical one. This would indicate that the soft X-ray emission comes from a region with its own intrinsic variability (Courvoisier et al., 1990), which can be identified with the actual birth place of a new knot in the base of the jet. The evolution of the spectrum of such a knot entrained by the underlying relativistic jet is initially dominated by Comptonization (X-ray flare), later on by synchrotron losses (mm–optical activity), and finally by adiabatic expansion (radio–flare). What is really needed for crucial tests of such models is a close flux monitoring for superluminal sources in X-rays as well as in UV, optical, IR and millimeter regimes.

Acknowledgements: The author would like to thank Dr. H. Lesch, Dr. S. Appl, M. Haehnelt and R. Khanna for general enlightening comments on the properties of magnetized plasmas. This work is supported by the Deutsche Forschungsgemeinschaft (SFB 328).

References

Appl, S.: 1990, Ph. D. thesis, University of Heidelberg

Blandford, R.D., Königl, A.: 1979, *Astrophys. J.* **232**, 34

Blandford, R.D., Payne, D.: 1982, *Mon. Not. R. astr. Soc.* **199**, 883

Blandford, R.D.: 1989, in *Theory of Accretion Disks*, eds. F. Meyer et al., Kluwer (Dordrecht), p. 35

Camenzind, M.: 1986, *Astron. Astrophys.* **156**, 137

Camenzind, M.: 1989, in *Accretion Disks and Magnetic Fields in Astrophysics*, ed. G. Belvedere, Kluwer (Dordrecht), p. 129

Camenzind, M.: 1990, in *Reviews of Modern Astronomy* **3**, ed. G. Klare (Springer–Verlag, Berlin), in press

Courvoisier, T.J.-L., Turner, M.J.L, Robson, E.I., Gear, W.K., Staubert, R., Blecha, A., Bouchet, P., Falomo, R., Valtonen, M., Teräsranta, H: 1987, *Astron. Astrophys.* **176**, 197

Courvoisier, T.J.-L., Robson, E.I., Blecha, A., Bouchet, P., Hughes, D.H., Krisciunas, K., Schwarz, H.E.: 1988, *Nature* **335**, 330

Courvoisier, T.J.-L., Camenzind, M.: 1989, *Astron. Astrophys.* **224**, 10

Courvoisier et al.: 1990, *Astron. Astrophys.* **234**, 73

Haehnelt, M., Camenzind, M.: 1991, to appear

Hughes, P.A., Aller, H.D., Aller, M.F.: 1989a, in *BL Lac Objects*, Lecture Notes in Physics (Springer–Verlag, Berlin) **334**, 30

Hughes, P.A., Aller, H.D., Aller, M.F.: 1989b, *Astrophys. J.* **341**, 54

Königl, A.: 1989, *Astrophys. J.* **342**, 208

Krichbaum, T.: 1990, Ph. D. thesis, Max-Planck-Institut für Radioastronomie, Bonn

Kunieda, H., et al.: 1990, *Nature* **345**, 786

Laor, A., Netzer, H.: 1989, *Mon. Not. R. astr. Soc.* **238**, 897; **242**, 560

Lesch, H.: 1990, this volume

Lesch, H., Appl, S., Camenzind, M.: 1989, *Astron. Astrophys.* **225**, 341

Lovelace, R.V.E., Wang, J.C.L., Sulkanen, M.E.: 1987, *Astrophys. J.* **315**, 504

Marscher, A.P.: 1987, in *Superluminal Radio Sources*, eds. J.A. Zensus, T.J. Pearson, Cambridge University Press, p. 280

Marscher, A.P., Gear, W.K.: 1985, *Astrophys. J.* **298**, 114

Pounds, K.A.: 1990, *Vistas in Astronomy* **33**, 83

Pudritz, R.E.: 1981, *Mon. Not. R. astr. Soc.* **195**, 881; 897

Rees, M.J.: 1984, *Ann. Rev. Astron. Astrophys.* **22**, 471

Rees, M.J., Begelman, M.C., Blandford, R.D., Phinney, E.S.: 1982, *Nature* **295**, 17

Ruzmaikin, A.A., Shukurov, A.M., Sokoloff, D.D.: 1988, *Magnetic Fields of Galaxies*, Kluwer (Dordrecht)

Sun, W.-H., Malkan, M.: 1989, *Astrophys. J.* **346**, 68

Svensson, R.: 1990, in *Proc. from NATO Advanced Res. Workshop on Physical Processes in Hot Cosmic Plasmas*, ed. W. Brinkmann, Kluwer (Dordrecht), p. 357

Taylor, J.B.: 1986, *Rev. Mod. Phys.* **58**, 741

Thorne, Kip S., Price, R.H., Macdonald, D.A.: 1986, *Black Holes: The Membrane Paradigm*, Yale University Press (New Haven)

Ulrich, M.-H.: 1989, in *Theory of Accretion Disks*, eds. F. Meyer et al., Kluwer (Dordrecht), p. 1

Wang, J.C.L., Sulkanen, M.E., Lovelace, R.V.E.: 1990 *Astrophys. J.* **355**, 38

Zensus, J.A., Baath, L.B., Cohen, M.H., Nicolson, G.D.: 1988, *Nature* **334**, 410

Particle Acceleration and Variability: Magnetic Reconnection in AGN

Harald Lesch

Landessternwarte Königstuhl, 6900 Heidelberg, Germany

Abstract: Magnetic reconnection is discussed as an important acceleration mechanism in active galactic nuclei. The dissipation of magnetic energy is directly connected with the activity in a magnetized accretion disk. Ascending magnetic flux is dissipated in the magnetosphere by reconnection. Thereby an electric field parallel to the magnetic field is induced which accelerates *only the electrons* to relativistic energies. This process is nonstationary, occurs on small scales and could provide the energetics and the observed time scales of variability in AGN.

1. Introduction

The reported fast variability of AGN in the nonthermal radio (Quirrenbach, 1990) ($\simeq 1\,day$) and optical regime (Wagner, 1990) ($\simeq 6\,hours$) sets severe constraints on the underlying acceleration process of the relativistic electrons. Since the frequency at which the variability occurs is constant, any kind of acceleration process has to increase the number of radiating electrons and this has to proceed on time scales of hours. So there is the necessity for fast and very effective acceleration processes, in which almost the whole electron energy distribution is involved. The observed time scales indicate that acceleration should occur in small localized regions within the central pc of the nucleus.

Evidences for localized dissipation regions are also given from global models for the origin of relativistic jets in AGN (Camenzind, 1990). Such models contain a rotating, supermassive black hole, surrounded by a magnetized accretion disk and a magnetosphere. The relativistic jet velocities are due to magnetic acceleration and the collimation of the outflows is produced by currents flowing along the jet axis. The requirement for a flowing current means $\nabla \cdot \mathbf{J} = 0$, where $\mathbf{J}$ is the current density. If $\mathbf{J}$ is resolved into a field aligned current $\mathbf{J}_{\parallel}$ and a current perpendicular to the magnetic field $\mathbf{J}_{\perp}$, one can relate $\mathbf{J}_{\parallel}$ to $\mathbf{J}_{\perp}$ by $\nabla \cdot \mathbf{J}_{\parallel} = -\nabla \cdot \mathbf{J}_{\perp}$; that is, $-\nabla \cdot \mathbf{J}_{\perp}$ acts as a source for the field aligned currents. In the scenario described above, this means that $\mathbf{J}_{\parallel}$ communicate between adjacent regions of the magnetosphere above the disk: the plasma in the magnetosphere is coupled electrodynamically with the plasma in the accretion disk. $\mathbf{J}_\parallel$ therefore represents the means by which the magnetic energy is transported from the generator (the disk with $-\nabla \cdot \mathbf{J}_\perp$) to its "storage" site. The generation of perpendicular currents in the accretion disk is due to the mechanical force of the differential shear acting on the disk magnetic field. Whenever the magnetic field in the disk (generated by a dynamo) is in equipartition with the amplifying forces (differential shear and turbulent motions) magnetic flux tubes, convective turbulent cells, etc. will ascend, which by driving parallel currents feed magnetic energy into the corona. In other words, if the driving forces cannot be counteracted by normal magnetic stresses in the disk the field lines are bent in the direction of the force and currents parallel to the magnetic field propagate into the corona (see Figure 1). To close the current circuit it is necessary to drive perpendicular currents somewhere in the corona. Moreover, in order for a current to flow, a resistor (load) within the current circuit is necessary, which may be provided by a perpendicular current. This leads to localized dissipative regions where magnetic energy is transferred into particle acceleration. Similar phenomena are known from the sun, where particle acceleration in the first phase of a flare (the injection phase) is associated with strong magnetic dissipation in small current sheets (Priest, 1982).

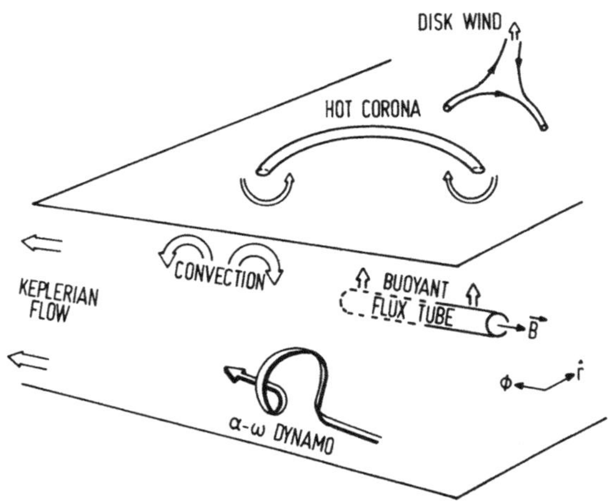

Fig. 1. Hydromagnetic phenomena in thin accretion disks

The question arises how the dissipation of the currents with density $J = env_d$ (with the drift velocity v_d, the charge e and the number density n) is provided on small scales. The rate Q of current dissipation is

$$Q = \frac{J^2}{\sigma}. \tag{1}$$

If the current density is constant, enhanced dissipation is equivalent to the reduction of the electrical conductivity σ

$$\sigma = \frac{\omega_{pe}^2}{4\pi\nu_{coll}}. \tag{2}$$

ν_{coll} is the collision frequency and $\omega_{pe} = 5.6 \times 10^4 s^{-1}\sqrt{n}$ denotes the electron plasma frequency.

For Coulomb-collisions one obtains (T is the plasma temperature)

$$\sigma = 6.3 \times 10^{10}\, T^{3/2}\, s^{-1}, \tag{3}$$

which means that dissipation is very weak. This can be understood as follows. When the conductivity is high, the magnetic field lines are "frozen into" the plasma. Field lines embedded within a volume element of plasma are carried along by the moving plasma itself. Any two plasma elements that are threaded by the same lines of force will remain threaded in this way, i.e. two plasma elements that are threaded by different lines of force can never be threaded together. Thus a high conductivity prevents the field lines from merging with one another. If reconnection occurs, it can happen only in those plasma regions where the electrical conductivity is drastically reduced below its classical Coulomb value.

The reduction of the electrical conductivity is provided by plasma instabilities. If the plasma is locally unstable microscopic instabilities excite waves which enhance the collision frequency by wave-particle interactions; they lead to *anomalous conductivity*. The plasma instabilities do not influence the global magnetic structure, but determine for a limited time the local plasma properties in the dissipation zone.

Magnetic reconnection or *magnetic merging* appears when the conductivity is reduced and currents are involved. It is the purpose of this contribution to discuss the properties of magnetic reconnection with respect to variable and strong electron acceleration to relativistic energies.

2. Particle acceleration and reconnection

Spatially varying magnetic fields with different polarity correspond to parallel, attracting currents (see Figure 2). When the plasma pressure between the oppositely fields $\pm B$ is insufficient to keep the fields apart (e.g. by pushing the different flux systems together (Figure 3)), the plasma squeezes from between and the two fields approach each other. The field gradient steepens and eventually the current density $\frac{c}{4\pi}\nabla \times B$ becomes so large that there is strong dissipation no matter how large the electrical conductivity is. If the current sheet is as thin as the thermal ion cyclotron radius the conducting electrons excite plasma microturbulence and the effective σ becomes anomalously low, greatly enhancing the dissipation and reconnection of the line of force. Then the concept of "frozen in" field lines is no longer valid and the plasma moves relativ to the field lines. When this occurs strong electric fields are induced along the reconnection length L, leading to the generation of high energy particles.

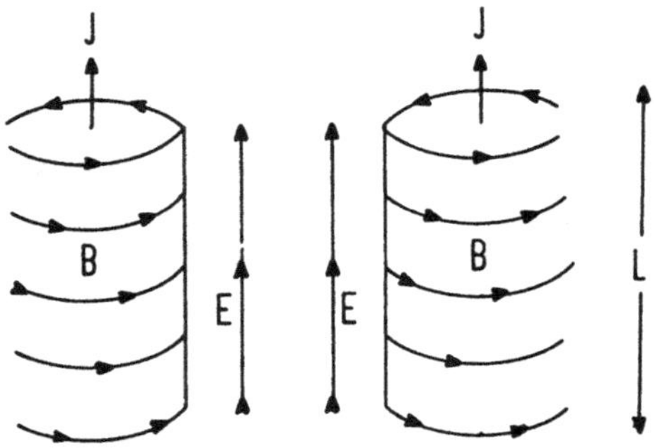

Fig. 2. A perspective view of two attracting current driven filaments, which could have their origin in unbalanced forces in a magneto-active accretion disk. When the magnetic gradient is strong enough strong dissipation sets in and an electric field E is induced which accelerates the particles. L denotes the characteristic length of the reconnection zone.

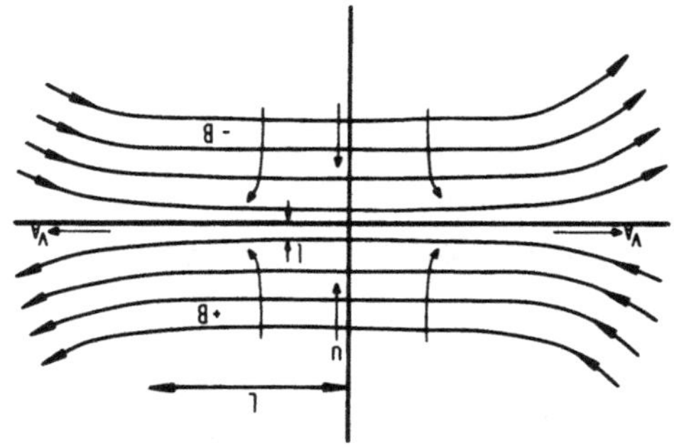

Fig. 3. A schematic drawing of two opposite magnetic fields pressed together so as to squeeze the plasma from between. The plasma is ejected from a thin layer of thickness $2l$ across the field changes from +B to −B. The two fields approach each other with the velocity u, the short arrows indicate the general direction of the plasma motion.

The electric field induced by magnetic reconnection follows from an application of Faraday's law. The magnetic field lines of opposite polarity merge at a neutral point, where $B = 0$. The merging speed u is typically a fraction of the Alfvén speed v_A, but for the present u is not explicitly determined.

If the pre-merging field is B_0, a particle of charge e and Larmor radius r_L finds the magnetic flux $\simeq \pi r_L^2 B_0$ decreasing through its orbit with a rate $(-\pi r_L^2 B_0/t_{rec})$. $t_{rec} = r_L/u$ is the reconnection time as seen from the orbit. The change of magnetic flux induces an electromotive force $emf \simeq \pi r_L B_0 u/c$ which is equal to the work done per unit charge by an electric field E; $emf \simeq 2\pi r_L E$. One gets (Biskamp, 1988)

$$E \simeq \frac{\pi u}{c} B_0. \tag{4}$$

This electric field accelerates the particles, i.e. increases the kinetic energy W of the particles and therefore increases r_L. Since in the acceleration region r_L must remain smaller than the size L of the reconnection zone, one gets an estimate for the maximum energy gain ($r_L \simeq \frac{W}{eB_0}$)

$$W_{max} \simeq eB_0 L. \tag{5}$$

The electric potential ϕ drop along L is $\phi = EL = \frac{\pi u}{c} B_0 L$. The total energy gain is

$$W \simeq e\phi = \frac{\pi e u B_0}{c} L.. \tag{6}$$

Reconnection is inherently variable and nonstationary, since the driving force, i.e. is the energy of the attracting currents, is dissipated until an equilibrium is reached and the merging process stops. In terms of acceleration by reconnection the flare duration time t_{flare} corresponds to a reconnection time for all particles along L: $t_{flare} \simeq \frac{L}{u}$. On the other hand, the flare duration time is also the time it takes the magnetic field B_0 to diffuse into the neutral region where $B_0 = 0$ along the neutral line. This diffusion time t_D is (Parker, 1979)

$$t_D \simeq \frac{L^2}{10\eta}. \tag{7}$$

$\eta = \frac{c^2}{4\pi\sigma}$ denotes the magnetic diffusivity of the plasma. By equating $t_D \simeq t_{flare}$ and using Eqs. (2 and 6) one gets

$$W \simeq 10\pi e B_0 c \frac{\nu_{coll}}{\omega_{pe}^2}. \tag{8}$$

Reconnection implies the presence of strong electric fields $j \simeq \sigma E$. In a resistive plasma following this simple form of Ohm's law, the current density depends only on the local values of σ and E, which implies that all electrons effectively randomize their directed energy gain. But the Coulomb cross-section is a strongly decreasing function of W. Therefore, electrons with energies above the thermal energy are almost collisionless and almost freely accelerated. This is called the *runaway effect* (Dreicer, 1959), an effect not restricted to Coulomb collisions. Any kind of turbulent resistivity would lead to a randomization of the electrons energy. The number of runaway electrons depends on the strength of the electric field. For $E \ll E_c \simeq m_e v_{the} \nu_{coll}/e$ (v_{the} is the electron thermal velocity) only a small fraction will be freely accelerated, while for $E \geq E_c$ almost the whole electron distribution is involved into the acceleration process and the current will be mainly carried by runaways. Ion runaway is much more difficult, so that reconnection primarly leads to high energetic electrons in contrast to diffusive shock acceleration. Therefore with $W \simeq \gamma m_e c^2$ one obtains ($\gamma \simeq \frac{B}{\sigma}$)

$$\gamma_{rec} \simeq 10\pi \nu_{coll} \frac{\omega_{ce}}{\omega_{pe}^2}. \tag{9}$$

$\omega_{ce} = \frac{eB}{m_e c}$ is the electron gyrofrequency.

The energy of the electrons does not explicitly depend on the reconnection speed u but it depends on the anomalous collision frequency ν_{coll} produced by plasma instabilities. Those instabilities have different excitation thresholds, i.e. they do not generally appear but need special conditions in order to produce an anomalous collision frequency. This is one reason why reconnection is a rather nonstationary process. A brief discussion of the important plasma instabilities is given in the next section.

3. Anomalous transport in current sheets

The basic plasma and magnetic field configuration is that of Figures 2 and 3. Since the ascending flux tubes contain currents, in the following only current driven instabilities are discussed, especially instabilities which are driven by a so-called *cross-field current*, which is equivalent to $J_\perp$.

For current-driven instabilities to be excited it is necessary that the drift velocity of the electrons v_D exceeds some threshold. According to Maxwell's equation

$$\nabla \times \mathbf{B} = \frac{4\pi}{c}\mathbf{J}, \tag{10}$$

one can associate the drift velocity with the magnetic field gradient B/l

$$v_D \simeq \frac{c}{4\pi e n}\frac{B}{l}. \tag{11}$$

Eq. (11) means that the larger the magnetic field gradient is the higher is the drift velocity. So one could expect that in the innermost current sheet plasma waves are present which need the highest drift velocity. Moreover, one has to distinguish between magnetized and unmagnetized instabilities. Namely, the former could not appear in the neutral sheet (with $B = 0$). Thus, these instabilities do not directly produce anomalous conductivity in the null region.

3.1. The Buneman instability (BI)

The BI is the classical electron-ion two-stream instability (Buneman, 1959). The turn-on condition for instability is that the drift velocity exceeds the electron thermal velocity. By using Eq. (10) one can show that the current sheet thickness for this instability is $l \simeq c/\omega_{pe}$. Because of the extremely thin current sheet it seems unlikely that the BI can be of any importance for the reconnection process.

3.2. The ion acoustic instability (IAI)

The IAI is also driven by the relative electron-ion drift v_D. The turn-on condition is that v_D is larger than the sound velocity $c_s = \sqrt{k_B T_e/m_p}$ and that $T_e > T_p$. It has been shown that the anomalous collision frequency associated with this instability is (Papadopoulos, 1979)

$$\nu_{coll}^{IAI} \simeq 10^{-2}\omega_{pe}. \tag{12}$$

Even though the turn-on condition for IAI is less stringent than for BI, a thin current sheet is still required to excite this mode $l \simeq few\ (c/\omega_{pe})$.

3.3. The lower-hybrid instability (LHI)

The LHI was first proposed as a source of anomalous conductivity for reconnection in the earth's magnetotail (Huba et al. 1978). Two properties in favor of this instability are (1) the mode can be excited in relatively broad current sheets since the necessary drift velocity is the thermal velocity of the ions $l \simeq (m_p/m_e)^{1/4} r_L^p$ (where r_L^p is the proton Larmor radius) and (2) the mode is insensitive to the temperature ratio T_e/T_p. Unfortunately, the LHI appears only if $B \neq 0$, that is not in the neutral region. The anomalous collision frequency associated with LHI is approximately equal to the lower-hybrid frequency

$$\nu_{coll}^{LHI} \simeq \omega_{LH} \simeq 4 \times 10^5 \, B. \tag{13}$$

Summarizing, the two "favored" instabilities are the ion acoustic instability and the lower-hybrid instability. The IAI can be excited in the null region but requires quite thin current sheets and is more easily excited in hot electron plasmas ($T_e > T_p$). On the other hand the LHI can be excited in broader regions but the waves are damped in the neutral region. In a dynamic situation (forced reconnection, where flux tubes are pressed together) the lower-hybrid turbulence may penetrate into the neutral region and is likely to exist over a substantial portion of the current sheet and can strongly affect the particle acceleration by determining the anomalous collision frequency.

4. Applications

If reconnection provides the energy of the relativistic electrons radiating in the optical and/or in the radio waveband, the necessary condition is $\gamma_c \simeq \gamma_{rec}$. γ_c is the synchrotron Lorentz factor at a specific frequency

$$\gamma_c = \sqrt{\frac{\nu_c 2\pi}{\omega_{ce}}} \simeq \begin{cases} 3 \cdot 10^3 \left[\frac{\nu_c}{3 \cdot 10^{14}\, Hz}\right]^{1/2} \left[\frac{B}{10\, Gauss}\right]^{-1/2} \\ 15 \quad \left[\frac{\nu_c}{5 \cdot 10^9\, Hz}\right]^{1/2} \left[\frac{B}{10\, Gauss}\right]^{-1/2} \end{cases} . \tag{14}$$

Equating $\gamma_c \simeq \gamma_{rec}$, one gets a relation for the magnetic field strength and the number density. Hereby it is assumed that the anomalous collision frequency is provided mainly by the lower-hybrid waves, as stated in the preceeding section.

For the optical regime this leads to

$$B_c^{opt} \simeq 0.4 \, n^{2/5} \tag{15}$$

and for the radio band

$$B_c^{radio} \simeq 0.05 \, n^{2/5}, \tag{16}$$

(see also Figure 4).

If $B \geq B_c$, then $\gamma_{rec} \geq \gamma_c$. Figure 4 shows this field regime, for densities probably given in the central parsec of an active galactic nucleus ($10^4 \, cm^{-3}$ (in the intercloud medium) - $10^{10}\,cm^{-3}$ (in the line emitting clouds)). In the context of acceleration models for relativistic jets by rotating magnetic fields, the required high field strengths ($1-1000$ Gauss) are typical within the central $10^{17}\,cm$ (Camenzind, 1990).

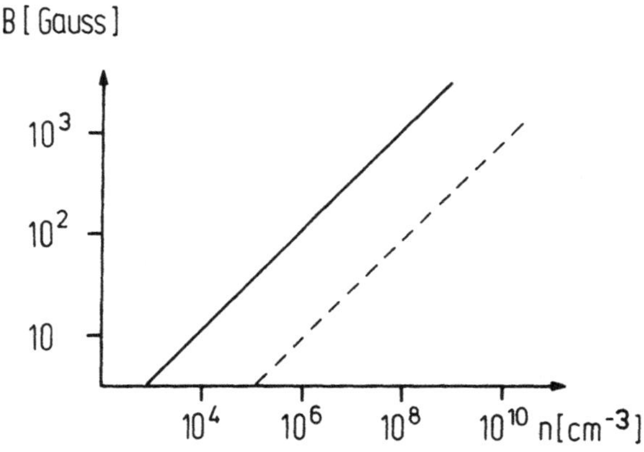

Fig. 4. The relation of the magnetic field strength B and the number density n. The lower-hybrid waves provide the anomalous conductivity. The solid (dashed) line is for optical (radio) synchrotron radiation. Above the lines reconnection provides energy for synchrotron radiating relativistic electrons.

Subsequently as an example, reconnection is applied to the optical variability as reported by Wagner (1990).

The observed luminosity L^* in the variable optical emission has to be of the order of $10^{46}\,ergs^{-1}$ in order to be brighter than the nonvariable optical emission. With an assumed beaming factor for the relativistic jet $\gamma_b \simeq 10$ and the relation $L^* \simeq \gamma_b^4 L_{int}$ one obtains an intrinsic luminosity $L_{int} \simeq 10^{42}\,ergs^{-1}$. The total intrinsic energy is then ($t_{int} \simeq t_{obs}/\gamma_b$ is the intrinsic time scale for variability, t_{obs} is the observed variability time scale)

$$E_{tot} \simeq L_{int} \cdot t_{obs}/\gamma_b \simeq 3 \cdot 10^{46} erg \left[\frac{t_{obs}}{3\,days}\right] \left[\frac{\gamma_b}{10}\right]^{-1} . \tag{17}$$

On the other hand one can estimate the total energy by assuming some effiency factor ϵ for the reconnection process which should occur in the volume V.

$$E_{tot} \simeq \epsilon \cdot V \cdot \frac{B^2}{8\pi} . \tag{18}$$

From Eq. (18) one determines the typical dimension of the reconnection zone

$$L \simeq 3 \cdot 10^{14}\,cm \left[\frac{B}{100\,Gauss}\right]^{-2/3} \left[\frac{\epsilon}{0.1}\right]^{-1/3} \left[\frac{E_{tot}}{3 \cdot 10^{46}\,ergs^{-1}}\right]^{1/3} . \tag{19}$$

The flaring time t_{flare} is given by

$$t_{flare} \simeq t_{int} \simeq \frac{l}{u}. \tag{20}$$

u is the reconnection speed and l denotes the thickness of the reconnection zone which is approximately $0.1 \cdot L$. The typical reconnection speed for a current sheet is $u \simeq 0.1 \cdot v_A$ (Parker, 1979). One obtains the intrinsic flaring time scale

$$t_{int} \simeq \frac{l}{u} \simeq 7\,hr \left[\frac{l}{3 \cdot 10^{13}\,cm}\right] \left[\frac{u}{0.1\,v_A}\right]^{-1} \left[\frac{B}{100\,Gauss}\right]^{-1} \left[\frac{n}{10^6\,cm^{-3}}\right]^{1/2}. \tag{21}$$

Similar estimates hold for the reported radio variability on the observed time scales of one day ($t_{int} \simeq 3\,hr\,(\gamma_b/10)$) (Quirrenbach, 1990).

The necessary field strengths for acceleration in the intercloud medium are relatively high. The nonstationary character of the optical and radio emission can be understood in terms of energy storage. Due to a mechanical force in the disk parallel currents store magnetic energy in the corona. When the field strength is sufficiently high, reconnection sets in. At this phase the electrons are effectively accelerated. At larger time scales when the magnetic field gradients become weaker and weaker, the conductivity rises again because the plasma instabilities cease. An enhanced conductivity means a reduced dissipation and the acceleration process becomes less efficient and therefore invisible.

5. Conclusions

The observed fast variability on time scales of hours in AGN points towards nonstationary and efficient acceleration process. In a large number of astrophysical systems the most favoured acceleration mechanism is diffusive shock acceleration (Drury, 1988). However, since the increase of energy is diffusive, shock acceleration is an inherently slow process. Moreover, since the electrons loose their energy by synchrotron and/or inverse Compton radiation, their attainable energies remain on a low level. In contrast to shock acceleration, the acceleration by a d.c. parallel electric field is more effective. To show how such fields are set up in an active galactic nucleus is one of the purposes of this contribution.

Therefore, it is argued that the observed variability in AGN is associated with flaring phenomena arising in the corona of a magnetized accretion disk. Such strongly magnetized disks are a necessary ingredient of models concerning the acceleration of relativistic jets. Since the disk and the corona are connected by ascending and rotating magnetic flux tubes, local distortions of the disk plasma drive currents into the corona, which store magnetic energy. When the current density (which corresponds to a magnetic field gradient) exceeds some critical value (i.e. if a large amount of magnetic energy is stored in a small volume) microscopic plasma processes set in which give rise to turbulent electrical conductivity and large parallel electric fields. These result in efficient electron acceleration.

Here, the acceleration in only one current sheet is considered. The results are very promising, both the energetics and the time scales of optical and radio outbursts can be explained in a satisfactory manner. However, if the whole corona is in a turbulent state a large number of current sheets could exist. If the mean free path of the electrons is larger than the acceleration length scale, they can reach much higher Lorentz factors than those given by Eq. (9). Such electrons are accelerated by a mean electric field along a scale length, which is given by the typical scale of the macroscopic turbulence (Biskamp, 1988).

Acknowledgements: I would like to thank S. Appl, P.L. Biermann, D. Biskamp, M. Camenzind, R. Schaaf and T. Schmutzler for helpful comments and criticism. This work was supported by the Deutsche Forschungsgemeinschaft (SFB 328).

References

Biskamp, D.: 1988, in *Hot Spots in Extragalactic Radio Sources*, Eds. Meisenheimer, K., Röser, H.-J., Springer, Berlin, 279
Buneman, O.: 1959, *Phys. Rev.* **115**, 503
Camenzind, M.: 1990, these proceedings
Dreicer, H.: 1959, *Phys. Rev.* **115**, 242
Drury, L.O.C.: 1988, in *Hot Spots in Extragalactic Radio Sources*, Eds. Meisenheimer, K., Röser, H.-J., Springer, Berlin, 279
Huba, J.D., Gladd, N.T., Papadopoulos, K.: 1978, *J. Geophys. Res.***83**, 5217
Papadopoulos, K.: 1979, in *Dynamics of the Magnetosphere*, Ed. Akasofu, S.I., Reidel, Dordrecht
Parker, E.: 1979, *Cosmical Magnetic Fields*, Clarendon Press, Oxford
Priest, E.R.: 1982, *Solar Magnetohydrodynamics*, Reidel, Dordrecht
Quirrenbach, A.: 1990, these proceedings
Wagner, S.: 1990, these proceedings

Electron-Ion Coupling in Compton-Heated Plasmas

Thomas Schmutzler [1] and Harald Lesch [2]

[1]Max-Planck-Institut für Kernphysik, Saupfercheckweg 1, 6900 Heidelberg 1, Germany
[2]Landessternwarte Königstuhl, 6900 Heidelberg, Germany

1. Introduction

The strong high energy photon flux of active galactic nuclei (AGNs like quasars, Seyfert galaxies and BL Lac objects) is a very effective heating source for the illuminated plasma in the inner regions of these objects (broad line and narrow line region – BLR and NLR). The heating processes due to photoionization and Compton scattering act directly on the electrons and are balanced by several cooling processes depending on plasma temperature and density. The main cooling is due to bremsstrahlung and collisional line excitation as long as $T > 10^6$ K and magnetic fields are negligible. Compton scattering also becomes a cooling mechanism (inverse Compton scattering) if the mean energy of the photon flux is less than the thermal plasma energy (Schmutzler, 1987).

Thus the equilibrium temperature of the electrons in thin plasmas may be determined only by the Compton effect. In this case the timescale of electron heating and cooling is much shorter than the Coulomb timescale which is usually responsible for the energy exchange between electrons and ions. Consequently a two-temperature plasma ($T_e \neq T_{ion}$) may be possible within the Coulomb timescale if the photon flux varies on shorter timescales. Typical variation timescales of quasars are in the range of months to years, whereas recent X-ray observations of Seyfert galaxies confirmed even shorter timescales down to a few hours (Pounds et al., 1987; Barstow and Stanger, 1988; Morini et al., 1987; Brodie et al., 1987; Maccagni et al., 1987; Courvoisier et al., 1987). Hence the electron temperature of the surrounding plasma may be tuned on equal timescales depending on the local photon flux. However, in the frame of two-phase models for BLR/NLR an equal-temperature plasma is assumed. Of course, this assumption is valid for a steady state situation, but a more sophisticated description must include time dependent processes as well (Krolik, McKee and Tarter, 1981).

2. The radiation induced sound instability

We have shown that the assumption of an equal-temperature plasma is still appropriate if the radiation field is locally anisotropic: Compton scattering acts in a twofold way: firstly, there is an energy transfer, and secondly a momentum transfer from the photons to the electrons. Here we have to note that Compton scattering is a statistical process which first of all acts on individual electrons. Only the average over a few photon-electron collision timescales (which also determines the Compton heating and cooling timescales) yields the well known net effect on the complete electron distribution. On smaller timescales the Compton acceleration of electrons drives a current along the direction of incoming photons which would build up an electric field. However, the high conductivity of the fully ionized plasma does not allow charge separation. Hence return currents are drawn locally and a filamentary structure appears. This situation is similar to that of a current-carrying plasma (a jet) penetrating another plasma which has high conductivity but no net current (Benford, 1987, 1988). In our special case the situation is much more comfortable since there is no problem of interpenetrating plasmas. The small scale filaments, carrying currents in the same direction, will attract each other and the possibility of return currents flowing in the surrounding medium may support this merging of current-carrying filaments in analogy to descriptions which have been established for the formation and propagation of current-carrying jets (Benford, 1978; Camenzind, 1986). However, the growth of the currents is restricted: firstly, the length of such a filament is of the order of the mean free path of an electron with respect to collisions with protons, and secondly, the self-generated magnetic field limits the size of the pinches. Thus the locally working electron-ion coupling has to operate within the life time of the current-carrying filaments.

Current-driven plasma instabilities are able to couple the the temperatures of electrons and ions close together on timescales much shorter than both the timescales of relaxation by electron-ion Coulomb collisions and the timescale of Compton heating. If the electron temperature exceeds the ion temperature the excited ion sound waves will lead to an instability (ion sound instability) as soon as the electron drift velocity exceeds the sound velocity. It is well known that the onset of an instability strongly influences the transport coefficients in a collisionless plasma by increasing the effective collision frequency due to resonant wave-particle interactions. Hence the drift velocity will be reduced and the free energy of the dragged electrons dissipates via plasma heating (Sagdeev and Galeev, 1969). In our scenario only the ions are heated since the electrons have reached their maximum temperature fixed by the photon field due to the Compton effect.

The drift velocity attains a value near the velocity of sound. Whenever the radiation energy density is enhanced (flux variability of the radiation source) the electrons are heated up and accelerated whereby the instability sets in and heats the ions up to $T_e \simeq T_{ion}$. Therefore the temperature relaxation is directly connected with the radiation variability and the excitation of ion sound turbulence.

3. The case of hot ions and cooling electrons

What happens if the electrons are cooled rapidly by inverse Compton scattering? If $T_{ion}/T_e > \sqrt{m_{ion}/m_e}$ is reached by this cooling process, the thermal velocity of the electrons is equal to the velocity of sound and the kinetic instability of hot ions arises. This instability (hot ion kinetic) is quite similar to the instability of a beam with a large thermal spread in a dense cold background plasma (Mikhailovski, 1974, p. 65). In this case the wave-particle interactions transfer the free energy of the current into heating of the cooling electrons on timescales which are ten times smaller than the that of the ISI (Davidson et al., 1970).

In summary, we have the following physical picture of the temperature coupling in an optically thin illuminated plasma: If the electrons are heated up and accelerated by an intense radiation flux the ion heating is governed by the ion sound instability. In the opposite case, $T_{ion} > T_e$, the electrons are heated by the kinetic instability of the hot ions. In both cases the temperature relaxation due to these instabilities is orders of magnitude faster than by classical Coulomb relaxation. The high temperature of the ions cannot be reduced by such collective plasma processes. This means that in principle a two-temperature plasma in active galactic nuclei is possible within the Coulomb timescale since the electrons suffer much stronger energy losses than the ions do.

4. Discussion and conclusions

We have presented a physical mechanism of rapid electron-ion coupling in fully ionized thin plasmas which are illuminated by an anisotropic radiation source. Hot thin plasmas which are illuminated by a strong photon flux are a part of common models of the BLR and NLR in AGNs (Krolik et al., 1981; Davidson and Netzer, 1979 and references therein). The knowledge of rapid coupling processes justifies the assumption of an equal temperature plasma for several quantitative time dependent calculations within the two-phase model of the BLR/NLR, e. g., phase transitions between the coronal gas component and line emitting clouds, rapid changes of the illuminating continuum sources or fast anisotropic Compton heating processes.

We conclude:

1. Heating up the electron component in an anisotropically X-ray illuminated thin plasma caused by variations of photon flux or by density perturbations does not lead to a two temperature plasma because
   a) the gas density is so large that the Coulomb relaxation couples the ion component on timescales of the electron temperature variations or
   b) the arising temperature difference between the electrons and the ion component together with the Compton induced drift of the electron distribution lead to the ion sound instability which couple electrons and ions on a timescale negligible in comparison to the electron heating timescale.

2. Cooling of the electrons due to inverse Compton scattering by an anisotropic radiation field is possible as long as $T_e/T_{ion} < \sqrt{m_e/m_{ion}}$. Then the hot ion kinetic instability heats the electrons faster than inverse Compton scattering can cool them. Thus a two-temperature plasma may form and the timescale for cooling down both the electrons and ions to a temperature at least two orders of magnitude less than the initial temperature is given by the Coulomb relaxation time unless other mechanisms (e. g., including magnetic fields) are more efficient.
3. All described processes operate on microscopic scales and do not contradict the common picture of plasma acceleration due to the radiation pressure. But we have argued that both plasma components are heated on similar timescales. Since the energy exchange works locally, the restriction to a cylindrical geometry of the illuminated plasma embedded in a highly conducting background supports the merging of current-carrying filaments, but it seems to be unnecessary. Therefore the electron-ion coupling may also work in a plasma which is illuminated by central point source.

A more detailed discussion of the relevant timescales and plasma processes is given by Schmutzler and Lesch, 1989.

Acknowledgements: This work was supported by the Deutsche Forschungsgemeinschaft (SFB 328).

References

Barstow, M.A., Stanger, V.J.: 1988, *Monthly Notices Roy. Astr. Soc. f 230*, 207
Benford, G.: 1978, *Monthly Notices Roy. Astr. Soc.* **183**, 29
Benford, G.: 1987, in: *"Astrophysical Jets and Their Engines"*, NATO ASI series, *ed. W. Kundt*, Reidel, Dordrecht
Benford, G.: 1988, *Astrophys. J.* **333**, 735
Brodie, J., Bowyer, S., Tennant, A.: 1987 *Astrophys. J.* **318**, 175
Camenzind, M.: 1986, *Astron. Astrophys.* **156**, 137
Courvoisier, T.J.-L., Turner, M.J.L., Robson, E.I., Gear, W.K., Staubert, R., Blecha, A., Bouchet, P., Falomo, R., Valtonen, M., Teräsranta, H.: 1987, *Astron. Astrophys.* **176**, 197
Davidson, K., Netzer, H.: 1979, *Rev. Mod. Phys.* **51**, 715
Davidson, R.C., Krall, N.A., Papadopoulos, K., Shanny, R.: 1970, *Phys. Rev. Lett.* **24**, 579
Krolik, J.H., McKee, C.F., Tarter, C.B.: 1981, *Astrophys. J.* **249**, 422
Maccagni, D., Garilli, B., Schild, R., Tarenghi, M.: 1987, *Astron. Astrophys.* **178**, 21
Mikhailovski, A.B.: 1974, *Theory of Plasma Instabilities*, Consultant, Bureau, New York
Morini, M., Lipuni, N.A., Molteni, D.: 1987, *Astrophys. J.* **317**, 145
Pounds, K.A., Stanger, V.J., Turner, T.J., King, A.R., Czerny, B.: 1987 *Monthly Notices Roy. Astr. Soc.* **224**, 443
Sagdeev, R.Z., Galeev, A.A.: 1969, *Nonlinear Plasma Theory*, Benjamin, New York
Schmutzler, T.: 1987, *Thesis*, University of Bonn
Schmutzler, T., Lesch, H.: 1989, *Astron. Astrophys.* **223**, 71

Variable Radio Sources and the Two–Fluid Models

G. Pelletier [1], Jacques Roland [2], Hélène Sol [3]

[1]Observatoire de Grenoble, CERMO BP68, 38402 St. Martin d'Hères Cedex, France
[2]Institut d'Astrophysique, 98 Bis Bd. Arago, 75014 Paris, France
[3]Observatoire de Meudon, 92190 Meudon, France

Abstract: Superluminal radio sources associated with extended double sources can be explained by a two–fluid model that consist of *i) a non–relativistic* e^-*–*p *jet* responsible for the formation of the extended lobes and *ii) a relativistic* e^-*–*e^+ *beam* responsible for superluminal radio sources. When the magnetic is greater than a critical value B_c, the beam can propagate along the magnetic field lines without dissipation. Natural consequences of the beam-jet interaction are, in the dissipation area, the acceleration of relativistic electrons by strong Langmuir turbulence, the possible formation of shocks, the emission of coherent emission and the generation of a local toroïdal magnetic field. Depending on the value of the magnetic field in the beam compared to the critical value, one can observe superluminal quasars, BL Lac objects or flickering radio sources.

1. On the velocity of jets powering extended lobes

1.1. Introduction

Models for the formation of extended lobes of extragalactic radio sources allow the determination of the characteristics of the jets powering them, and consequently, one can obtain important constraints for the formation of jets in active nuclei of galaxies. We will distinguish two simple extreme cases of radio galaxies and we will not study in this article intermediate cases. So, we will discuss the two following cases

i) the Cygnus A type radio sources,

ii) the 3C 449 type radio sources.

Cygnus A type radio sources are characterized by a high integrated radio luminosity, namely $L_r \simeq 10^{44} - 10^{45}$ erg s^{-1}, and the existence of hot spots in the extended lobes (throughout this paper, the numerical values are given assuming $H_o \simeq 100$ km s^{-1} Mpc^{-1} for the Hubble constant). Hot spots and the extended lobes are due to the interaction of a supersonic jet with an hot intergalactic medium.

3C 449 type radio sources are characterized by a low integrated radio luminosity, i.e. $L_r \simeq 10^{41} - 10^{42}$ erg s^{-1}, two bright radio jets which Mach number $M \simeq 1$, no hot spots in the extended lobes and no violent interaction with the intergalactic medium, except for head–tail radio sources associated with clusters of galaxies.

1.2. Cygnus A type radio sources

In Pelletier and Roland (1986), we shown that it was a priori possible to explain the formation of Cygnus A hot spots by the interaction of a non–relativistic supersonic jet with the external medium. Assuming the magnetic field quasi–perpendicular to the jet, we found that

i) the speed of the jet is $v_j \simeq 0.35$ c,

ii) the dominant pressure inside the hot spots is the pressure of the non–relativistic thermal protons,

iii) the ratio of the pressure of the relativistic protons to the relativistic electrons is smaller than unity,

iv) the ratio β of the particle pressure to the magnetic pressure is $\beta \simeq 2 - 3$, indicating that the Alfvenic Mach number in the jet is $M_A \simeq 6$.

Consequently, in Pelletier and Roland (1988), we developed a spectral theory of magnetized mixed shocks to take into account the finite value of the Alfvenic Mach number.

Then, from the knowledge of the synchrotron spectrum of Cygnus A hot spots (Muxlow et al. 1988), we made a self-consistent determination of the physical parameters inside the hot spots and of the jets powering them (Roland et al. 1988). We shown that

i) with a velocity $v_j \simeq 0.35$ c and an ejected mass $\dot{M} \simeq 0.5\ M_\odot$ yr^{-1}, the integrated synchrotron luminosity of the hot spots is about 4 % of the kinetic power of the jet,

ii) the low energy cut–off of the relativistic electrons is about 300 MeV, inducing that the Inverse Compton emission of the hot spots begins in the U.V. range,

iii) from the knowledge of the high energy cut–off of the relativistic electrons, one can deduce the level of the MHD turbulence is $\eta_{turb} \simeq 10^{-3} - 10^{-4}$ and thus models of mixed shocks are very good to explain the acceleration of relativistic electrons which radiate from the radio to the optical.

A similar result was obtained for 3C 20, 3C 33, 3C 111, 3C 123 and Pic A hot spots by Meisenheimer et al. (1989).

1.3. 3C 449 type radio sources

Even if a self-consistent model does not exist to obtain the physical parameters of 3C 449 radio jets, empirical models have been developed which constraint the speed of bright radio jets.

Head-tail radio sources, which are of 3C 449 type sources, have jets which velocities are $v_j \simeq 1000 - 3000$ km s^{-1}. These velocities are similar to the speeds of the galaxies moving through the intergalactic medium.

From 3C 449 radio map (Perley et al. 1979) and from the opening angle of the jets, one can deduce that they must have a low Mach number, i.e. M$\simeq$ 1, inducing an important role for the acceleration of the relativistic electrons by the MHD turbulence, because the efficiency of this kind of acceleration is inversely proportional to the cube of the Mach number (Pelletier and Zaninetti 1984).

An empirical model for bright radio jets and its application to 3C 31 and NGC 315, has been developed by Bicknell (1984, 1986-a and 1986-b). Bicknell (1986-b) found a typical range of velocities from 1600 to 3800 km s^{-1} for NGC 315 radio jets. These velocities correspond to the velocities of head-tail radio sources jets.

Recent observations of low luminosity sources with two bright radio jets have been done by Fanti et al. (1987). The properties of the jets have been obtained by Morganti et al. (1987), Parma et al. (1987). They found that bright radio jets have velocities smaller than 10^4 km s^{-1}.

2. Superluminal radio sources and the two-fluid models

2.1. Superluminal radio sources

A large fraction of superluminal radio sources is observed in nuclei of strong double radio sources with hot spots in the extended lobes.

However, there exists probably one example of a superluminal radio source which shares some characteristics with the 3C 449 type radio sources, it is 1928+738. Indeed, if we except the emission of the nucleus which is Doppler boosted and then anisotropic, the integrated synchrotron luminosity of the source is quite low, i.e. it is about 10^{42} erg s^{-1} and VLA observations show two bright radio jets powering the extended lobes (Johnston et al. 1987). See the radio maps in Figure 1. As the two arc sec. scale bright jets have the same luminosities, none of them seems to be Doppler boosted which suggests that they have non-relativistic velocities, as for other low luminosity sources with two bright jets. The difference between the spectral indices of the extended lobes indicates a probable interaction of the jets with the intergalactic medium similar to head-tail radio sources. 3C 120 is also a superluminal radio source which has a low luminosity, but its morphology is more complicated than those of 1928+738 and 3C 449.

Superluminal expansions, the Doppler beaming effect and the absence of the inverse Compton catastrophe can be explained if the relativistic e$^-$ and e$^+$, responsible of the synchrotron emission, are ejected with a mean Lorentz factor $\bar{\gamma} \leq 10$.

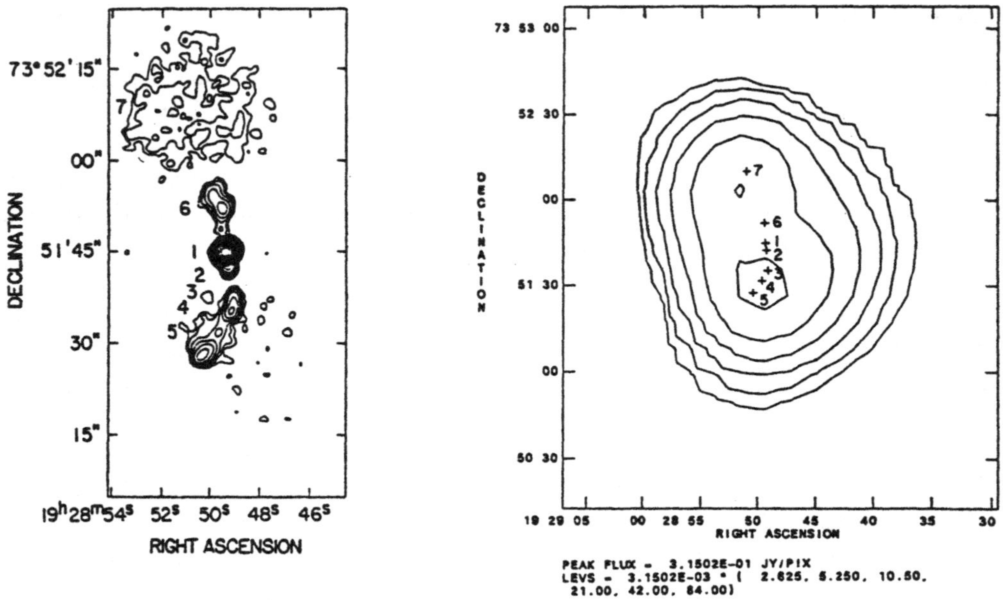

Fig. 1. 1928+738 from Johnston et al. (1987)

Indeed, assuming relativistic ejections from nuclei of galaxies, observations of superluminal radio sources imply that the bulk Lorentz factor γ is related to the Hubble constant by $\gamma H_o \leq 900$ (Cohen 1990). However, observations donot imply that the thermal component of the plasma is also ejected relativistically.

2.2. The two–fluid models (Cygnus A type radio sources)

Superluminal radio sources associated with Cygnus A type radio sources can be explained if the jets in their nuclei contain two fluids (Sol et al. 1989, Pelletier and Roland 1989), i.e.

i) a non–relativistic jet of e^-–p plasma which is responsible for the formation of the hot spots and the extended lobes. With a velocity of $v_{\rm j} \simeq 0.35$ c and an ejected mass $\dot{M} \simeq 0.5\ M_\odot\ {\rm yr}^{-1}$, its kinetic power is $P_{\rm kj} \simeq 1.8\ 10^{45}$ erg s^{-1} and it contains most of the mass and the kinetic power of the ejected plasma.

ii) a relativistic beam of e^-–e^+ plasma, which is responsible for the formation of superluminal radio sources in the nucleus. With a mass loss $\dot{M}_{\rm b} \simeq 3\ 10^{-6} M_\odot\ {\rm yr}^{-1}$ and a mean Lorentz factor $\gamma_{\rm b} \simeq 10$, the kinetic power of the beam is $P_{\rm kj} \simeq 1.8\ 10^{42}$ erg s^{-1}.

iii) the magnetic field B is parallel to the flow in the beam and in the mixing layer (see more details in Figure 2).

Sol et al. (1989) suppose that the beam pervades the jet and Pelletier and Roland (1989) suppose that the beam moves in a channel through the jet. The physical parameters of the beam and of the jet can be found in Pelletier and Roland (1990). The two–fluid models are supported by the detection of two different velocities in the VLBI jet of NGC 1275 (Marr et al. 1989) and by the limb brightening of the VLBI jet of M 87 (Reid et al. 1989), even though these two sources are not superluminal.

Pelletier et al. (1988) showed that if the magnetic field is greater than some critical value

$$B_c = (4\pi n_{\rm th} m_e c^2)^{1/2} \ , \tag{1}$$

a relativistic e^-–e^+ beam can propagate along the magnetic field lines that are parallel to a non–relativistic e^-–p jet ($n_{\rm th}$ is the density of the thermal plasma in the mixing layer).

2.3. Further consequences of the two–fluid models

a) *Formation of the Jets*

It has been shown in Pelletier and Roland (1990) that VLBI polarization observations of superluminal quasars can be understood if, close to the nucleus, the magnetic field is helicoïdal in the jet and becomes toroïdal when the jet opens (Figure 2).

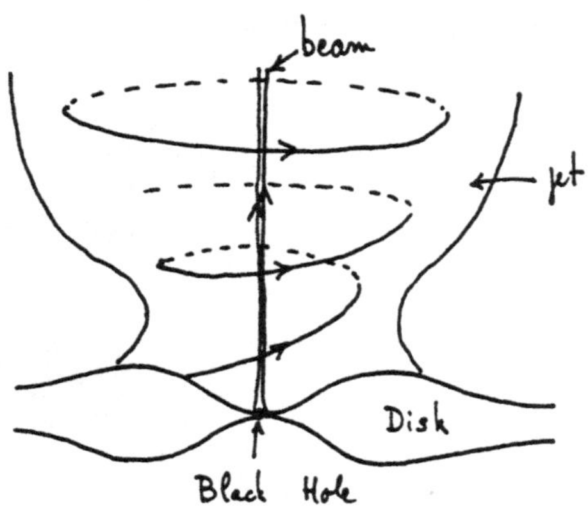

Fig. 2. The magnetic field configuration in the jet and in the beam

The magnetic field configuration in the jet and in the beam, and the formation of the jet can probably be explained by the MHD extraction process (e.g., Blandford and Payne 1982, Draine 1983, Pudritz and Norman 1983, Uchida and Shibata 1985). The e^-–e^+ beam is created close to the black hole, but a model for the formation of the beam has to be made.

b) *Formation of Asymmetries in Strong Radio Sources*

A number of asymmetries in strong radio sources can be explained by the two–fluid model, remarking that velocity of the jet is $v_{\rm j} \simeq 0.35$ c and without supposing any relativistic effect.

i) Relationship between VLBI jets and large scale jets

VLBI jets and large scale jets are often observed on the same side. As indicated in Pelletier and Roland (1990), the relationship between VLBI jets and large–scale jets can be understood if there is an asymmetry between the two beams created in the nucleus.

ii) Sources with asymmetrical hot spots: e.g. 3C 263

Sources similar to 3C 263 have asymmetrical hot spots and extended lobes (see Figure 3, the radio maps from Browne 1987, and from Leahy et al. 1989). One lobe contains a strong hot spot and the other one without hot spot is weaker. A number of sources similar to 3C 263 can be found in Leahy et al. (1989).

The advance speed, $v_{\rm sep}$, of the hot spots is

$$v_{\rm sep} \propto 1/n_{\rm ext}^2 \;, \tag{2}$$

where $n_{\rm ext}$ is the density of the hot intergalactic medium. With $n_{\rm ext} \simeq 6\ 10^{-3}$ e$^-$ cm^{-3}, one obtains for Cygnus A hot spots $v_{\rm sep} \simeq 0.05$ c (Muxlow et al. 1988). Thus, when the external density is $n_{\rm ext} \geq 10^{-4}\ e^-\ {\rm cm}^{-3}$, we have $v_{\rm sep} \simeq 0.35\ {\rm c} \simeq v_{\rm j}$ and the strong shock at the end of the jet desappears. The relativistic electrons are accelerated by the bow shock only. So if we suppose two symmetric jets, but an asymmetry in the external medium, on the side where $n_{\rm ext} > 10^{-4}\ e^-\ {\rm cm}^{-3}$ an hot spot can be formed and on the other side where $n_{\rm ext} \leq 10^{-4}\ e^-\ {\rm cm}^{-3}$ one does not observe an hot spot and the extended lobe is more distant from the optical galaxy than the lobe containing the hot spot.

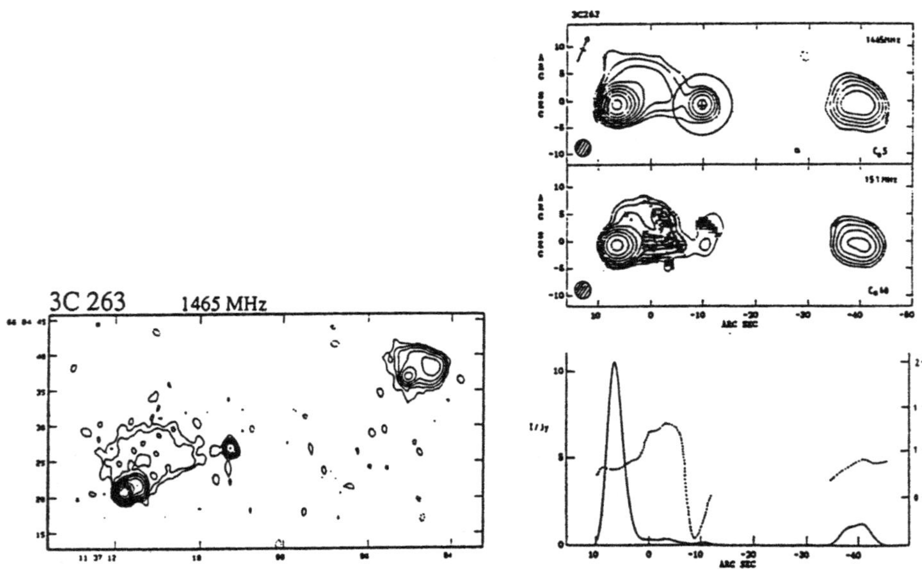

Fig. 3. 3C 263 from Browne (1987) and from Leahy et al. (1989)

iii) The extreme case of asymmetry: 3C 273

Radio observations from 3C 273 (Foley and Davis 1985, Davis et al. 1985 and Flatters and Conway 1985) suggest that this radio source is intrinsically asymmetrical, i.e. it is a one–sided source. As already suggested by Shklovsky (1982),

3C 273 could be understood if the 14 arc sec. radio component is mainly a pure e^-–e^+ beam. Indeed, radio polarization observations (Conway 1988) suggest an extremely low level of the thermal electron density in both the extended lobe and the intergalactic medium surrounding the quasar.

2.4. Remarks concerning 3C 449 type radio sources

The two–fluid model has not yet been applied to sources similar to 3C 449. However, it is especially suitable to explain the superluminal radio souce 3C 120 (Sol et al. 1989). 3C 120 is a low luminosity source compared to Cygnus A but is not simple as 3C 449 or head–tail radio galaxies, it is an intermediate case between these two extreme simple cases, i.e. between Cygnus A and 3C 449. It will be interesting to verify if the two–fluid model can explain the source 1928+738. If it does, it will be nice to determine the ratio $P_{\rm kj}/P_{\rm kb}$ for this source. Up to now, there is no theory to explain the formation of 3C 449 type jets and to understand the difference between their speed and the speed of Cygnus A type jets.

3. Variability in nuclei of galaxies

3.1. Introduction

Compact radio sources are frequently variable sources. One observes either flux and polarization variations and/or structure variations, in the case of superluminal radio sources for instance. Various examples of variable sources and references concerning their observational properties can be found in this Workshop and in "Parsec–scale Radio Jets" (Cambridge University Press, A. Zensus and T. Pearson Editors). The variability of some sources implies brightness temperatures greater than 10^{12} K, but the lack of the inverse Compton catastrophe for these sources point out an anisotropic emission of the nucleus, which can be due either to relativistic ejections or/and to coherent emission. Observational properties of some variable sources indicate rapid and local acceleration processes which can be shock waves.

The most interesting consequences of the two–fluid models to understand variable sources are the possibility to

i) produce shocks,

ii) accelerate relativistic electrons via strong Langmuir turbulence,

iii) emit coherent emission in the radio range, such as coherent Compton radiation of the relativistic electrons scattered by Langmuir waves when the relativistic beam interacts with the jet (any fluctuation of the Langmuir wave background generated by the beam dissipation leads to some variability),

iv) allow the occurence of intermittent events when the relativistic beam interacts with the jet.

In the simplest case, if we have a continuous jet and a sporadic emission of e^-–e^+ clouds, one can expect to observe a variable nucleus. Moreover, the relativistic motion of the e^-–e^+ clouds produces a strong Doppler beaming effect which amplifies the observed phenomena.

3.2. Different processes induced by the beam–jet interaction

a) *Beam pervading the jet*

If the magnetic field in the beam becomes smaller than the critical value B_c, the beam dissipates through strong Langmuir turbulence which accelerate relativistic electrons and a radio knot appears in the dissipation area (Sol et al. 1989).

As the kinetic power of the beam is much smaller than the kinetic power of the jet, when it dissipates it does not destroy the jet. However, depending of the various physical parameters of the beam and the jet, different situations can happen (Sol et al. 1989). Let us summarize them; one can have

i) regular stationary supersonic or subsonic flows,

ii) regular stationary transonic solutions, then either a subsonic flow can be transformed into a supersonic jet after the dissipation area, or the supersonic jet becomes smoothly subsonic (this happens when the sound speed increases more rapidly than the plasma velocity and thus the heating of the plasma is more efficient than its entrainment),

iii) no regular stationary flow exists and a shock develops,

iv) a generation of a poloïdal current in the thermal plasma, and thus a generation of a local toroïdal magnetic field.

When a pervading beam dissipates in a jet, one can have also emission of coherent radiation, which emitted frequencies are in the radio range (Baker et al. 1988).

b) *Beam moving in a channel through the jet*

In the mixing layer, if the magnetic field is smaller than the critical value, all the phenomena previously described can happen. Moreover, in front of the beam, its interaction with the jet induces relativistic shocks.

3.3. Different kinds of variable radio sources

The beam–jet interaction induces many kinds of variabilities similar to those observed. It is therefore tempting to try a classification of the different variable sources depending on the criterium of stability of the beam propagation in the jet. This suggets a classification of the different kinds of variable radio sources according the value of the magnetic field in the beam close to the nucleus, namely B_{nu}, compared to the value of the critical magnetic field. Three different cases may happen

i) $B_{nu} > B_c$: superluminal quasars

The e^-–e^+ clouds move in the jet without dissipation. The apparent speeds of the VLBI radio knots are $v_{ap} \simeq$ 1–10 c. VLBI polarization observations (see details in "Parsec–scale Radio Jets") show the magnetic field parallel to the direction of propagation, indicating no relaxation of the e^-–e^+ clouds with the jet. The beam can propagate on scales as great as few kpc and dissipates via strong Langmuir turbulence when the magnetic field becomes smaller than the critical value.

ii) $B_{nu} \gtrsim B_c$: BL Lac radio sources

The e^-–e^+ clouds move in the jet and start rapidly to interact with it. The apparent speed of the VLBI knots is $v_{ap} \simeq$ 1–2 c. VLBI polarization observations (see details in "Parsec–scale Radio Jets") show the magnetic field non parallel to the direction of propagation in the knots, suggesting a dissipation of the e^-–e^+ clouds with the jet. The beam propagates on scales smaller than few 100 pc.

iii) $B_{nu} < B_c$: flickering radio sources

The e^-–e^+ clouds interact immediately with the jet. The beam propagates on scales smaller than 1 pc.

4. Further observational tests for the two–fluid models

A number of observations can be made to test the two–fluid models. They are for instance

i) VLBI observations of 3C 449 type radio sources,

ii) VLBI and arc sec. observations (with MERLIN and the VLA) of flickering radio sources,

iii) VLBI polarization observations with smaller bandwiths and at 2 cm (see details in Pelletier and Roland, 1990) and observations of the variations of the polarization percentage and of the P.A. with time.

These observations will allow

i) to detect in which sources one has formation of shocks and/or emission of coherent radiation,

ii) the determination of the magnetic field and of the thermal density, two key parameters of the model,

iii) the detection of the dissipation area of the beam in the jet, which will provide a direct test of the model (indeed, the beam dissipation occurs when the magnetic field reaches its critical value),

iv) to determine for which sources the beam pervades the jet or the beam moves in a channel through the jet.

References

Baker, D.N., Borovsky, J.E., Benford, G., Eleik, J.A. (1988): Astrophys. J. **326**, 110

Bicknell, G.V. (1984): Astrophys. J. **286**, 68

Bicknell, G.V. (1986-a): Astrophys. J. **300**, 591

Bicknell, G.V. (1986-b): Astrophys. J. **305**, 109

Blandford, R.D., Payne, D.G. (1982): Monthly Notices Roy. Astron. Soc. **199**, 883

Browne, I.W.A. (1987): Superluminal Radio Sources, Cambridge University Press, A. Zensus and T. Pearson Editors

Cohen, M.H. (1990): Astrophysical Ages and Dating Methods, Editions Frontières, E. Flam, M. Cassé, J. Audouze and J. Tran Than Van Editors

Conway, R.G. (1988): Magnetic Fields and Extragalactic Objects, Editions de Physique, E. Asseo and D. Grésillon Editors

Davis, R.J., Muxlow, T.W.B., Conway, R.G. (1985): Nature, **318**, 343

Draine, B.T. (1983): Astrophys. J. **270**, 519

Fanti, C., Fanti, R., de Ruiter, H.R., Parma, P. (1987): Astron. Astrophys. Suppl. **69**, 57

Flatters, C., Conway, R.G. (1985), Nature, **314**, 425
Foley, A.R., Davis, R.J. (1985): Monthly Notices Roy. Astron. Soc. **216**, 679
Johnston, K.J., Simon, R.S., Eckart, A., Biermann, P., Schalinski, C., Witzel, A., Strom, R.G. (1987): Astrophys. J. **313**, L85
Leahy, J.P., Muxlow, T.W.B., Stephens, P.W. (1989): Monthly Notices Roy. Astron. Soc. **239**, 401
Marr, J.M., Backer, D.C., Wright, M.C.H., Readhead, A.C.S., Moore, R. (1989): Astrophys. J. **337**, 671
Meisenheimer, K., Röser, H.J., Hiltner, P.R., Yates, M.G., Longair, M.S., Chini, R., Perley, R.A. (1989): Astron. Astrophys. **219**, 63
Morganti, R., Fanti, C., Fanti, R., Parma, P., de Ruiter, H.R. (1987): Astron. Astrophys. **183**, 203
Muxlow, T.W.B., Pelletier, G., Roland, J. (1988): Astron. Astrophys. **206**, 237
Parma, P., Fanti, C., Fanti, R., Morganti, R., de Ruiter, H.R. (1987): Astron. Astrophys. **181**, 244
Pelletier, G., Zaninetti, L. (1984): Astron. Astrophys. **136**, 313
Pelletier, G., Roland, J. (1986): Astron. Astrophys. **163**, 9
Pelletier, G., Roland, J. (1988): Astron. Astrophys. **196**, 71
Pelletier, G., Sol, H., Asseo, E. (1988): Phys. Rev. A. **38**, 2552
Pelletier, G., Roland, J. (1989): Astron. Astrophys. **224**, 24
Pelletier, G., Roland, J. (1990): Parsec–scale Radio Jets, Cambridge University Press, A. Zensus and T. Pearson Editors
Perley, R.A., Willis, A.G., Scott, J.S. (1979): Nature **281**, 437
Pudritz, R.E., Norman, C.A. (1983): Astrophys. J. **274**, 677
Reid, M.J., Biretta, J.A., Junor, W., Muxlow, T.W.B., Spencer, R.E. (1989): Astrophys. J. **336**, 112
Roland, J., Pelletier, G., Muxlow, T.W.B. (1988): Astron. Astrophys. **207**, 16
Shklovsky, I.S. (1982): Extragalactic Radio Sources, Reidel Publishing Company, D. Heeschen and C. Wade Editors
Sol, H., Pelletier, G., Asseo, E. (1989): Monthly Notices Roy. Astron. Soc. **237**, 411
Uchida, Y., Shibata, K. (1985): Publ. Astron. Soc. Japan, **37**, 515

Infrared Variability of Active Nuclei

Martin Ward

Royal Greenwich Observatory, Madingley Rd., Cambridge, U.K.

Abstract: AGN are known to vary in flux at near infrared wavelengths. Unfortunately because of limited accuracy and poor time coverage, the interpretation of the results is often ambiguous. Recent statistical surveys and detailed studies of individual AGN have improved the situation. With the introduction of the large format infrared arrays, we are poised to commence new programmes that will help disentangle the variable and non-variable components of the infrared continuum.

1. Introduction

It is often claimed that temporal studies of infrared flux variations from Seyferts and quasars, are a powerful means of distinguishing between the presence of different physical processes. A specific aim is to use the variability characteristics to test models of thermal and non-thermal emission, on the grounds that thermal components can only vary relatively slowly, whilst non-thermal components can in principle vary on much shorter time-scales. Such arguments are valid, ***but*** only if regard is given to the following additional points; the length of the monitoring baseline and sampling frequency, the amplitude and point to point trends, and ideally, coordinated observations at other frequencies.

Because of the relatively greater difficulty in obtaining infrared monitoring data, and the fact that small telescopes with less over-subscription of time are not generally used, it has taken longer to arrive at definite conclusions compared to the situation for variability at optical wavelengths. The early pioneering work by Penston *et al* (1974) and Rieke and Lebofsky (1979), established that some AGN did vary at wavelengths of a few microns, but the limited accuracy and sampling frequency meant that no firm conclusions concerning the origin of these variations could be drawn.

However, thanks to a number of comprehensive studies the identification and interpretation of infrared variations is now substantially improved. Cutri *et al* (1985) undertook a combined optical and infrared monitoring study of quasars, and obtained the important result that the amplitude of variability is greatest at blue optical wavelengths and decreases toward longer wavelengths. Indeed, in their small sample of 7 objects only one (3C273) was observed to vary at 2.2 microns, whereas *all* varied at optical wavelengths. This suggested that the origin of the optical and infrared continua were distinct.

Some years later Neugebauer *et al* (1989) analysed a long-baseline set of infrared photometry on a much larger sample of quasars and arrived at a number of important conclusions; roughly half of the quasars in their sample showed little sign of near infrared variability over a time span of many years. For those that do vary, the mean amplitude is less than a factor 2.5, and most vary by a few 10's of percent. Both of these properties are in strong contrast to what is found from long-term optical monitoring programmes. Their work further suggests that during the variations the shifts in the infrared fluxes are *grey*, i.e. the color remains constant. More luminous quasars tend to have less chance of varying, compared to lower luminosity examples. The sub-group with the highest chance of varying are those quasars with flat radio spectra (compact radio sources).

2. Interpretation of infrared variability

A simple model that can explain most of the observations discussed above has been described by Sanders *et al* (1989). They suggest that a *Blazer* type component is present in the radio quasars such as 3C273, whereas in radio-quiet quasars this component is weak or absent. There is additional evidence for this from polarisation data (Impey, Malkan and Tapia 1989). Changes in this component are responsible for the short-term variations such as the flares and flickering reported by Courvoisier *et al* (1988). By analogy with the properties of BL Lacs, this hypothetical Blazar component could exhibit large amplitude variations on very short timescales. Impey *et al* (1982) report near infrared variability in 3C446 of a factor 2 in one day! It is relevant to note here that BL Lac objects are the *only* type of AGN to show significant variations in the mid-far infrared IRAS wavelengths (Edelson and Malkan 1987, and Sembay, Hanson and Coe 1987). The far infrared variability of OJ287, 3C345 and BL Lac suggest a typical doubling time of order two months. For an isotropic source this implies a minimum brightness temperature of If $T_b > 106$K, which is taken as further evidence that these sources must be non-thermal. The overall continuum energy distributions of radio loud quasars are also consistent with the existence of a Blazar component, contrary to the situation for radio-quiet examples.

In radio-quiet quasars like most in the PG sample, the longer term trends could be predominantly the time averaged response of a thermal component, to the established higher degree of optical/ultraviolet continuum variability which provides the heating source. There is a complication in that there would be a *heating echo* analogous to the *light echo* of the BLR in response to changes in the ionizing continuum. Also, even in radio-quiet quasars there may still be some contribution from a non-thermal infrared continuum. Just as we have now recognised both constant and variable components in the optical/uv continuum (due to a hot accretion disc, and an underlying non-thermal component) it seems very probable that the same situation exists for the infrared continuum.

The thermal component will be effectively constant over the short timescales on which the non-thermal component can vary. So, to infer the spectral properties of the non-thermal component in its ***high*** and *low* states, we must subtract the thermal component. It is therefore an intricate problem requiring good sampling of the IR light curve over extended periods. The two best examples to date in which this has been achieved are 3C273 (Courvoisier *et al* 1987), and F9 (Clavel, Wamsteker, and Glass 1989). These two AGN provide excellent examples of the detailed variability behaviour of what is very likely a non-thermal component in a radio-loud AGN, and a radio-quiet example in which the long term trend of changing flux is interpreted as thermal variability. At the time of writing there are *no* cases of short-term, large amplitude infrared variations in radio-quiet AGN, and *no* cases of any infrared variability in Seyfert 2s. Should either of these statements be contradicted by future observations, we would need to revise significant parts of the above explanations.

Very little genuinely simultaneous infrared/X-ray monitoring of AGN has been carried out. This is a pity since one of the best correlations between continuum fluxes is of L_x *vs.* near-infrared luminosity. The tightness of this correlation has led to the suggestion that the infrared and X-ray continua are closely related. However, at least in one case we can now rule out models which link the continua by means of Compton scattering or synchrotron emission. This conclusion resulted from a multi-frequency monitoring campaign on the low luminosity Seyfert 1 nucleus in NGC 4051 (Done *et al* 1990). They found the 2 micron stellar subtracted continuum flux to be constant to within the measurement uncertainty of 8%, whilst the X-ray flux varied by about a factor of 2. Clearly this type of experiment needs be repeated on more AGN, including higher luminosity examples.

3. Alternative models and the importance of supernovae

In the above discussion we have assumed the *standard model* of Seyfert and quasar type AGN, namely one in which the stellar component plays a minor role in the energy output. There are other models which suggest that supernovae may be responsible for many of the characteristics of Seyferts, if not quasars (Terlevich 1990). Such events would clearly lead to variability, which might best be observed in the infrared if the explosions occurred in dusty and hence heavily reddened regions. Using the new large format infrared arrays it is now feasible to make sensitive searches for supernovae events in AGN. If this model were correct there would also be the possibility of observing an outburst sightly displaced from the nucleus (as defined by the peak of the light distribution). This would then pose serious problems for the standard model in which fast variability must occur relatively close to the black hole.

Whether or not supernovae are important in the context of Seyferts is controversial, but their presence in the *starburst* galaxies is accepted almost by definition.

For example, in NGC 253 and M82 a family of compact radio sources observed in their central regions, are believed to be young supernova remnants, see Antonucci and Ulvestad (1988). Although the rate of supernova explosions will depend on various factors, and may be much less than one per year, monitoring of a sample of these galaxies over a few years ***should*** provide an unambiguous result. This programme is ideally suited to the use of infrared array detectors, because the supposed SNRs observed in the radio are distributed over a large number of arcseconds, and we know that the inner regions of this type of galaxy are extremely dusty. Failure to detect supernovae in this way would require us to critically re-examine the simple starburst model.

4. Conclusions

Despite the lack of infrared monitoring data of an accuracy and time coverage approaching that available at optical wavlengths, a number of overall characteristics can be identified. The infrared continua of AGN vary far less than their optical continua. This is partly due to the diluting effect of the contribution from the constant stellar component, but could also indicate that the non-stellar optical and infrared components have distinct origins. The probability of an AGN being variable in the infrared is strongly correlated with its radio properties. BL Lac objects are highly variable at all frequencies, and the presence of such a component in some AGN could explain the difference in variability characteristics between radio-loud (RL) and radio-quiet (RQ) objects. In the RQ objects the infrared variability is consistent with that due to dust heating and cooling effects. In RL objects the infrared variability timescales can be short, and are best fitted by non-thermal models.

Supernovae are widely assumed to be important in starburst galaxies, and in some schemes are invoked to explain Seyfert activity. Future infrared monitoring programmes with arrays, will offer a good opportunity for testing these models.

References

Antonucci, R.J, & Ulvestad, J.S., 1988. ***Astrophys. J.***, **330**, L100

Clavel, J., Wamsteker, W., & Glass, I.S., 1989. ***Astrophys. J.***, **337**, 236

Courvoisier, T.J-L, Turner, M.J.L., Robson, E.I., Gear, W.K., Staubert, R., Blecha, A., Bouchet, P., Falomo, R., Valtonen, M., & Terasranta 1987. ***Astron. Astrophys.***, **176**, 197

Courvoisier, T.J-L, Robson, E.I., Blecha, A., Bouchet, P., Hughes, D.H., Krisciunas, K., & Schwarz, H.E., 1988. ***Nature***, **335**, 330

Cutri, R.M., Wisniewski, W.Z., Rieke, G.H., & Lebofsky, 1985. ***Astrophys. J.***, **296**, 423

Done, C., Ward, M.J., Fabian., A.C., Kunieda, H., Tsuruta, S., Lawrence, A., Smith, M.G., & Wamsteker, W. 1990. ***Mon. Not. R. astr. Soc.***, **243**, 713

Edelson, R.A., & Malkan, M.A., 1987. ***Astrophys. J.***, **323**, 516

Impey, C.D., Brand, P.W.J.L., Wolstencroft, R.D., Williams, P.M., 1982. *Mon. Not. R. astr. Soc.*, **200**, 19

Impey, C.D., Malkan, M.A., & Tapia, S., 1989. *Astrophys. J.*, **347**, 111

Neugebauer, G., Soifer, B.T., Matthews, K., & Elias, J.H., 1989. *Astron. J.*, **97**, 957

Penston, M.V., Penston, M.J., Selmes, R.A., Becklin, E.E., & Neugebauer, G., 1974. *Mon. Not. R. astr. Soc.*, **169**, 357

Rieke, G.H., & Lebofsky, M.J., 1979. *Astrophys. J.*, **227**, 710

Sanders, D.B., Phinney, E.S., Neugebauer, G., Soifer, B.T., Matthews, K. & Green, R.F., 1989. *Astrophys. J.*, **347**, 29

Sembay, S., Hanson, C.G., & Coe, M.J., 1987. *Mon. Not. R. astr. Soc.*, **226**, 137

Terlevich, R. *Windows on Galaxies (ed. Fabbiano, G. et al.) 87-96 (Kluwer, Netherlands, 1990).*

Continuum Variability in Quasars and Seyfert Galaxies

Thierry J.-L. Courvoisier

Geneva Observatory, CH-1290 Sauverny, Switzerland

Abstract: Multiwavelength observations of AGN give several pieces of information which help advance our understanding of the physics of the core regions. Very good correlations at zero lag are found between the UV and visible domains, suggesting that the variability cannot be due to variations in thin accretion discs. A possible significant correlation is found between the UV and the radio emission of 3C 273, which might be due to links between accretion and non thermal activity. Although no correlation between the X-ray flux and other components can be found, analysis including spectral information suggests that the X-ray emission may be due to Comptonised EUV flux, at least in the galaxy NGC 5548.

1. Introductory remarks

The question of the response of the line emission to the ionising continuum fluctuations has been much discussed at this conference. The question is well understood, even if the answers are complex. In simple words, there are two elements playing a role, the fluctuating ionising continuum and the responding line emitting gas. Both elements are well identified. The difficulties are caused by the complex structure of the responding medium and possibly by the fact that this structure is a function of time. It seems, however, that the timescale of the variations in the responding medium is much larger than the continuum variability timescale. When dealing with the relationships between the continuum emission components emitting in different regions of the electromagnetic spectrum one is met with considerable difficulties because one doesn't know a priori which component is responding to which. In addition, when one such relationship is identified (see below), one finds that the responding medium varies with a similar timescale as the source component.

There is no doubt, however, that the different components contributing to the continuum energy distributions of AGN are inter-related. If this was not the case, there would be no reason for the gross similarities observed in objects of very different luminosities. The mere fact that, in a very rough approximation, the flux

$f_\nu \sim \nu^{-1}$ means that all components have approximately the same luminosity and is a characteristic of the AGN phenomenon at least in Seyfert galaxies and Quasars. If no interrelationship was present between the different components, one would expect to find objects with the X-ray luminosity of, say, NGC 5548 and the infrared luminosity of, say, 3C 273. Since no such objects are known, we can conclude that the different emission components of an object "know" the emission level of the other components.

2. Correlation analysis of the continuum components

Our first expectation (or at least mine) when observing repeatedly the continuum emission of the bright quasar 3C 273 in different wavebands was that the flux of the various emission components would be characterised by different timescales and that cross correlations of the components would indicate several well defined lags. These timescales and lags would then be expected to provide direct information on the geometry of the source. The reality has proved considerably more complex (Courvoisier et al. 1987, 1990). I summarise here the results, details and data can be found in the above papers.

The data on the 2–10 keV X-ray flux show no correlation with the UV-optical flux, nor with the IR or mm fluxes. Unfortunately the X-ray coverage is far too sparse to characterise its variability properly, we can only deduce that X-ray flux variations with an amplitude of factors 2 or 3 are not infrequent and happen on a timescale typically less than 0.5 year. Even though the X-ray sampling is poor, one could expect to see in the correlation analysis whether the X-ray flux is well correlated with any other component, since the other components are much better sampled. Our negative result shows that any correlation is more subtle than can be seen with our kind of data.

Two qualitative conclusions can be drawn from this analysis. The first is that if the UV emission is a signature from the inner regions of an accretion disc, and thus a measure of the accretion flow, the X-ray flux is **not** directly linked with this accretion flow. The second is that the X-rays are **not** a self-Compton branch of the mm synchrotron emission in a constant magnetic field. If this were the case, both the mm flux and the X-ray emission would depend on the parameters of the electron population and vary simultaneously.

Excellent correlations are found between the 1250Å flux and the the visible flux in 3C 273 (Courvoisier and Clavel 1990 and work in preparation) and between the U and 5800Å fluxes (Fig. 1). This means that all the blue bump is varying simultaneously, the lags being less than few tenths of a year. This result is not caused by the very rapid and high amplitude flares which were discarded for the correlation analysis. It is caused by the slower trends which have a characteristic timescale of approximately 0.5 years. This lack of lag is very difficult to understand with models based on thermal emission from optically thick gas. The distance between emission regions with temperatures such that they cool primarily at 1200Å or at 6000Å is of the order of the sizes of these regions. Thin accretion discs give a geometrical model for these distances (Courvoisier and Clavel 1990), but more general cases can be estimated by simply considering black body spheres. In the case of a luminosity of $10^{46} ergs/s$ for a 6000K black body, the radius of such a sphere is of the order of $10^{17} cm$, so that the UV emitting sphere would be at least this distance from the sphere emitting in the V–band. For the 2 spheres to vary with a time difference of less than, say, 0.2 year, the communication speed would have to be larger than half the speed of light, a very unlikely situation indeed.

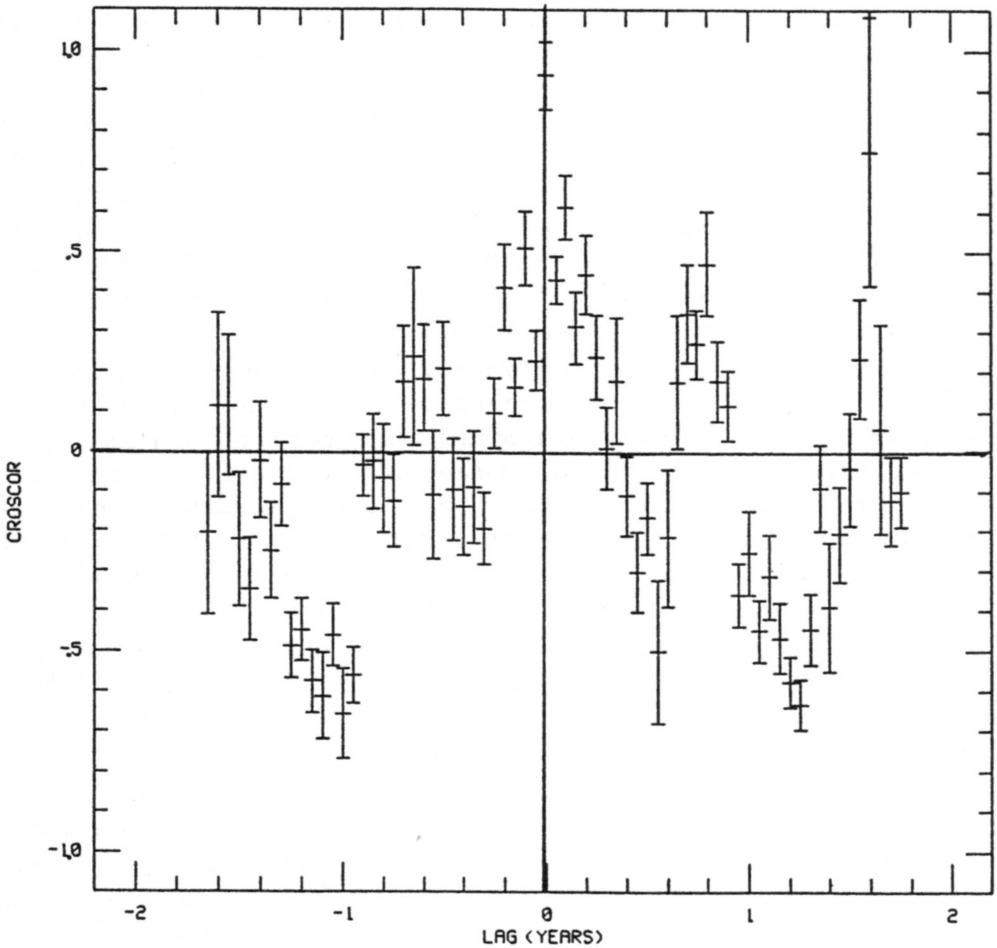

Fig. 1. Cross correlation coefficient between the U and the 5800Å flux in the QSO 3C 273

In the case of 3C 273, the visible-UV continuum energy distribution changed in such a way that the difference between the high and low states can be represented by a single black body curve of $T \sim 30\,000\,K$. This led to the sugestion that the only variable part is the hot inner regions of the disc, leaving the outer regions constant, at least on the timescales investigated here. The same reasoning can possibly be applied to the sources which show a hardening of the continuum when the flux increases.

Low luminosity Seyfert galaxies for which sufficient UV and visible continuum data are available show the same phenomenon: the UV and visible fluxes vary simultaneously, whereas expected lags based on thin disc models are orders of magnitude longer than observed upper limits (Courvoisier and Clavel 1990). The problems linked with interpretations of the visible to UV emission in terms of thin accretion discs are therefore not limited to intrinsically bright objects, probably accreting close to the Eddington rate, but is generic to Seyfert galaxies and Quasars. Possible solutions may be found in the properties of thick accretion discs, which could be unstable, but not sufficiently so to be completely disrupted (see the contribution by W. Glatzel in these procedings). Clearly, models in which part of the visible-UV flux is reflected by the disc (or any other structure) may be able to explain the observed timescales. The flux involved in the variable component is, however, a large fraction of the visible-UV flux (as indicated by the amplitude of the variations which is often a factor 2) and thus an important fraction of the bolometric flux. Successful models of the UV to visible flux involving reflection should therefore provide a way to hide the primary source.

The near IR flux in 3C 273 doesn't follow the UV or visible trends. The near IR (J,H,K,L bands) is characterised by long periods of constant flux with short intervals of violent activity. The best observed such flare occurred in the spring of 1988 (Courvoisier et al. 1988) and was most probably associated with the appearance of a new component in the VLBI jet. This activity was most probably of synchrotron origin, as indicated by the spectral behaviour and the high polarisation observed during the flare. These flares are also observed in the visible bands, although with an amplitude that decreases towards shorter wavelengths.

There are then at least two very different physical origins for the visible variability of 3C 273, one linked with the 0.5 year, probably thermal, timescale (Ulrich et al. 1988) and the other one linked with the very fast but low duty cycle IR flaring related to the synchrotron emission and the jet.

A further significant correlation is observed between the UV and the radio emission (Courvoisier et al. 1990). This result is surprising in that it relates two components which are thought to be of very different physical origins, namely thermal and synchrotron. The possible explanations for this correlation are 1) Comptonisation or self Comptonisation, but the lag which is indicated by the data suggests that the UV flux leads the radio flux by about 2 tenths of a year, which is not expected if the UV is a comptonised radio flux. 2) A link between the accretion related UV-visible emission and the non thermal radio emission. Such a link must be present at some level if the UV is emitted by accretion related processes and if accretion is the prime energy source in AGN, both of which are expected to be true. 3) A statistical fluke which will disappear when data from the coming years is gathered and analysed.

3. UV–X comptonisation

One possible origin for the X-ray emission is Comptonised lower energy radiation. This process naturally produces a power law spectral energy distribution, which can be modified by different reprocessing, like for example reflection effects. The possible low energy photon sources are the synchrotron emission, the blue bump or the soft excess emission (which may or may not be the high frequency tail of the blue bump). It is very unlikely that dust emission contributes significantely to the soft photon flux, as the dimensions of the emission region are much larger than the X-ray source.

Walter and Courvoisier (1990) have analysed the EXOSAT data available on the Seyfert NGC 5548. This source, which was observed several times, showed a well defined soft exces that cannot be explained by partial or hot absorbers, large amplitude variability and significant 2–10 keV power law slope changes. The EXOSAT LE data were assumed to be a fraction of the soft photon source which is then Comptonised in a hot medium of optical depth (τ) of few tenths and covering a few percent of the soft source (covering factor η). In this case one obtains a linear relationship between the spectral index of the Comptonised component (in this case the 2–10 keV X-rays) and the logarithm of the soft to Comptonised photon fluxes. The observed data satisfy this linear relationship very well.

The correlation obtained with this model between the spectral index and the logarithm of the soft to Comptonised photon fluxes is much more significant than the correlation of the same spectral index with the 2–10 keV flux. The correlation coefficients are greater than 0.9 (α vs ln(soft/Comptonised)) and approximately 0.5 (α vs flux) respectively, leading to probabilities of less than 0.1% in our analysis that the two quantities are not correlated and less than 15% for α and the flux.

A preliminary study of all AGN observed several times with EXOSAT shows that most data fit this model. It must be stressed, however, that only in the cases where the soft photon source can be well identified (i.e. where the soft exces cannot be fitted by partial or hot absorber models but represents a genuine additional photon component) does the model provide non trivial information. This model is very similar to the wind and shock model described for 3C 273 by Courvoisier and Camenzind (1989). The only difference resides in the fact that the UV flux of 3C 273 was taken to be representative of the soft photon source, rather than the weak soft excess.

The model described here provides a natural explanation for the lack of direct correlation between the soft and the hard fluxes if the product $\eta \cdot \tau$ varies in time, i.e. if the physical conditions in the reprocessing medium (here the hot Comptonising electrons) vary in time. This is to be expected since the reprocessing medium, here a hot mildly relativistic plasma, must be heated, for example by shocks. Note that in any case energy must be provided to the Comptonising medium in addition to the soft photon flux. This is in contrast to the dust and to the line emission gas, where the reprocessing medium reacts only to the incoming photon flux, without requiring an additional, possibly mechanical, energy source.

4. Conclusion

Reprocessing is a very natural explanation for the similarity in the flux of the X-ray, visible–UV and IR components of the continuum emission in Quasars and Seyfert galaxies. The differences from object to object are due to geometrical aspects. In this scenario, the prime photon source is the visible–UV component extending into the EUV, which is being reprocessed in dust to produce the IR flux, in shocks to produce the X-rays by Comptonisation and also in the broad and narrow line emission clouds.

There remains the question of the radio to far IR component, undoubtedly of synchrotron origin. This component is a major difference between the radio loud and the radio quiet objects, since it is quite possible that this component is altogether absent in radio quiet objects (Barvainis 1990). The strength of the magnetic field present in the core regions of AGN may be able to explain this difference, since synchrotron emission of single electrons depends on B^2. In such a picture, radio quiet objects have weak magnetic fields and radio loud objects have strong magnetic fields (magnetic AGN). The differences between classes of radio loud objects or classes of radio quiet objects are then due to orientation effects (Barthel 1990).

The study of cross correlations between the fluxes of the continuum emission components is insufficient to reveal links between them that are caused by a variable medium, as is observed between the EUV and the X-rays. In order to observe these connections, it is necessary to consider not only the total flux in any one component, but also its spectral shape parameters. This requires a more detailed analysis than had been expected, it also gives a considerably larger parameter space to study.

References

Barthel, P.D. (1989), Ap.J. **336**, 606
Barvainis R. (1990), Ap.J. **353**, 419
Courvoisier T.J.-L. et al. (1987), A& A **176**, 197
Courvoisier, T.J.-L. and Cmenzind,M. (1989), A& A **224**, 10
Courvoisier T.J.-L. et al. (1990), A& A **234**, 73
Courvoisier T.J.-L. and Clavel J. (1990), to appear in the proceedings of Structure and Emission Processes of Accretion Discs, Editions Frontieres, Paris, Eds. Bertout, Collin-Souffrin and Lasota.
Courvoisier, T.J.-L. et al. (1988), Nature **335**, 330
Ulrich, M.-H., Courvoisier, T.J.-L. and Wamsteker,W. (1988) A& A **204**, 21
Walter, R. and Courvoisier T.J.-L. (1990), A& A **233**, 40

The Variability of Ly-alpha in 3C273

Paul T. O'Brien & T. J. Harries

Department of Physics and Astronomy, UCL, Gower Street, London,
WC1E 6BT

Abstract

Evidence is presented for correlated variability between the ultraviolet continuum and the Ly-alpha emission line in the luminous quasar 3C273. Using IUE data covering a period of approximately 1000 days, Ly-alpha varied by 15 percent whilst the continuum varied by a factor of two. A time series analysis gives a lag for Ly-alpha relative to the ultraviolet continuum of 74 +/- 35 days. The broad flat-topped shape of the cross-correlation function, and the small amplitude of the Ly-alpha variability, indicate that the derived lag is a measure of the inner radius of a geometrically thick BLR.

The Electromagnetic Spectrum of the Radio-Quiet Quasar 1821+643 and Comparison with 3C 273

Marie-Hélène Ulrich

European Southern Observatory, Karl-Schwarzschild-Str. 2, 8046 Garching bei München, Germany

The quasar 1821+643 whose ecliptic latitude is 86° is a good candidate for simultaneous observations at a number of frequencies during the ROSAT survey. It is therefore timely to summarize the data available just before the ROSAT surveying of this quasar.

1. General characteristics

					f_ν	Jy
Source	z	$N_{\rm H}$ *	$E_{\rm B-V}$	$M_{\rm v}$	6 cm	3 cm
1821+643	0.297	$4.1 \cdot 10^{20}$ cm^{-2}	$E_{\rm B-V} = 0.085$	$M = -27.2$	–	< 0.5
3C 273	0.158	$3 \cdot 10^{20}$ cm^{-2}	$E_{\rm B-V} = 0.050$	$M = -27.0$	43	–

* Stark *et al.* 1989; Courvoisier *et al.* 1987.

1821+643 is radio quiet. Its absolute and apparent fluxes in the X-ray, UV, optical and infrared ranges are comparable to those of 3C 273 (the fluxes of the two objects differ by factors 2 to 4 depending on the energy ranges and the dates of the observations). It is therefore of interest to compare 3C 273, a powerful core dominated radio source, and 1821+643, a bright radio quiet quasar.

2. X-ray variability

The most interesting feature of the electromagnetic spectrum of 1821+643 is the flux variation by a factor 7 of the low energy X-ray component (energy < 2 keV). This variation occurred between 1980, the date of the HEAO–A2 experiment (Pravdo and Marshall 1984) during which the object was in a bright state, and 1985, an epoch of several EXOSAT observations. The flux at 5 keV decreased by a factor 1.8 during the 2 dates (Warwick *et al.* 1989). In comparison, no large variations have been found in the soft X-ray spectrum of 3C 273 but variations by up to a factor ~ 2.8 have been detected in the hard X-ray spectrum (Courvoisier *et al.* 1987; Turner *et al.* 1989).

3. UV variability

Until now, no UV observations of 1821+643 have been performed contemporaneous with the X-ray observations. Examination of the IUE archive, and the adding of our own observations of 24 July 1990, reveal no flux variations by more than 10%. We recall that the error on the IUE flux typically is 5% on the well exposed parts of the spectra (Fig. 1).

The only spectrum taken on 10 August 1987, SWP 31523, is of definitely lower quality than the other spectra with an error on the flux of $\sim 15\%$. There is therefore no significant variation between 29 July and 10 August 1989. The lack of UV variability is also found in a preprint by Halpern *et al.* which we received after presenting this article in Heidelberg and which is based on the same data except for our latest spectra of 24 July 1990.

The spectrum in Fig. 2 of 24 July 1990 is representative of the spectra in the IUE archives.

4. Comparison with the electromagnetic spectrum of 3C 273 and conclusion

The energy distribution of 1821+643 and 3C 273 is shown in Fig. 3. The X-ray flux of 1821+643 is that of 1985; the infrared data are from Neugebauer *et al.* (1986). The data of 3C 273 are those of May 1985. At this date the UV flux happened to be the highest observed for 3C 273, while both the IR and X-ray fluxes have been observed at higher intensities at other epochs (Courvoisier *et al.* 1987).

No correlation has yet been found between the UV and the medium energy X-ray flux in 3C 273. It would be surprising to find such a correlation in 1821+643 in view of the fact that large variations have been found in the X-range and practically none in the UV. However 1821+643 has been less frequently observed than 3C 273, and therefore new observations could reveal as yet unobserved states of this radio quiet quasar.

References

Courvoisier, T.J.-L., *et al.* (1987): Astron. Astrophys. **176**, 197
Neugebauer, G., *et al.* (1986): Astrophys. J. **308**, 815
Pravdo, S.H., Marshall, F.E. (1984): Astrophys. J. **281**, 570
Stark, A.A., *et al.* (1989): in preparation
Turner, M.J.L., *et al.* (1989): 23rd ESLAB Symposium, Vol. 2, p. 769, ESA publ. SP-296
Warwick, R.S., Barstow, M.A., Yagoob, T. (1989): Monthly Notices Roy. Astron. Soc. **238**, 917

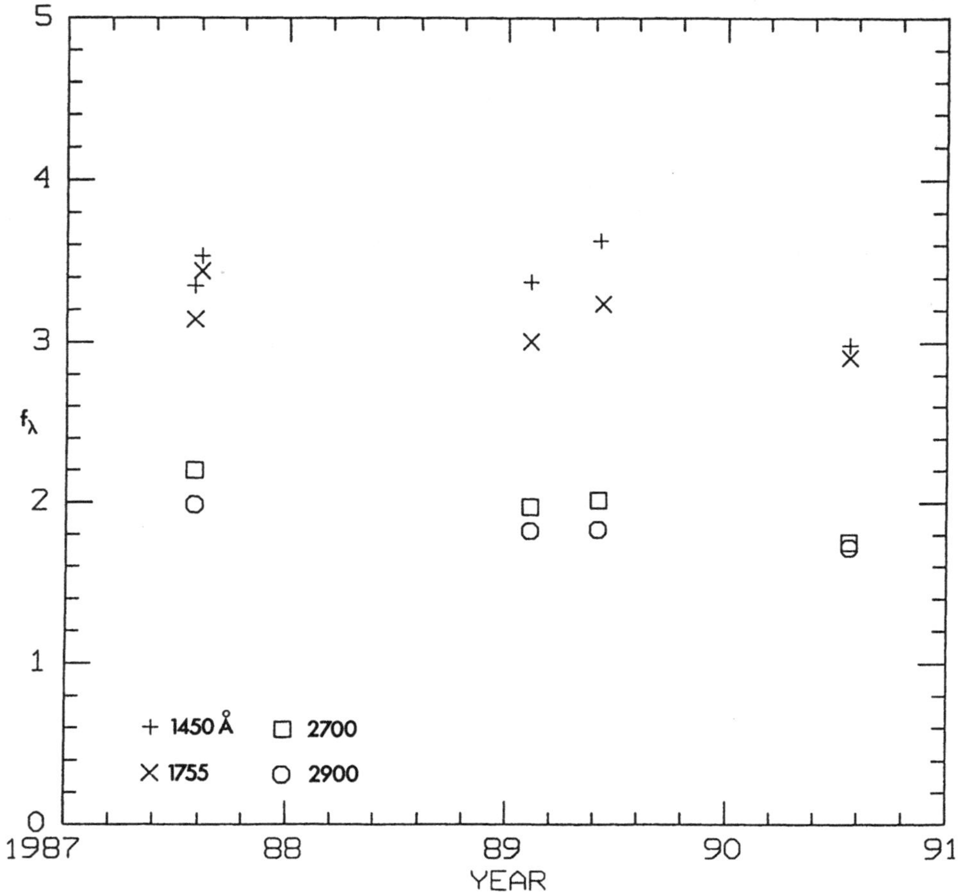

Fig. 1. The light curve of 1821+643 in 4 ultraviolet continuum bands. The difference observed between the fluxes on 29 July and 10 August 1989 is not significant (for details, see text). Ordinates in 10^{-14} erg cm^{-2} s^{-1} Å^{-1}.

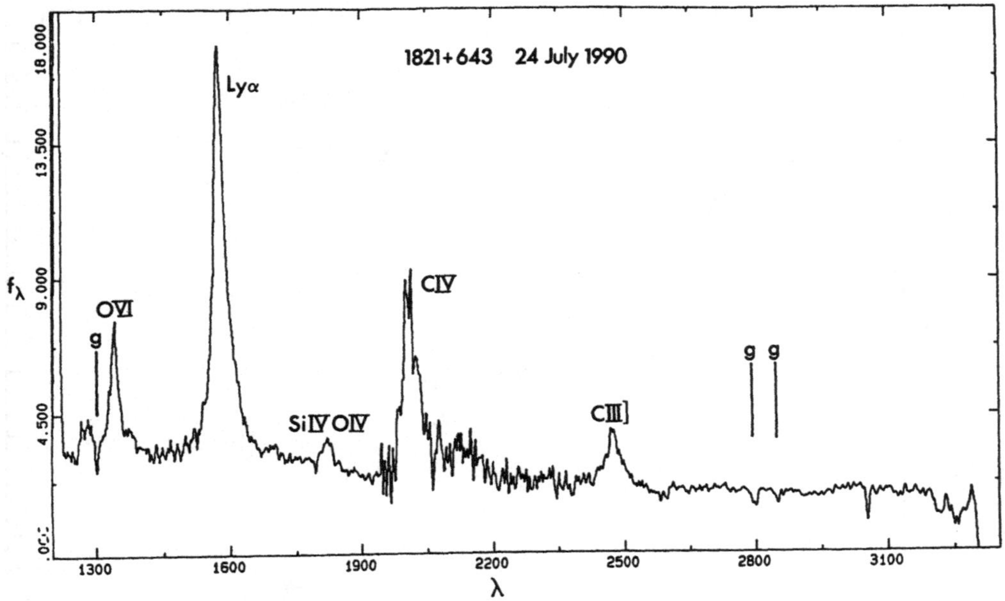

Fig. 2. The UV spectrum of 1821+643 on 24 July 1990. Ordinates in 10^{-14} erg cm^{-2} s^{-1} $Å^{-1}$.

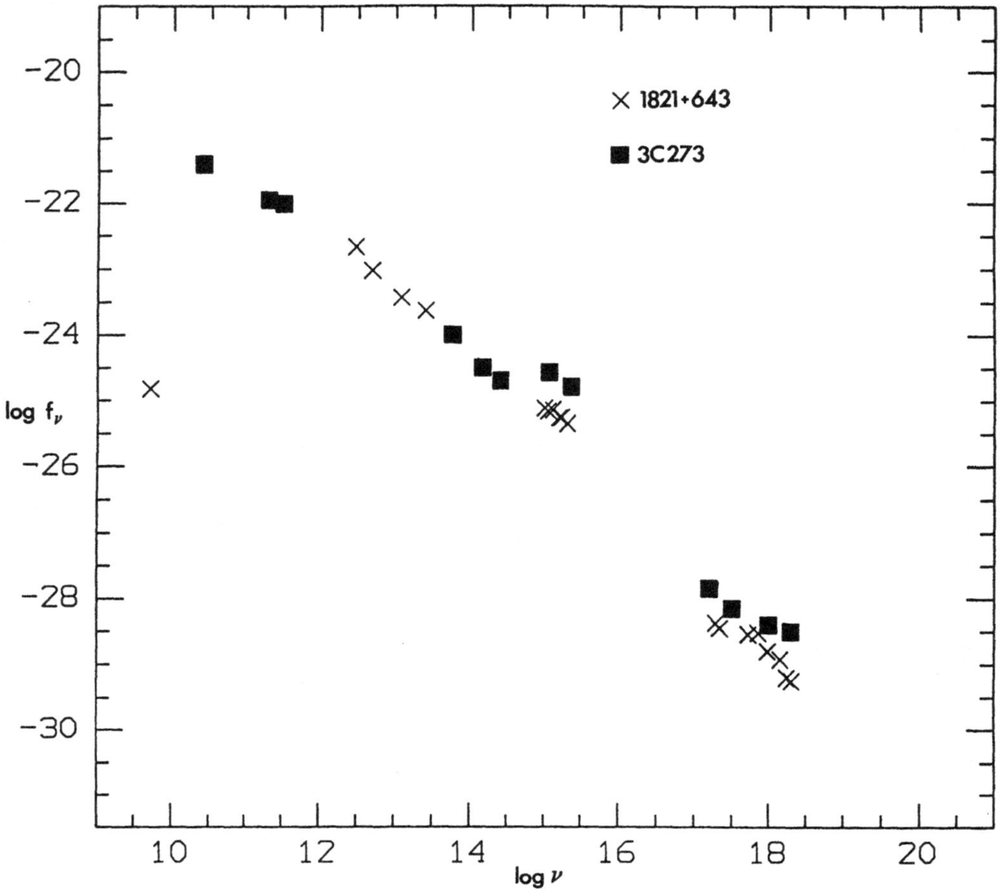

Fig. 3. The electromagnetic spectrum of 1821+643 and of 3C 273 in 1985. (The IUE data have been corrected for galactic absorption.) Ordinates in erg cm^{-2} s^{-1} Hz^{-1}.

Variability of Active Galactic Nuclei: a Theorist's View.

Marek A. Abramowicz

SISSA and ICTP, Strada Costiera, Trieste, Italy,
Nordita, Blegdamsvej 17, DK-2100 København Ø, Denmark

Observations indicate that the central engines of active galactic nuclei consist of an optically thick accretion disk orbiting a central black hole. Typically mass of the hole is in the range between $10^6\,M_\odot$ and $10^9\,M_\odot$, the accretion rate is between $10^{-2}\,\dot{M}_{Edd}$ and $10\,\dot{M}_{Edd}$, and the viscosity parameter α between 10^{-3} and 10^{-2}. Viscous stresses act against shear motion and release the gravitational energy of accreted material in the form of thermal radiation. Theory predicts that most of the radiation originates close to the central hole, in the range of radii,

$$3\,r_G \lesssim r \lesssim 10\,r_G, \tag{1}$$

and comes out as optical, UV and soft X-ray photons. Non-thermal processes in optically thin corona above the disk (or in a jet) may produce hard X-ray and γ-ray photons.

The most important timescales in the region (1) are the orbital (dynamical) timescale, $t_d \sim 1/\Omega$, where Ω is the local angular velocity of rotation,

$$\Omega \approx 7 \times 10^{-3}\,\frac{10^7\,M_\odot}{M}\left(\frac{r_G}{r}\right)^{3/2}\ \mathrm{Hz}, \tag{2}$$

and the thermal timescale $t_{th} \sim t_d/\alpha$. For the typical values of the accretion disk parameters mentioned above,

$$10^{-3}\,\mathrm{Hz} \lesssim t_d \lesssim 10^{-5}\,\mathrm{Hz}, \quad 10^{-5}\,\mathrm{Hz} \lesssim t_{th} \lesssim 10^{-7}\,\mathrm{Hz}. \tag{3}$$

These can be roughly compared with the observational data shown in Figure 1 which schematically represents a power spectrum, $P(\omega)$, of X-ray variability of an active nucleus. The variability power spectrum measures the strength of the X-ray time-dependent signal, $u(t)$, at a particular variability frequency ω,

$$P(\omega) = \lim_{\delta\omega \to 0} \frac{\langle u^2(t,\omega,\delta\omega)\rangle}{\delta\omega}. \tag{4}$$

In the frequency range corresponding to dynamical timescales, which is well covered by observational data, the variability power spectra of all observed active galactic nuclei are featureless and can be well approximated by a power law,

$$P(\omega) \sim \omega^{-\beta}, \tag{5}$$

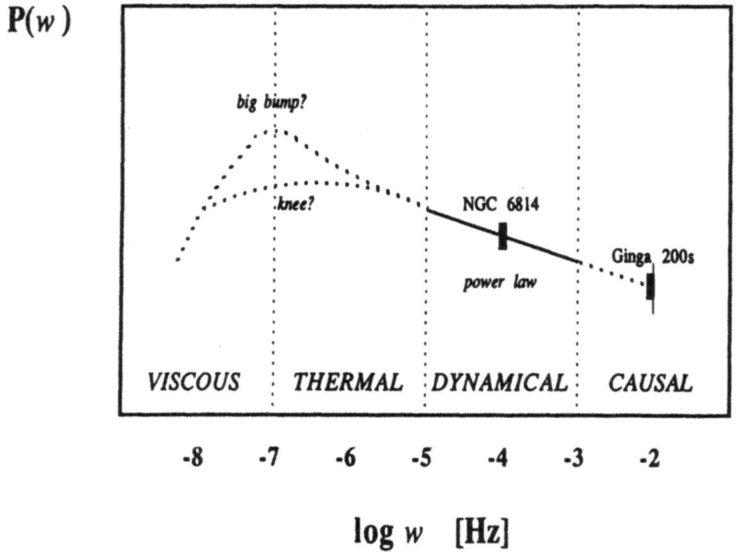

Fig. 1. Schematic summary of the observational data on the variability power spectrum of active galactic nuclei. The data was discussed in several contributions to this Meeting.

with $1 \lesssim \beta \lesssim 2$. The only exception is a spectral peak, corresponding most probably to an orbital frequency, observed in NGC 6814. In the frequency range $\omega \lesssim 10^{-7}$ Hz, corresponding to thermal timescales, it was not yet possible to collect a sufficient amount of the X-ray variability data. There is, however, some evidence for either a "knee" or a "big bump" at these frequencies. It was noticed by myself and my students E. Szuszkiewicz and F. Wallinder [1] that when a proper scaling factor is adopted, typical variability power spectra for active galactic nuclei are strikingly similar to those of some galactic X-ray sources, in particular the "big bump" corresponds to a similar feature due to quasi-periodic oscillations (QPO). Impressed by these similarities in the observational data we suggested that both the featureless variability pattern at dynamical frequencies and the possible quasi-periodic variability at thermal frequencies may reflect some *very basic* properties of accretion disks around compact objects and thus should be explained in terms of these basic properties. Our approach should be contrasted with several studies in which particular models have been proposed to explain particular observational facts. These models often postulate some *ad hoc* hypothesis, for example, that there is a large-scale magnetic field with a particular (assumed *ad hoc*) topology, that the disk is stably divided into two different fluids, one of electrons and the other one of ions, which do not recombine, that some unknown mechanism ejects relativistic particles with a particular (assumed *ad hoc*) energy distribution into an optically thin corona, that there is a network of colliding local shocks (with *ad hoc* assumed properties), *etc.* In my opinion these studies, although based on serious physics and solid calculations, do not really explain *why* active galactic nuclei have the observed properties, because they ignore the most important part of the story – the physics of the primary energy source. In this spirit I will address in my lecture the following two problems:

(a) All accretion disks relevant for active nuclei rotate in the region (1) with relativistic orbital velocities. Does this influence variability patterns with dynamical timescales?
(b) All accretion flows relevant for active nuclei are transonic close to region (1). I have shown [2] that transonic advection of heat is a very efficient global cooling mechanism, which stabilizes the otherwise thermally unstable innermost part of accretion disks, very close to region (1). Does the competition of local thermal instability and global advective stabilization produce a quasi-periodic variability with (roughly) thermal timescales?

My discussion of the first problem will be based on my work with G. Bao, A. Lanza and X.-H. Zhang [3, 4, 5], and that of the second problem will be based on the idea of "slim accretion disks" proposed by myself and J.-P. Lasota a few years ago [6]. The idea was then developed by both of us in collaboration with E. Szuszkiewicz and B. Czerny [7, 8] and later by myself in collaboration with the Kyoto group, S. Kato, R. Matsumoto and F. Honma, [9, 10]. Most recently time-dependent models of slim disks have been calculated by F. Honma, R. Matsumoto and S. Kato [11] and by J.-P. Lasota and D. Pelat [12] who used a different physical approach and obtained results different from those of the Kyoto group.

(a) The importance of relativistic rotation

It is unlikely that the innermost region of an accretion disk (1) is smooth and steady; rather one expects dissipation induces there many types of instabilities of a thermal, viscous or hydrodynamical nature, as well as shocks, magnetic flares, convective cells, lumps of locally denser matter, *etc.* Because the dissipative processes important for accretion physics are not well understood, there is not yet any theory describing these phenomena. However, it is sufficient for our present purposes to notice that the disipative processes will cause nonstationary, non-axially symmetric fluctuations of the disk surface brightness. These fluctuations can be described in terms of a time dependent distribution of "spots" with statistical properties given by:

$$\begin{pmatrix}\text{surface number density}\\ \text{of spots}\end{pmatrix} \equiv n, \qquad \begin{pmatrix}\text{characteristic size}\\ \text{of a single spot}\end{pmatrix} \equiv \delta r, \tag{6}$$

$$\begin{pmatrix}\text{characteristic lifetime}\\ \text{of a single spot}\end{pmatrix} \equiv \tau, \qquad \begin{pmatrix}\text{characteristic brightness}\\ \text{of a single spot}\end{pmatrix} \equiv I_0. \tag{7}$$

The quantities n, δr, τ, and I_0 are functions of location, time, and energy of the X-ray photons. In all of the models for variability considered so far, these functions have been approximated by radial scalings,

$$n \sim r^{1-\alpha_n}, \quad \delta r \sim r^{\alpha_r}, \quad \tau \sim r^{\alpha_\tau}, \quad I_0 \sim r^{\alpha_I}. \tag{8}$$

All of the proposed models for the observed short-term X-ray variability of AGNs postulate values of α_i, $i = n, r, \tau, I$, ***ad hoc*** either explicitly or implicitly. This is, of course, equivalent to making ***ad hoc*** assumptions about the characteristic time-scales and length-scales of variability and from these assumptions one can derive observational predictions of the models *directly*. Discussion of very detailed physics which some authors provide in this context only confuses the matter, because it is totally irrelevant: if one model assumes (say) variability due to local shocks, and another one assumes variability due to radiation transfer instabilities, but they both postulate ***ad hoc*** the same time-scales and length-scales, the variability pattern which these models predict will be exactly the same and it has nothing to do with the physics of shocks or radiation transfer.

Rotation of the disk is, of course, a common and basic feature for all of the models, but it has not been realized before that the rotation affects variability even when the orbital speed v is only mildly relativistic. If the disk is not seen exactly pole–on, *i.e.* if the viewing angle $i \neq 0^o$, relativistic rotation causes a periodic Doppler beaming of the signal from each individual spot, each time the spot moves towards the X-ray receiver. Thus, the single spot produces a single line in the variability power spectrum

$$p(\omega, \Omega, \tau, I_0) = I_0 v^2 \sin^2 i \left[\frac{1}{\tau(\omega - \Omega)^2} \left(\sin \frac{(\omega - \Omega)\tau}{2} \right)^2 + ... \right], \tag{9}$$

and the variability power spectrum of the whole ensemble of the spots is then given by

$$P(\omega) = \int p(\omega, \Omega(r), \tau(r), I_0(r)) n(r) dr \sim \sin^2 i \omega^{-\beta},$$

$$\beta = 1 + \frac{2\alpha_I + \alpha_n - \alpha_\tau - 4}{\alpha_\Omega}. \tag{10}$$

When the distribution of the statistical properties of the spots is uniform, ($\alpha_n = \alpha_r = \alpha_\tau = \alpha_I = 0$) and the rotation of the disk is Keplerian ($\alpha_\Omega = -\partial \ln \Omega / \partial \ln r = 2/3$), then $\beta = 5/3 = 1.7$, which is in the observational range.

Figure 2 shows X-ray theoretical light curves and variability power spectra for a thin Keplerian disk orbiting a Schwarzschild black hole of mass $10^7 M_\odot$. The model includes 100 very small spots (which contribute most of the luminosity) randomly distributed in region (1) and having a uniform distribution of statistical properties. The lifetime of each spot is much longer than the corresponding orbital period. The upper panels in Figure 2 correspond to a viewing angle $i = 20^o$, the lower panels to a viewing angle $i = 60^o$.

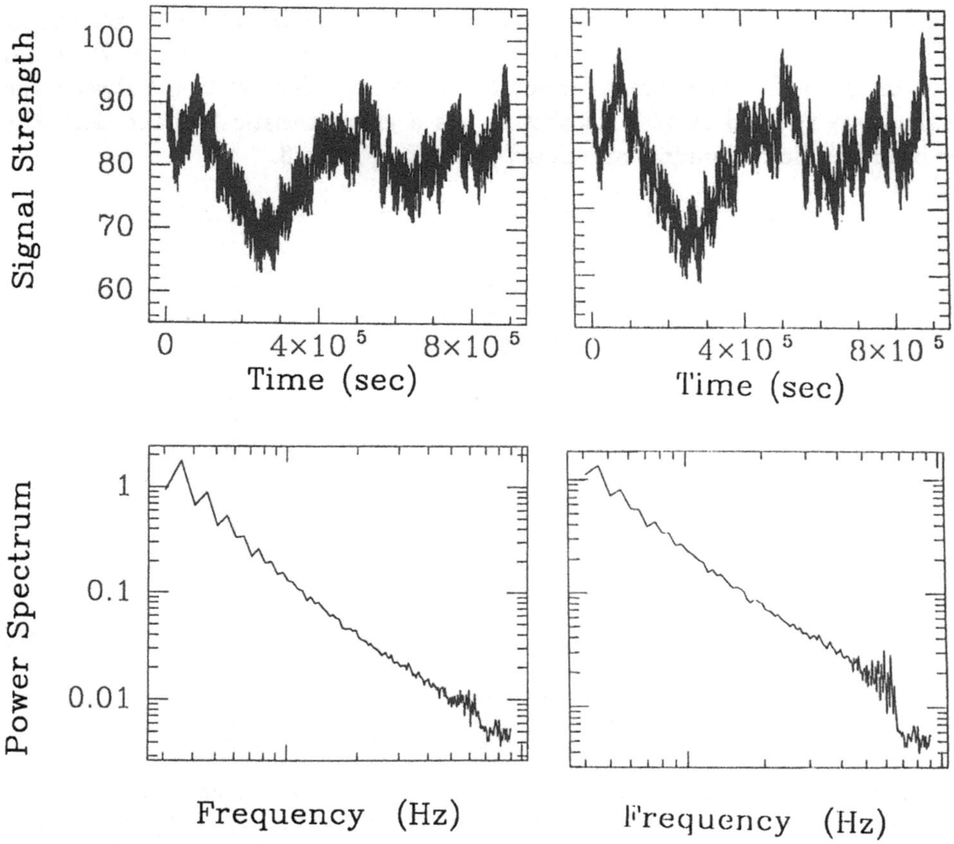

Fig. 2. Theoretical X-ray light curves and variability power spectra due to a distribution of small, long-lived spots on the surface of an accretion disk seen at different viewing angles (left panel $i = 20^o$, right panel $i = 60^o$) . It is clear that relativistic rotation strongly influences variability patterns.

(b) The importance of transonic advective cooling

Close to the central black hole the accretion flow is transonic and for this reason strong nonlocal cooling by advection is more important than radiative cooling. In addition the horizontal pressure gradient and the deviation from Keplerian flow are very important. These effects are ignored in the standard Shakura-Sunyaev model and its relativistic version due to Novikov and Thorne and this makes these well known models of accretion disks totally inadequate close to the black hole. *Slim accretion disk models*, on the other hand, include all of these effects [5 - 11].

I will now describe a sequence of slim accretion disk models at some fixed radius r. The mass of the central black hole M, the form of the opacity $\kappa = \kappa(T, \rho)$ and the viscosity coefficient α are fixed for the sequence. The relation between the accretion rate $\dot{M}$ and surface density Σ has a characteristic S-shape with the three branches (lower, middle an upper) shown in Figure 3.

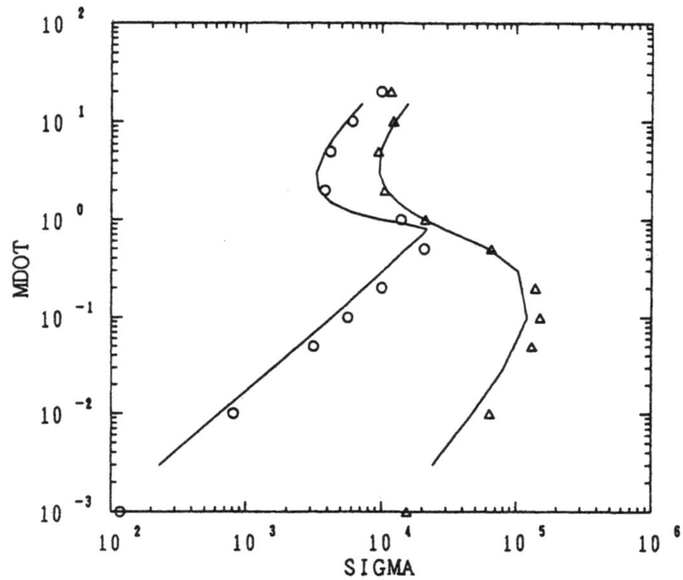

Fig. 3. Sequences for slim disk models for $r = 3\,r_G$ and $r = 5\,r_G$ and for $\alpha = 10^{-2}$, $M = 10\,M_\odot$ computed in Trieste [7] (triangles and circles) and in Kyoto [9] (solid lines). The small difference for low accretion rates is due to different opacities being used: in the Trieste models $\kappa(T, \rho)$ was taken from the Cox tables, in the Kyoto models $\kappa = \kappa_{es} =$ *const* was assumed.

On the *lower branch* the gas pressure P_{gas} is greater than the radiation pressure P_{rad} and the opacity is dominated by electron scattering. The cooling is provided by the vertical radiative flux. Accretion is stable against local thermal and viscous perturbations, as follow from the positive slope of the $\dot{M}(\Sigma)$ curve. On the *middle branch* the opacity and cooling mechanisms are the same as on the lower branch, but here $P_{gas} \ll P_{rad}$. Accretion is thermally and viscously unstable here reflecting the negative slope of the $\dot{M}(\Sigma)$ curve. The thermal instability is due to an insufficient dependence of the rate of radiative cooling Q^- on the vertical thickness of the flow, H. For radiative cooling $Q^- \sim H$, while for viscous heating $Q^+ \sim H^2$, and so when overheating occurs it causes expansion and the expansion then causes overheating so that a thermal runaway arises. On the *upper branch* the accretion flow cannot be described by the Shakura-Sunyaev model. Here $P_{gas} \ll P_{rad}$ as on the middle branch, but cooling is provided mainly by horizontal advection of heat and the relativistic Roche lobe overflow mechanism which I will now describe.

Close to the sonic point one of the equipotential surfaces (shown in Figure 4 by the solid lines) crosses itself. This self-crossing surface is called the Roche lobe and the place of crossing is called the cusp. If the surface of the disk overflows

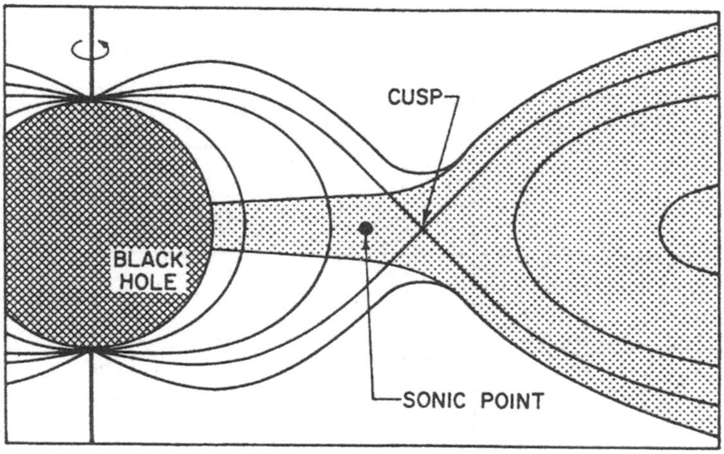

Fig. 4. Thermal stabilization of the innermost part of accretion disk due to relativistic Roche lobe overflow.

the Roche lobe, then hydrostatic equilibrium is not possible close to the cusp, and dynamical mass loss takes place. The matter flowing out of the disk through the cusp carries away heat and so the Roche lobe overflow is always connected with cooling. The cooling connected with the overflow is a global effect, while the advective cooling far away from the cusp is a local one. Far away from the cusp the advective heat transport may locally heat up or cool down the disk at some fixed location according to whether the local entropy gradient is positive or negative there. However, close enough to the cusp the advective flux always globally cools the whole region of the disk which is affected by the mass loss. The region of the disk cooled by the Roche lobe overflow is thermally stable because for the "Roche cooling" $Q^- \sim H^3$, which means that this cooling mechanism always dominates over viscous heating ($Q^+ \sim H^2$) during vertical expansion. For this reason models on the upper branch of the slim disk sequence are thermally stable.

On the upper branch, the slim disks have comparable pressure scale-heights in both the vertical and radial directions (I will denote these by h). When the disk expands due to overheating its density decreases as h^{-2}. If hydrostatic equilibrium is maintained, $dP/dz \sim \rho/h^2$, which means that pressure decreases as h^{-3} during expansion. Because the viscous heating rate $Q^+ \sim \alpha P h$, one concludes that $Q^+ \sim h^{-2}$ during expansion. The radiative cooling rate is approximately given by $Q^- \sim P_{rad}/\Sigma \sim h^{-2}$.

Thus, when only viscous heating and radiative cooling of the disk are considered $Q^+ \sim Q^- \sim h^{-2}$, *i.e.* the disk is in neutral thermal equilibrium. Let us now consider what happens in a more general situation when $Q^+ \sim h^{-a}$, $Q^- \sim h^{-b}$, where a and b are some positive numbers (perhaps both close to 2) and in addition $a < b$. Clearly, the last condition introduces thermal instability: the disk will

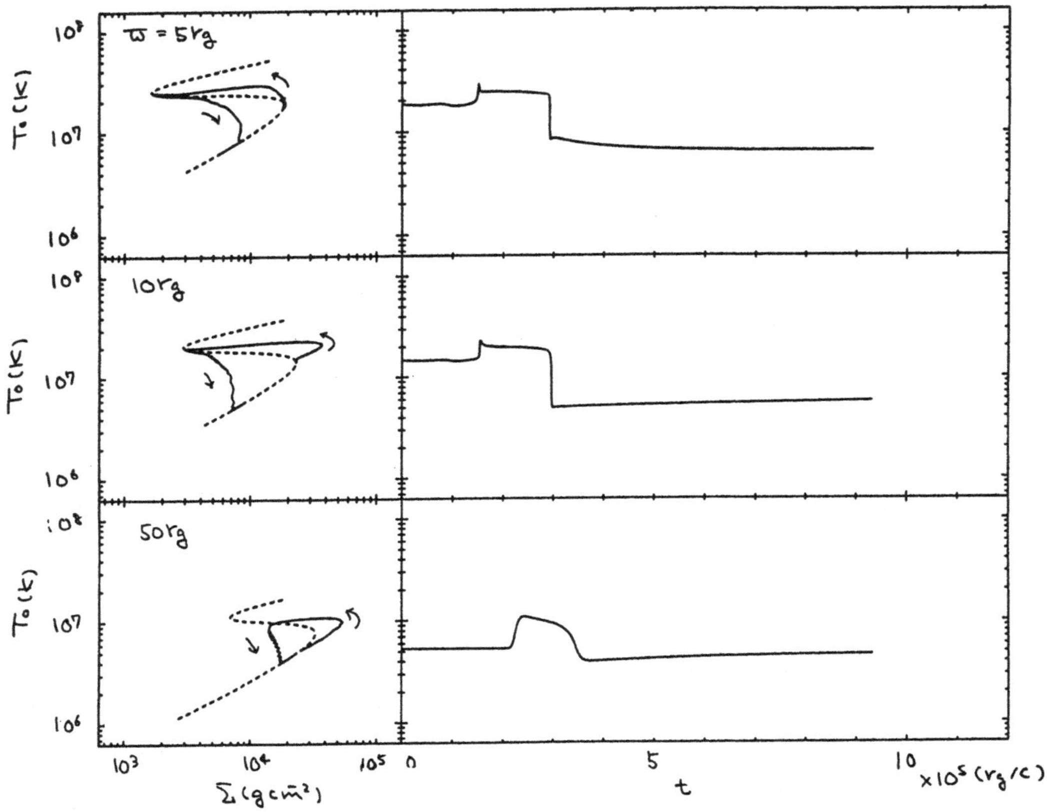

Fig. 5. Thermal limit cycles of slim accretion disks computed by the Kyoto group. The left panels show the disk evolution path on the $T - \Sigma$ plane, the right panels show the time evolution of T at three different radii. This Figure was taken from a recent, very important, work "Nonlinear Oscillation of Thermally Unstable Accretion Disks around a Relativistic Compact Object", by F. Honma, R. Matsumoto and S. Kato, which for the first time unequivocally proved the existence of the slim disk limit cycle postulated a few years ago by myself and J.-P. Lasota.

expand and heat up. However, when its expanded surface finally reaches the Roche lobe a new cooling mechanism turns on whose efficiency scales with the overflow as Δh^3 and therefore quickly dominates the viscous heating. Thus, the size of the disk will be almost exactly equal to that of the Roche lobe and it will be in a stable thermal state, as long as the mass supply from outside compensates the loss through the cusp. If it does not, then a stationary stable state is not possible and instead the disk will experience a *limit cycle*, expanding and contracting on a roughly thermal timescale. The expanded "hot" state corresponds to the upper branch of the S-curve, while the contracted "cool" state corresponds to the lower branch. (See Figure 5.)

This situation resembles (formally) the well-known limit cycle mechanism which gives quasi periodic dwarf nova outbursts. It works because: (1^0) the relation between the accretion rate and surface density is characteristically S-shaped (2^0) the lower and upper branches of the S-curve correspond to stable stationary equilibria, while the lower one corresponds to an unstable stationary equilibria, (3^0) the accretion rate fixed by the eternal conditions lies in the unstable range.

(c) Astrophysical conclusions

(1) Short term X-ray variability ($10^{-5}\text{Hz} \lesssim \omega \lesssim 10^{-3}\text{Hz}$) of Active Galactic Nuclei is strongly influenced by the relativistic rotation of the accretion disks.

(2) The normal branch quasi periodic oscillations with timescales $\omega \sim 1$ Hz observed in several low mass X-ray binaries, quasi periodic variability with timescales $\omega \sim 10^{-1}$ Hz observed in a few Galactic black hole candidates and variability with timescales $10^{-7}\,\text{Hz} \lesssim \omega \lesssim 10^{-6}\,\text{Hz}$ suspected in several Active Galactic Nuclei all have exactly the same physical nature: limit cycle behaviour driven by competition between thermal instability of the accretion flow and the stabilizing effect of global advective cooling close to the transonic part of the accretion flow.

References

1. M.A. Abramowicz, E. Szuszkiewicz, and F. Wallinder, in *Theory of Accretion Disks*, eds. F. Meyer, W.J. Duschl, J. Frank and E. Meyer-Hofmeister, Kulwer Academic Publishers, Dordrecht (1989).
2. M.A. Abramowicz,*Nature*, **294**, 235, (1981).
3. M.A. Abramowicz, G. Bao, A. Lanza and X.-H. Zhang, in *Proceedings of the 23rd ESLAB Symposium*, eds. J. Hunt and B. Battrick, ESA Publication Division, Noordwijk, (1989).
4. M.A. Abramowicz, G. Bao, A. Lanza and X.-H. Zhang, *Astrophys. J. Letters*, submitted.
5. Zhang X.-H. and G. Bao, *Astron. Astophys.*, in press, (1990).
6. Abramowicz M.A., Lasota J.-P. and Xu C., in *Quasars* eds. G. Swarup and V.K. Kapahi, D.Reidel Publishing Company, Dordrecht, (1986)
7. Abramowicz M.A., Czerny B., Lasota J.-P. and Szuszkiewicz E., *Astrophys. J.*, **332**, 646, (1988).
8. Szuszkiewicz E., *Month. Not. Roy. astr. Soc.*, **244**, 377, (1990).
9. Abramowicz M.A., Kato S. and Matsumoto R., *Publ. Astron. Soc, Japan*, **41**, 1215, (1989).
10. Honma F., Matsumoto R., Kato S. and Abramowicz M.A., *Publ. Astron. Soc, Japan*, submitted.

11. Honma F., Matsumoto R., Kato S., *Publ. Astron. Soc, Japan*, submitted.
12. Lasota J.-P. and Pelat D., a poster presented at the 6th IAP Meeting, Paris, July 1990.

The Stability of Thick Disks

Wolfgang Glatzel

Universitäts - Sternwarte Göttingen, Geismarlandstraße 11, 3400 Göttingen, Germany

Abstract: The properties of instabilities occurring in thick disks are reviewed. Their origin and mechanism is discussed and a summary of the influence of various model parameters and stabilising effects is given.

1. Introduction

Thick disks orbiting massive black holes are an essential ingredient in models for the central engine of active galactic nuclei. For small deviations from constant specific angular momentum distribution they exhibit steep funnels which makes them particularly attractive for the collimation and generation of jets. If there are considerable (specific) angular momentum gradients, however, the opening angle of the nozzle becomes wider and well collimated jets cannot be produced. The range of admissible rotation laws is determined by stability considerations. Centrifugal stability requires the specific angular momentum h to increase with radius (Rayleigh's criterion, $\frac{\partial h}{\partial r} \geq 0$). However, also centrifugally stable accretion tori, in particular those having constant specific angular momentum distribution have been found unstable on dynamical timescales when subject to nonaxisymmetric perturbations (Papaloizou & Pringle, 1984). Since this discovery many authors have studied various aspects of the instability (e.g. Drury, 1985, Goldreich & Narayan, 1985, Kato, 1987 and Papaloizou & Pringle, 1985, 1987). Apart from investigations concerning the origin and mechanism of the instability and its dependence on the parameters of the accretion torus their nonlinear evolution and final result has been of major interest. A crucial question with respect to the viability of thick disk models for AGN's is, if and how the models have to be changed in order to avoid instabilities. Studies in this direction have been made by Blaes (1987) and Glatzel (1989).

The present knowledge about the origin and mechanism of the instability will be reviewed in Sects. 2 and 3, its dependence on flow parameters is described in Sect. 4 and stabilising effects are discussed in Sect. 5.

2. The origin of the instability

Assuming a polytropic equation of state the origin of the instabilities occurring in accretion tori may be uncovered by considering the evolution of the energy $\tilde{E}$ of a perturbation defined as

$$\tilde{E} = \frac{1}{2}\bar{\rho}\tilde{\mathbf{v}}^2 + \frac{1}{2}\frac{\tilde{p}^2}{\bar{\rho}a_0^2} \tag{1}$$

$\tilde{\mathbf{v}}$ and $\tilde{p}$ denote the Eulerian perturbations of velocity and pressure, a_0 and $\bar{\rho}$ are the stationary sound speed and density stratifications. $\tilde{E}$ is the acoustic energy of sound waves and consists of two terms describing kinetic and compression energy respectively. Moreover, $\tilde{E}$ is a quadratic positive definite quantity in the perturbations and can be used as a norm to measure the deviation from the stationary state. Its evolution is derived from the linear perturbation equation as

$$\frac{\partial}{\partial t}\langle\tilde{E}\rangle = -\bar{\rho}r\frac{\partial\Omega}{\partial r}\langle\tilde{v}_r\tilde{v}_\varphi\rangle - \nabla(\langle\tilde{p}\tilde{\mathbf{v}}\rangle) \tag{2}$$

where $\langle\ \ \rangle$ denotes an angular average. The first term on the r.h.s. of (2) is associated with the Reynolds stress and represents the energy transfer from the stationary flow to the perturbation. The second term describes an acoustic energy flux and corresponds to the redistribution of perturbation energy by sound waves. Integrating (2) over the volume occupied by the fluid this term can be transformed into a surface integral and vanishes for appropriately chosen boundary conditions, in particular if a closed system is considered. Thus the time derivative of the integrated perturbation energy vanishes too for $\frac{\partial\Omega}{\partial r} = 0$, i.e. if there is no differential rotation. In other words, shear is necessary for an instability ($\frac{\partial}{\partial t}\langle\tilde{E}\rangle \neq 0$) and may therefore be regarded as its origin. Equation (2) has been derived for cylindrical configurations by Glatzel (1987a) and was generalized to threedimensional tori by Kojima (1989).

Except for the limit of a slender torus (Blaes, 1985) even the linear stability of a threedimensional accretion torus can be studied by a numerical analysis only (Frank & Robertson 1988, Kojima 1989) and requires a considerable amount of computing time on large memory supercomputers. Therefore the instability has been studied in simplified twodimensional models such as infinitely long cylinders or even plane shear layers. This idealisation is justified, since shear causes the instability and it is verified a posteriori that its behaviour is qualitatively identical in the simplified models and in threedimensional accretion tori. Moreover, the simplified models occasionally allow for analytical solutions of the perturbation problem and a deeper physical insight into the problem. In the following we shall discuss the properties and mechanism of the instability on the basis of a plane shear layer model.

3. The mechanism of the instability

A standard approach to the study of stability is a local analysis, where the coefficients of the perturbation equations are kept constant and boundary conditions are disregarded. The problem then allows for an analytical solution and leads to a dispersion relation describing the dependence of the frequency ω on the local wavenumber k. However, none of the instabilities found when solving the full set of equations together with appropriate boundary conditions shows up in a local analysis. Accordingly, a global analysis is necessary to study these instabilities and the differential equations describing the perturbation problem have to be integrated together with suitable boundary conditions.

Due to the failure of a local interpretation two complementary pictures have been developed to gain some insight in the mechanism of global shear instability occurring in accretion tori. Since one of them is thoroughly discussed by Narayan et al. (1987) and Narayan & Goodman (1989), we shall restrict ourselves in the following to a brief description of the alternative.

We consider a linear compressible shear layer (see also Glatzel, 1988), where a compressible medium with constant sound speed a_0 is confined between two infinite parallel planes moving with relative velocity $2V_0$ parallel to the planes. The velocity of the fluid is assumed to vary linearly in space and to match the velocities of the confining boundary planes.

For convenience we introduce a coordinate system where the x - axis is taken in flow direction, the z - axis perpendicular to the boundary planes and the origin is placed on the midplane between the boundaries. Lengths are measured in units of half of the thickness of the shear layer, z_0, velocities in units of the constant sound speed. The Mach number M - being the only dimensionless number characterising the flow - is defined as the ratio of flow velocity and sound speed at the edge of the shear layer: $M = \frac{V_0}{a_0}$. With these definitions the shear layer extends in z from $z = -1$ to $z = +1$ and the velocity may be written as $\mathbf{V} = (Mz, 0, 0)$. In the following the stability properties of the shear layer will be discussed as a function of the Mach number or, equivalently, in terms of the flow velocity at its edges, i.e. the shear strength.

The perturbations in the isentropic compressible configurations described have the physical properties of sound waves. In general the system allows for perturbations proportional to $\exp(ik(\omega t + x))$, i.e. for sound waves travelling in x - direction with wavenumber k and pattern speed ω. Perpendicular to the boundaries they have the character of standing sound waves and may be ordered according to n, their number of nodes in z.

For $M = 0$, i.e. for a medium at rest, the perturbation problem is easily solved analytically and the spectrum of these sonic modes is given by:

$$\omega = \pm\left[1 + \left(\frac{\pi}{2}\frac{n}{k}\right)^2\right]^{1/2} \quad ; \quad n = 0, 1, 2, \ldots \tag{3}$$

Due to the symmetries of the configuration we have - for given wavenumber k - a twofold infinite set of discrete neutrally stable sonic modes travelling in positive and negative x - direction respectively.

Considering now the case of finite shear strength we follow the evolution of the sonic spectrum for fixed wavenumber k when M is increased continuously from $M = 0$ to some finite value. For illustration the pattern speed ω for the neutrally stable sonic modes having $n \leq 12$ and $k = 1$ is plotted as a function of M in Fig. 1. Indicated by dashed lines is the range of the flow speed ($-M \leq V_x \leq M$), i.e. the dashed lines correspond to the extrema of the flow speed at the boundaries.

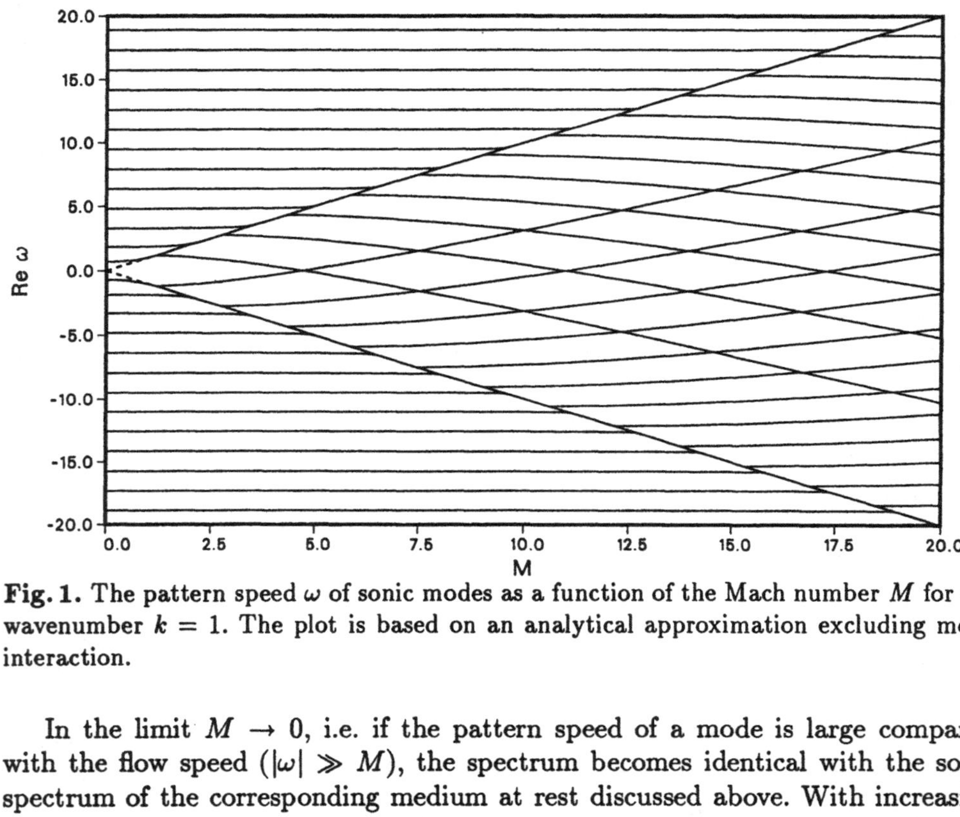

Fig. 1. The pattern speed ω of sonic modes as a function of the Mach number M for the wavenumber $k = 1$. The plot is based on an analytical approximation excluding mode interaction.

In the limit $M \rightarrow 0$, i.e. if the pattern speed of a mode is large compared with the flow speed ($|\omega| \gg M$), the spectrum becomes identical with the sonic spectrum of the corresponding medium at rest discussed above. With increasing M, i.e. increasing shear strength, the pattern speed of a mode is distorted if it gets comparable with the flow velocity, where the pattern speed of the mode having $n = 0$ is increased by the shear. Always remaining outside the flow velocity range ($|\omega| > M$) it approaches the flow velocity at the boundaries for large values of M. Contrary to the mode having $n = 0$ the pattern speed of all modes having $n \geq 1$ is decreased by the shear and enters the flow speed range, i.e. the mode gets a critical layer within the flow. For higher Mach numbers the pattern speed even reverses its sign and approaches the velocity of the boundary ($|\omega| < M$) moving in the direction opposite to the mode in the medium at rest. Thus multiple mode crossings are generated, where crossings involving the mode with $n = 0$ occur outside ($|\omega_{crossing}| > M$), all other mode crossings inside ($|\omega_{crossing}| < M$) the flow velocity range.

In order to clarify the situation and to enable a separation of different effects Fig. 1 is not based on the solution of the full problem but on an analytical approximation which does not allow for mode interaction. The discontinuity of modes with $n \geq 1$ appearing at $|\omega| \approx M$ is an artefact of the approximation in which different expansions of the exact dispersion relation are used for $|\omega| < M$ and $|\omega| > M$.

The counterpart of Fig.1 based on the exact dispersion relation is shown in Fig. 2. Apart from slight modifications due to the approximate treatment used for Fig. 1, Figs. 1 and 2 are identical except for the mode crossings. This is not surprising since mode interaction is essential around resonances or mode crossings and was artificially excluded in the approximation.

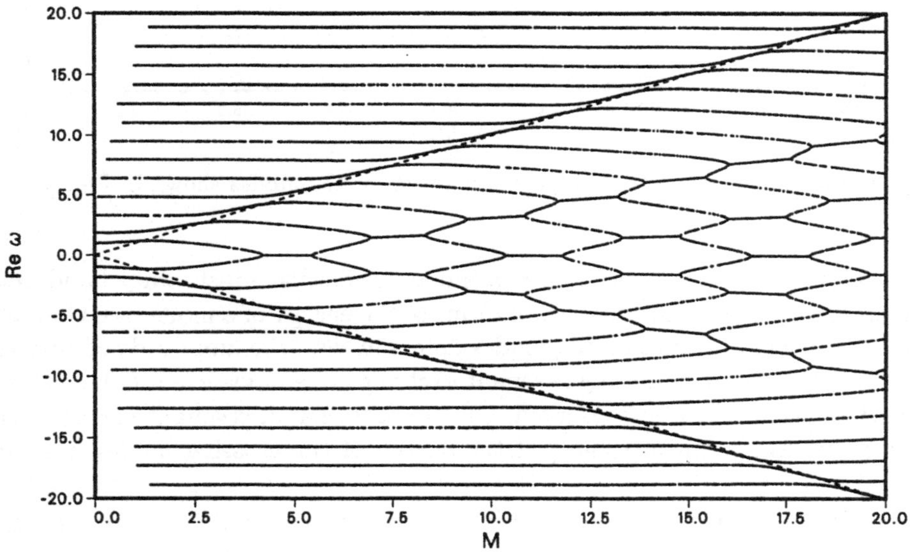

Fig. 2. Same as Fig. 1 but according to the exact dispersion relation. Note the unfolding of mode crossings into avoided crossings and instability bands.

According to the general theory (see, e.g. Glatzel, 1987b) mode crossings of neutrally stable modes A and B unfold - except for the very rare case of vanishing mode interaction - in two different ways. Either they remain neutrally stable and mode A before crossing is connected to mode B after crossing and vice versa. This case is called avoided crossing and occurs in our problem at all crossings involving the mode having $n = 0$ (see Figs. 1 and 2). The second possibility is the formation of a complex conjugate pair of modes - an instability band - around the original crossing by merging of modes A and B close to the crossing. Its pattern speed is the mean of the uncoupled crossing modes. This case is realised here at all crossings where the pattern speed lies within the flow speed range (see Figs. 1, 2 and 3). (The imaginary part of the pattern speed of the modes shown in Fig. 2 is plotted in Fig. 3.)

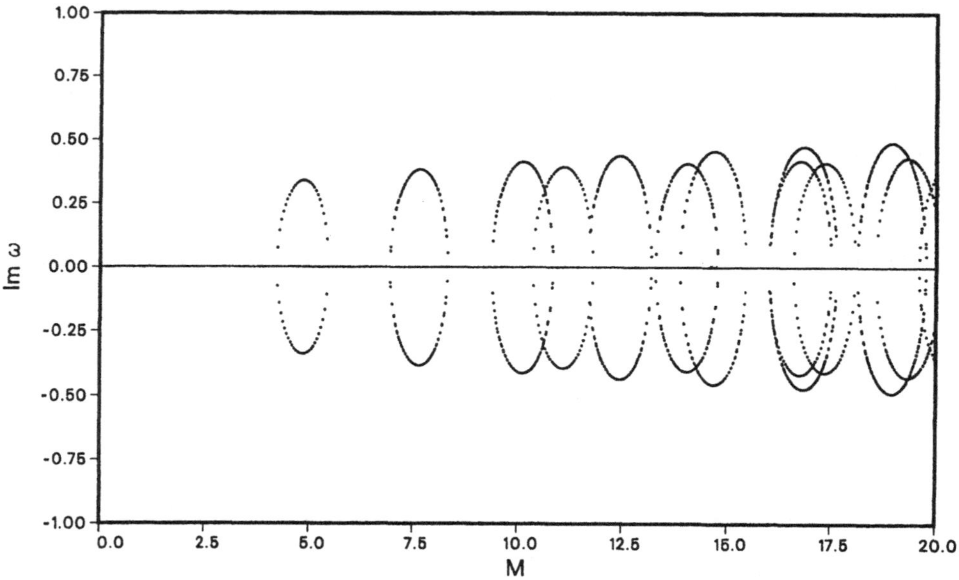

Fig. 3. The imaginary part of the pattern speed for the modes shown in Fig. 2.

Physically the two different results of mode interaction - instability band and avoided crossing - may be interpreted in terms of the energy of a mode (see Glatzel, 1987b). If the energy of the crossing modes differs in sign, the amplitude of both of them can grow without bound by resonant exchange of energy, even if the energy of the entire system is to be conserved. On the other hand, resonant exchange of energy cannot lead to an instability if the energy of the crossing modes has the same sign.

We emphasize that the energy used in this section must not be confused with the positive definite gauge invariant energy defined in Sect. 2. Here the "energy" of a mode is defined as energy difference of the shear flow with the mode superimposed and the stationary configuration, which is not a scalar gauge invariant quantity. Even its sign depends on the frame of reference chosen. Hence this "energy" is also referred to as pseudoenergy.

The existence of negative energy modes also provides a physical explanation for the instability of sonic modes when energy loss by acoustic radiation is considered: The amplitude of a negative energy mode grows by loss of energy. However, instability by energy loss due to acoustic radiation does not necessarily imply instability by energy loss in general as claimed by several authors. For instance, energy loss by viscous dissipation does not lead to an instability. (Viscous instabilities occurring in the flows considered here have a different physical origin (see Glatzel, 1989).) The reason for this apparent contradiction is that we are dealing with a pseudoenergy and different pseudoenergies, in particular different signs may be appropriate in different physical situations. In the case of energy loss by acoustic radiation a sonic mode is correctly characterised by a negative, for energy loss by viscous dissipation by a positive pseudoenergy.

Qualitatively the mechanism for shear instability presented here is not restricted to sonic modes. A similar behaviour is also found for surface waves (see, e.g. Blaes & Glatzel, 1986 or Goldreich et al., 1986), internal gravity modes (Ghosh & Abramowicz, 1991 and Glatzel, 1990) as well as for fast and slow magnetosonic modes (Glatzel, 1991). Whenever the pattern speed of a mode becomes comparable to the flow speed it is distorted by the shear. Thus multiple mode crossings are generated which unfold either into avoided crossings or instability bands. Independent of the physical nature of a mode the properties of mode crossings and resonance instabilites can be understood physically in terms of the energy of a mode. However, the direction of distortion of the pattern speed of a mode by the shear depends on the particular mode considered and only an appropriate distortion pattern generates the mode crossings necessary for resonance instabilities to occur. For an understanding of the shear instabilities it is therefore not sufficient to explain the properties of mode crossings. At least equally important is, how and why shear distorts the pattern speed of a mode. Since this question has been entirely ignored so far, a physical explanation is still missing.

4. The influence of model parameters

Apart from the investigation of the mechanism of the instability its dependence on various model parameters has been of major interest.

One of the most important parameters of accretion tori - in particular with respect to their ability to collimate jets - is the rotation law. Since shear is the origin of the instability most of them are not strongly affected by different angular momentum distributions (see, e.g. Hanawa 1987, 1988). Only the growth rate of the surface wave instability band sensitively depends on angular momentum gradients and is stabilised for rotation laws $\Omega \sim r^{-q}$ having $q < \sqrt{3}$ (Glatzel, 1987a, Goldreich, Goodman & Narayan, 1986, Papaloizou & Pringle, 1985). As shown by Sekiya & Miyama (1988) this result does not strictly hold for any configuration. Although the growth rate of this particular instability band is strongly reduced for $q < \sqrt{3}$, it does not vanish for $q > 3/2$ (Keplerian rotation) in an incompressible system.

The second parameter - also important for the collimation of jets - is the thickness of the torus. Its influence on the growth rate of the instabilities has been investigated for a twodimensional cylindrical configuration by Glatzel (1987a). As the outer boundary of a cylindrical shell tends to infinity the instability bands merge to form continuously unstable modes. This continuous instability is equivalent to the instability due to energy loss by acoustic radiation discussed in Sect. 3. In this sense the transition from a finite torus to an infinite thick disk causes a transition in the character of the instabilities from resonant to radiation type. The radiation type instabilities have somewhat lower growth rates than their resonant counterparts but do not vanish for infinite twodimensional systems. Threedimensional wide systems have been studied by Frank & Robertson (1988). Their results seem to indicate that the maximum growth rate for a threedimensional wide configuration is lower than for a twodimensional one. However, their calculations do not allow for a conclusion concerning infinitely thick disks.

Entropy gradients have been considered under various aspects. With respect to sonic and surface wave modes their influence has been found to be weak (Frank & Robertson, 1988, Kojima et al., 1989, Kubotani et al., 1990). On the other hand, entropy gradients together with gravity introduce the spectrum of internal gravity modes which provide an additional source of shear instability through mode crossings and resonances among the gravity modes (Ghosh & Abramowicz, 1991, Glatzel, 1990). According to Sect. 3 the distortion of the pattern speed of a particular mode, i.e. also the instabilities associated with it, occur typically at flow velocities which are of the same order as the pattern speed of the mode considered in the corresponding medium at rest. As a consequence, sonic instabilities occur in supersonic shear flows (such as accretion tori), whereas g - mode instabilities are found for low shear rates. Since we are dealing with high shear rates in accretion tori, this means that only low order g - mode resonance instabilities may be relevant here.

Similar to the g - mode spectrum caused by entropy gradients the spectra of fast and slow magnetosonic modes are generated by magnetic fields. Apart from mathematical subtleties the reaction of fast magnetosonic modes on a shear flow is the same as for sonic modes. Once the sound speed in the non magnetic case is identified with the sum of sound speed and Alfvèn speed in the magnetic shear layer, almost identical results are obtained (Glatzel, 1991). Preliminary results concerning the currently investigated slow magnetosonic modes seem to indicate a behaviour which reminds of g - modes.

5. Stabilising effects

Since standard accretion tori have been found to be violently unstable (Papaloizou & Pringle, 1984), the question has been asked, if this instability entirely rules out thick disks as essential parts in models for active galactic nuclei. Accordingly some studies have been made in order to figure out how the standard models have to be modified and which physical effects have to be taken into account in order to avoid or damp out shear instabilities.

In this context Blaes (1987) has studied the influence of accretion in a twodimensional cylindrical configuration whose outer boundary lies at infinity. A pseudo - Newtonian potential has been adopted in order to allow for relativistic effects and accretion onto the central object occurs through a sonic surface which determines the inner boundary of the system. The angular momentum distribution was assumed to be constant, since this is the only case where accretion and viscosity are decoupled and can be treated separately. In particular, viscosity needs not to be taken into account in the stability analysis which simplifies the problem considerably. Blaes found that even the most unstable modes are stabilised by rather small accretion rates. The stabilising effect of accretion is attributed to the existence of the sonic surface and interpreted by its property to absorb action. However, the significance of the twodimensional calculations for threedimensional models is not clear, since the sonic "line" in the threedimensional configuration is artificially extended to a sonic "surface" in the twodimensional model. Thus the stabilising influence of a sonic surface is certainly overestimated when extrapolated to a sonic line.

A stabilising effect closely related to accretion is viscous dissipation and it is straightforward to include it in a stability analysis. However, when shear viscosity is taken into account, the order of the differential equations to be integrated increases by a factor of 2 and yields an Orr - Sommerfeld type problem which causes considerable mathematical and numerical difficulties. As a first step in the direction of a viscous stability analysis of accretion tori the viscous stability of the plane linear compressible shear layer has been investigated (Glatzel, 1989). For low viscosities an additional viscous instability of the resonance type could be identified but all instabilities are damped for sufficiently high viscosities or sufficiently low Reynoldsnumbers respectively. (The Reynoldsnumber of the plane shear layer is defined as the ratio $Re = \frac{z_0 V_0}{\nu}$ where ν denotes a constant kinematic shear viscosity.) Critical Reynoldsnumbers below which the flow is stable with respect to a given instability have been calculated and are plotted in Fig. 4 as a function of the Mach number for the sonic and viscous resonance instabilities indicated. (Viscous instabilities are labelled (n, V).) We note that the critical Reynoldsnumber is defined and has to be determined separately for each resonance instability.

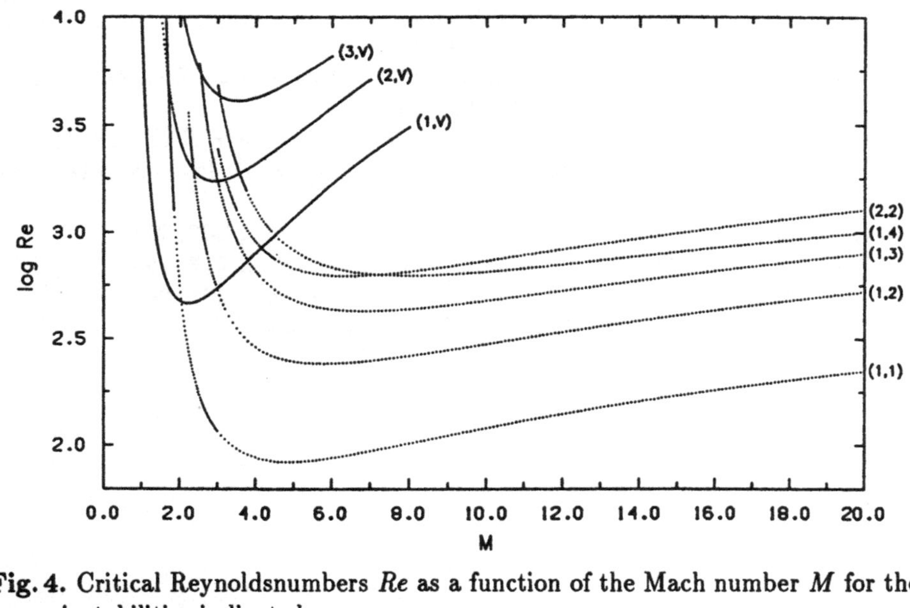

Fig. 4. Critical Reynoldsnumbers *Re* as a function of the Mach number *M* for the resonance instabilities indicated.

From Fig. 4 we deduce that the minimum critical Reynoldsnumber below which the flow is stable with respect to all instabilities depends on the Mach number *M* and is of the order of $Re \approx 10^2 \ldots 10^3$. This indicates that viscous dissipation can overcome violent shear instabilities and models for active galactic nuclei involving thick disks may still be viable. We emphasize, however, that this result is based on the analysis of a highly idealized shear flow and needs confirmation by a viscous stability analysis of less idealized models.

References

Blaes, O.M., 1985. *Mon. Not. R. astr. Soc.*, **216**, 553.
Blaes, O.M., 1987. *Mon. Not. R. astr. Soc.*, **227**, 975.
Blaes, O.M. & Glatzel, W., 1986. *Mon. Not. R. astr. Soc.*, **220**, 253.
Drury, L.O'C., 1985. *Mon. Not. R. astr. Soc.*, **217**, 821.
Frank, J. & Robertson, J.A., 1988. *Mon. Not. R. astr. Soc.*, **232**, 1.
Ghosh, P. & Abramowicz, M.A., 1991. *Astrophys. J.*, in press.
Glatzel, W., 1987a. *Mon. Not. R. astr. Soc.*, **225**, 227.
Glatzel, W., 1987b. *Mon. Not. R. astr. Soc.*, **228**, 77.
Glatzel, W., 1988. *Mon. Not. R. astr. Soc.*, **231**, 795.
Glatzel, W., 1989. *J. Fluid Mech.*, **202**, 515.
Glatzel, W., 1990. *Mon. Not. R. astr. Soc.*, **242**, 338.
Glatzel, W., 1991. in preparation.
Goldreich, P. & Narayan, R., 1985. *Mon. Not. R. astr. Soc.*, **213**, 7P.
Goldreich, P., Goodman, J. & Narayan, R., 1986. *Mon. Not. R. astr. Soc.*, **221**, 339.
Hanawa, T., 1987. *Astron. Astrophys.*, **185**, 160.
Hanawa, T., 1988. *Astron. Astrophys.*, **196**, 152.
Kato, S., 1987. *Publ. Astron. Soc. Japan*, **39**, 645.
Kojima, Y., 1989. *Mon. Not. R. astr. Soc.*, **236**, 589.
Kojima, Y., Miyama, S.M. & Kubotani, H., 1989. *Mon. Not. R. astr. Soc.*, **238**, 753.
Kubotani, H., Miyama, S.M. & Sekiya, M., 1990. preprint.
Narayan, R., Goldreich, P. & Goodman, J., 1987. *Mon. Not. R. astr. Soc.*, **228**, 1.
Narayan, R. & Goodman, J., 1989. in *Theory of Accretion Disks*, Proc. NATO ARW, eds. Meyer, F. et al., Kluwer, Dordrecht.
Papaloizou, J.C.B. & Pringle, J.E., 1984. *Mon. Not. R. astr. Soc.*, **208**, 721.
Papaloizou, J.C.B. & Pringle, J.E., 1985. *Mon. Not. R. astr. Soc.*, **213**, 799.
Papaloizou, J.C.B. & Pringle, J.E., 1987. *Mon. Not. R. astr. Soc.*, **225**, 267.
Sekiya, M. & Miyama, S.M., 1988. *Mon. Not. R. astr. Soc.*, **234**, 107.

Brightness and Color Variations of Accretion Disks: Implications for the Parameters

Rainer Wehrse and Herbert Störzer

Institut für Theoretische Astrophysik, Im Neuenheimer Feld 561, 6900 Heidelberg, Germany

Abstract: For the interpretation of long-term continuum variations of AGNs detailed non-grey models for stationary accretion disks are used to estimate the brightness and color variations as functions of the accretion rate. Possible consequences of multiple solutions of the structure equations are briefly discussed.

1. Introduction

Although no definite proof has yet been presented there seems to be general agreement that in the wavelength range from about 300 Å to a few microns most of the photons emerging from active galactic nuclei (AGNs) are produced in accretion disks surrounding massive black holes (c.f. e.g. Ulrich, 1989). Since the physical mechanisms for the various modes of observed AGN variability are not yet well understood it seems worthwhile to investigate in which way and to which extent the disks may be responsible for the brightness and spectral changes. In this paper we want to use stationary accretion disk models which have been calculated with the emphasis on realistic vertical structures and radiative transfer to study variations which occur as a consequence of changes in the mass accretion rate. It is evident that these investigations can only refer to time-scales of about ten years or longer since for shorter periods the disks cannot reach a steady state. On the other hand, it is known for disks in cataclysmic systems (Adam et al., 1988) and it is expected also for AGN disks that in the optically thin parts multiple solutions of the system equations are possible and the matter there may flip from one state to the other giving rise to spectral variations. Although such variations should show up predominantly in the lines we restrict the discussion here to the continuum since reliable calculations of line profiles are not yet available.

In Section 2 we describe the model construction and the resulting disk structures. The consequences for the interpretation of brightness and color changes are discussed in Section 3.

2. Construction of model disks and resulting temperature, pressure, and radiation field structures

Since the assumptions used in the disk calculations and the details of the resulting structures are presented elsewhere (Shaviv and Wehrse, 1990 a,b) we give here only a brief description.

Our models are based on the following assumptions:

- stationarity;
- axial symmetry;
- Newtonian physics;
- energy generation according to α model of Shakura and Sunyaev (1973);
- Keplerian rotation;
- negligible radial gradients;
- gravity provided by the central object;
- local thermodynamic equilibrium;
- no magnetic fields.

The assumption of Newtonian physics can certainly be regarded as a first approximation only; however, since we are mainly interested here in differential effects and we are not dealing particularly with those parts of the disks that are closest to the black hole we expect that the data given here are reliable.

Due to these assumptions we have to solve simultaneously the following set of equations:

(i) the hydrostatic equation

$$\frac{d(P_g + P_r)}{dz} = -\frac{GM_* z}{r^3}\rho \tag{1}$$

with P_g = gas pressure, P_r = radiation pressure, z = vertical coordinate, r = radial coordinate, G = gravitational constant, M_* = mass of the central object, ρ = density. Since the radiation pressure itself is not needed we calculate directly

$$\frac{dP_r}{dz} = \frac{1}{c}\int \chi_\lambda F_\lambda d\lambda \tag{2}$$

(c = velocity of light, χ = extinction coefficient, λ = wavelength, F = radiative flux);

(ii) energy equation

$$\int \kappa_\lambda (B_\lambda - J_\lambda) d\lambda = E - divF_{\mathrm{conv}} \tag{3}$$

with κ = absorption coefficient, J = mean intensity of the radiation field, $E = qP_g$ = local energy generation rate, q = efficiency factor depending on α, F_{conv} = convective flux;

(iii) radiative transfer equation

$$\pm\frac{dI^\pm}{dz} = \pm\chi(S - I^\pm) \tag{4}$$

with $I^{\pm}$ = wavelength dependent specific intensities in the positive and negative z-direction, S = source function which comprises scattering and true absorption contributions. Note that this radiative transfer equation (as the slightly more general one dimensional equation $\mu dI/dz = \chi(I - S)$) makes only sense for the continuum, since for lines the velocity fields have to be taken into account and the therefore the azimuthal dependence of the radiation field cannot be neglected;

(iv) equation for the radial dependence of the effective temperature $T_{\rm eff}$

$$T_{\rm eff}(r) = \left(\frac{3GM_*\dot{M}}{8\pi r^3 \sigma} \left[1 - (\frac{r_*}{r})^{1/2} \right] \right)^{1/4} \tag{5}$$

with $\dot{M}$ = accretion rate, σ = Stefan's constant, r_* = effective radius of the central object. The equation implies that the orbital velocity should vanish at the central object. If the rotation of the central object can be estimated the equation should be modified accordingly.

(v) constraint equation for the determination of the local disk height $z_0 = z_0(r)$

$$q \int_0^{z_0} P_g dz - \sigma T_{\rm eff}^4 = 0; \tag{6}$$

(vi) expressions for the absorption and scattering coefficients, for the thermodynamical functions (as e.g. adiabatic gradients, specific heats) and for the equation of state.

The equations are subject to the following boundary conditions:

$$P_g(z_0) = 10^{-4}\ \text{dyn/cm}^2; \tag{7a}$$

$$I^-(z_0) = 0; \tag{7b}$$

$$I^+(0) = I^-(0). \tag{7c}$$

The numerical solution of these equations proceeds essentially in the same way as for disks in cataclysmic systems (see Shaviv and Wehrse, 1990a). However, due to the dominance of radiation pressure over gas pressure the requirements for the numerical solvers are much more demanding, in particular the integration of the hydrostatic equation (eq. 2) with the energy equation (eq. 3) is very delicate since the r.h.s. of eq.2 is usually a small difference of two large numbers and the system is highly non-linear so that the choice of a slightly too large step-width can result in a completely incorrect solution.

The resulting models of galactic disks are characterized by a very small ratio of heights to radial extensions (except in the very inner parts where additional effects have to be considered, see above) and very large optical depths. The most striking difference between these galactic disks and cataclysmic ones refers to the gas pressure distribution: Whereas P_g increases monotonically from z_0 to $z = 0$ in the cataclysmic case, in most cases of galactic disks it first increases fast with decreasing z, then stays essentially constant for quite some z-range and then decreases strongly to the central plane. This behaviour was first found by Adam

(1989) and is evidently caused by the strong radiation pressure and its non-linear interactions. The calculations imply that the ratio of gas pressure and radiation pressure cannot be parameterized in a simple way but it will be interesting to understand such pressure distributions by means of analytical approaches. The local energy generation rates and the temperature gradients are generally quite small due to low gravities but -as a consequence of the high optical depths- the temperatures at the symmetry plane are quite high so that in some cases nuclear reactions may start. The emergent spectra are usually quite smooth. In most cases we find an increase in the flux up to the Lyman limit (the Balmer jump and the higher discontinuities are very weak and may be in absorption or in emission) followed by a minimum after which the emergent radiation field gets stronger again up to the HeII jump. For still smaller wavelengths we find essentially no flux any more.

3. Relevance for variability studies

The models can be used to estimate the changes in the brightness of the continuum and in the colors that occur when the mass loss rates change. In Fig. 1 we give a corresponding diagram for λ = 1300Å and λ = 3000Å. It is seen that with increasing accretion rates the colors become blueer as a consequence of the fact that at all given radii the effective temperatures increase. Since this means that a higher and higher fraction of the emergent radiation occurs at smaller wavelengths the brightness at a fixed frequency increases less than linearly with the accretion rate. In the models used in the diagram we have assumed $\alpha = 1.0$ and $\dot{M} = 10^8 M_\odot$ but tests showed that a reduction of the α parameter to 0.5 gave only minute changes and a mass accretion rate to $\dot{M} = 10^7 M_\odot$ hardly modified the diagram.

Due to the rather small velocity of sound in these objects the varaiations above can only occur on time scales of about ten years or more, as mentioned above. However, in several disk models for cataclysmic systems we found multiple and non-continuous solutions (Adam et al., 1988) which may lead to strongly inhomogeneous temperature distributions in the optically thin parts of the disks. Since the flipping from one state to the other should occur due to local disturbances the time scales for them may be very short. Although the local energy generation rates in AGN disks are usually smaller than in cataclysmic ones similar effects should also occur in disks around supermassive objects. They would mainly show up in the line profiles as increases or decreases in the emission components but might also be noticeable in regions of high continuous opacity. More specific statements however, have to wait for detailed hydrodynamic calculations.

Acknowledgements: This research was supported by a grant from the G.I.F., the German-Israeli Foundation for Scientific Research and Development (Contract I-94-142.7/88) and by the Deutsche Forschungsgemeinschaft (SFB 328).

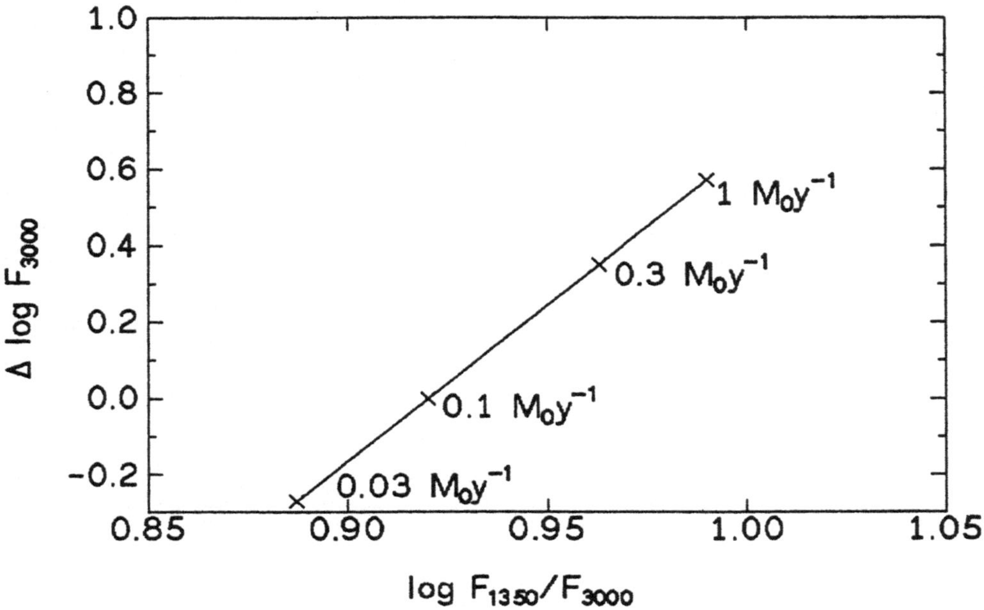

Fig. 1. The change of the flux at 3000 Å and of the 1350/3000 Å color of an accretion disk with the indicated accretion rates, a central object of $10^8 M_\odot$, and an α parameter of 1.

References

Adam, J., Störzer, H., Shaviv, G., Wehrse, R.: Astron. Astrophys. **193**, L1 (1988)

Adam, J.: PhD thesis, Univ. Heidelberg (1989)

Shakura, N.I., Sunyaev, R.A.: Astron. Astrophys. **24** 337 (1973)

Ulrich, M.H.: in: Theory of Accretion Disks, Meyer, F., Duschl, W.J., Frank, J., Meyer-Hofmeister, E., eds., Kluwer Publ. Comp. (1989)

Shaviv, G., Wehrse, R.: Astron. Astrophys., in press (1990a)

Shaviv, G., Wehrse, R.,: in preparation (1990b)

Structure and Variability in Broad Absorption Lines of Quasar Spectra

V. L. Afanasjev [1], E. A. Nazarov [1], Hilmar Lorenz [2]

[1]Special Astrophysical Observatory, Zelenchukskaja, USSR
[2]Sternwarte Babelsberg, Potsdam-Babelsberg, Germany

Abstract: We present spectroscopic observations of two Broad Absorption Line (BAL) QSOs 0752 + 617 and 0932 + 501, representing different classes in the complexity of structures in the absorption troughs. Our aim is the determination of column densities and the analysis of the absorption line fine structures. Applying a Gauss-analysis of the absorption line profiles we show that the troughs may be interpreted as a set of several individual components. The column densities of absorbing atoms in these proposed clouds are estimated. Physical conditions in the clouds are likely to change monotoniously depending on their velocity. We compare spectra of three different seasons and find no apparent changes in the C IV and Si IV absorbtion troughs.

1. Introduction

BAL QSOs show a substantial amount of resonance line absorption blueshift with respect to the emission redshift. Turnshek (1986) divided the spectra of BAL QSO into two typical classes. The first class exhibits smooth P-Cygni-like profiles, where the other class has several complexes of individual absorption lines and only slight evidence of a general smooth structure. We selected for the present study two objects with significant differences in the absorption troughs which represent extreme cases in the above mentioned sequence. Both quasars were discovered on Tautenburg objective prism AGN survey plates.

A still small amount of data in the literature suggest that BAL QSOs are optically variable. An essential point for the understanding of the geometry of the absorption region is a possible variability in the profiles of the broad absorption lines, due to bulk motion of the absorbing material. The long baseline of our observations, taking into account the data by Notni et al. (1979), allows us to set limits on the variability of absorption features.

2. Observations

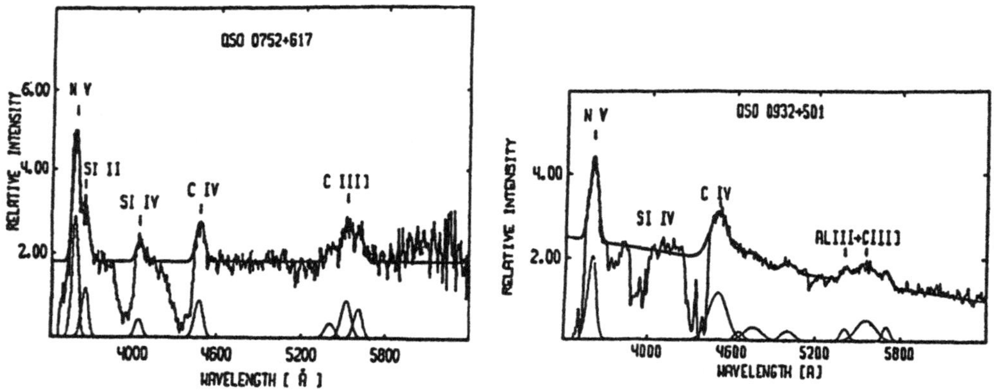

Fig. 1. Spectra of the objects 0752 + 617 (left) and 0932 + 501 (right). The spectra are smoothed over 3 channels, which corresponds to a spectral resolution of about 10 Å. Emission lines are represented by Gaussians (below). The sum of the continuum and the Gaussian profiles of the emission lines are plotted as the solid line in the figure.

The observations reported here were obtained in 1984 and 1986 using the Photon Counting Multichannal Spectrophotometer (Drabek et al., 1986) attached to the SP-124 Spectrograph at the Nasmyth focus of the 6-m telescope of the USSR Academy of Sciences. In its 1024 channal mode this SIT TV-type scanner has a typical resolution (FWHM) of 4 Å at a dispersion of 1.7 Å/channel. The observation and reduction procedure was standard as described by Afanasjev et al. (1986).

The spectra of the objects 0752 + 617 and 0932 + 501 obtained on March 15 1986, are plotted in Fig. 1.

3. Emission lines

The distribution of the BAL QSO emission line properties differ considerably from those of non-BAL QSOs (Hartig and Baldwin, 1986). Trunshek (1986) indicated a correlation between the structure of the BAL-profiles and the characteristics of the observed emission lines in the objects.

Table 1 contains the characteristics of the emission lines observed in both objects. The correlation between the absorption and emission line structures indicated by Trunshek (1986) is clearly visible. In 0932 + 501 the BAL-profile is less smooth and has a pronounced splited up absorption structure. With respect to the second object which shows a smoother BAL-profile the peak emission line height to continuum ratio in 0932 + 501 is smaller and the emission line width broader.

Our two objects follow thus the general effects observed by different authors. Table 1 compares our results with that given by Junkarinen et al. (1987) for a larger sample.

Table 1. Emission line properties of the BAL QSOs 0932 + 501 and 0752 + 617

	FWHM	$I/I_{(CIV)}$			
QSO	C IV λ1549 [kms^{-1}]	N V λ1240 [Å]	Si II λ1264 [Å]	Al III λ1860 [Å]	C III] λ1908 [Å]
0932 + 501 0752 + 617	6000 2700	1.7 3.3	- 1.3	0.3 0.5	0.4 1.0
Junkarinen et al. (1987)	< 4330 > (2300 – 6100)	< 1.8 > (1.1 – 3.5)	< 0.5 > (0.2 – 1.1)	< 0.2 > -	< 0.7 > (0.4 – 1.1)

A important peculiarity of the emission spectrum of BAL-quasars is the presence of strong iron lines (Hartwig and Baldwin, 1986). Their intensity is correlated with the intensity of Al III line observed in both objects. Due to this, the emission line at $\lambda = 5005$ Åin 0932 + 501 may probably be connected with Fe II λ1716 Å, λ1724 Å, λ1726 Å from UV 37, UV 38 and UV 39, respectively.

4. Absorption lines

The broad absorption lines were analysed after considering the effects of the continuum and the emission lines. Because the redshifts of the two objects are in close proximity to each other, there arises three doublets: Si IV λ1393.76 Å, λ1402.77 Å, C IV λ1548.20 Å, λ1550.77 Å, Al III λ1854.72 Å, λ1862.79 Åand the absorption near λ3700 Å, identified as C II λ1334.53 Å. Analysing the profiles we used a Gauss-technique, directed by the rule that the number of components describing the observed absorption troughs should be minimal. The results are presented in Fig. 2 where the observed and the synthetic profiles of C IV and Si IV are plotted in the comoving coordinate system in the QSO reference frame. In order to give an impression how the individuell absorption systems match the profile we plotted them together in the C IV-line.

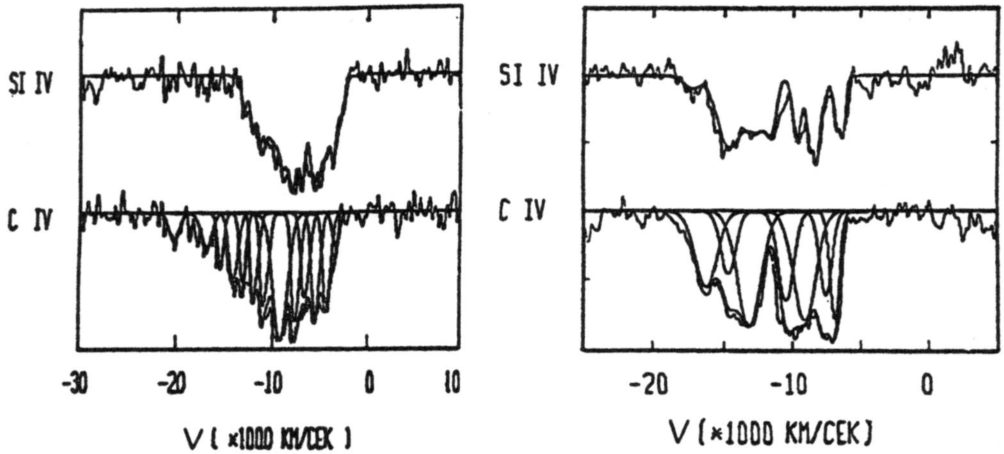

Fig. 2. Gauss-analysis of the Si IV and C IV BAL-profiles in the spectra of the quasars 0752 + 617 (left) and 0932 + 501 (right). The origin of the velocity coordinates is determined by the red wing of the C IV emission and corresponds to $z = 1.892$ (0752 + 617) and $z = 1.901$ (0932 + 501). The observed profiles are represented by a minimum of 13 in the case of the 0752 + 617 and 7 components for 0932 + 501.

Our spectral resolution allowed us to distinguish only between unresolved absorbers in the doublet. In all the cases we observed single absorption lines, which gives an estimation of the lower limit of FWHM $= 250\ kms^{-1}$ for separate absorption lines. Thus, only the doublet Si IV is reliably resolved. The general behaviour of the C IV and Si IV troughs is well described by a set of seven and fourteen Gaussians, respectively.

The results of Gauss-analysis were then used to calculate the column densities of absorbing atoms in the clouds. Using the procedure given by Afanasjev et al. (1989) the column densities N of the absorbing atoms of C IV, Si IV, Al III and C II are detemeined. We included into the discussion the results for the Mg II line, determined on the basis of a spectrum observed by Wampler (1983).

Comparing the absorption profiles of different observational seasons we found no variation larger than 5% in the intensity distribution of the C IV and Si IV absorbtion lines in both objects. Additional to that there are indications that the reported variations (Foltz and Turnshek, 1987) in the absorption profiles are due to continuum variations of the ionizing source. We will further investigate this effect in detail especially for the only positive detection in the above mentioned object QSO 1303+308. Thus we conclude, that up to now there is no direct indication in the variation of the BAL-profiles or variation in the velocity of individual absorbing components.

5. Discussion

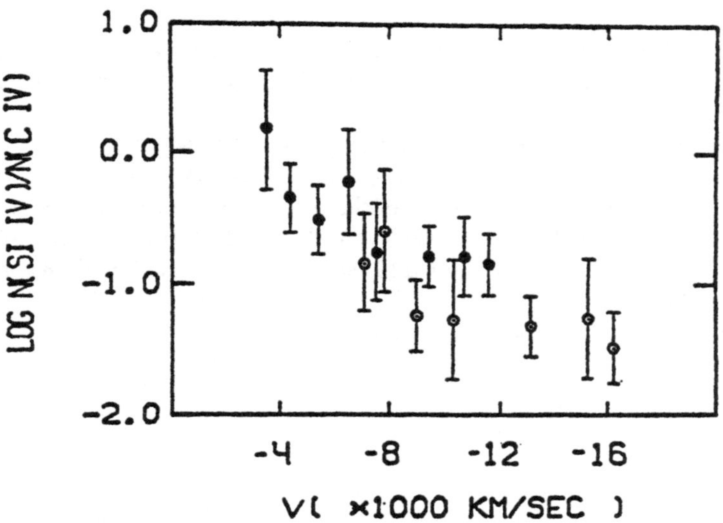

Fig. 3. Number of the observed Si IV column density in the individual clouds relative to the C IV column density as a function of the cloud velocity for both objects. • corresponds to 0932 + 501 and ∘ to 0752 + 617.

Analysing the data shown as a summary in Fig. 3 we found, that the Si IV density decreases with ingreasing cloud velocities, whereas Al III and Mg II are an order of magnitude larger than at lower velocities. This is supported by the fact, that in BAL-quasar spectra low-ionized absorption lines of Al III and Mg II are most often not observed, or observed as weak ones at high velocities (see Wampler, 1983 and Hartig and Baldwin, 1985).

Column density determinations for clouds in BAL-regions show levels of ionization that are difficult to understand using solar abandances and simple models for the ionisation. In particular from our observations there are indications for enhanced carbon abundances relative to solar values. The observation of iron emission may be interpreted as an enhanced abundance too.

Our observations of the emission line profiles and the residual intensities in the BAL troughs necessitates a small global covering factor. However, in order to have a reliable model for the acceleration of the BAL-clouds, a large local covering factor is required. The observed densities imply upper limits on the distances between the emission and absorption line regions. The constraint that the BAL-region lies beyond the broad Ly_α emitting region and in particular our observations which show no changes in the structure of the BAL-profile,as well as in the velocity of individual BAL components put a lower limit on the distance of the BAL-region to the central source of several pc.

References

Afanasjev, V. L., Nazarov, E. A. and Lorenz, H.: 1989, Pisma Astron. J., **15**, 195.
Drabek, S. V., Kopylov, I. M., Somov, N. N. and Somova, T. A.: 1986, Astrofiz. Issled., **22**, 64.
Folz, C. B. and Turnshek, D. A.: 1987, Astrophys. J., **317**, 341.
Hartig, V.P. and Baldwin, J.A.: 1986, Astrophys. J., **302**, 64.
Junkarinen, V. T., Burbidge, E. M. and Smith, H. E.: 1987, Astrophys. J., **317**, 460.
Notni, P., Karachentsev, I. D. and Afanasjev, V. L.: 1979, Astron. Nachr., **300**, 121.
Turnshek, D.A.: 1986, IAU Symposium 119, eds. G. Swarup and V.K. Kapahi, Reidel, Dordrecht, 317.
Wampler, E.J.: 1983, Astron. and Astrophys., **122**, 54.

Are the Broad Emission Lines of Quasars Affected by Gravitational Microlensing?

Joachim Wambsganss and Peter Schneider

Max-Planck-Institut für Physik und Astrophysik, Institut für Astrophysik, Karl-Schwarzschild-Str.1, 8046 Garching, Germany

Abstract

Gravitational microlensing by stars in galaxies can affect the continuum flux observed from quasars, but it is usually assumed that the size of the broad emission line region (BLR) is too large to be considerably magnified (or demagnified) by stellar-mass lenses. Although this is true, this does not mean that the radiation from the BLR is unaffected by microlensing. If the BLR has intrinsic structure (such as a large-scale velocity field), parts of it can be stronger magnified than others, leading to an influence on the resulting line profile. Whereas the changes of the line profile by microlensing will be too small to be detectable in a single quasar spectrum, a comparison of the same line in different images of a multiply imaged quasar can yield differences, which (with consideration of a possible difference in the light-travel time between these images) can be attributed to the action of microlensing in one (or more) of the images.

Using two simple models for the geometry and kinematics of the BLR (Keplerian rotation and gravitational infall), we study the resulting line profiles with and without microlensing[1]. Whereas our approach is quite general, most of the results presented here apply to image A of the quasar 2237+0305, in which microlensing of the continuum flux has recently been verified (Irwin *et al.*, 1989). The reason for the large interest in QSO2237+0305 is mainly due to the fact that the lensing galaxy has a very small redshift, which makes all geometric factors for lensing rather favourable.

Our main results can be summarized as follows: (1) There is an observable influence of microlensing on the broad emission lines of quasars, if the motion of the clouds in the BLR is not random or chaotic, but somehow ordered. (2) BLRs with large-scale rotation are much more sensitive to microlensing than infall

[1] This is a summary of the full article published in Astronomy and Astrophysics as (Schneider and Wambsganss, 1990)

models; in particular, even the (mean) redshift of the line can be changed in the case of rotation. (3) For all microlensing models of QSO2237+0305 A we have considered, there is an appreciable difference of the lensed and unlensed line profiles. Whereas this difference is larger for the case of Keplerian rotation, even for the infall model it is at a level which should be observable by sensitive spectroscopy of the individual images of this multiply imaged quasar. In the case of a rotating BLR, one might see redshift differences of a single broad emission line which amount to up to 200km/s, and fractional deviations of the lensed from the unlensed profile of typically 10%. In Fig.1, some example profiles are shown with the microlensing parameters of 2237+0305A for Keplerian rotation (top) and gravitational infall (bottom); ξ is a normalized measure of wavelength. In Fig.2, similar profiles are shown corresponding to some other microlensing parameters.

If such differences in line profiles of multiply imaged quasars are observed — and there are some indications for it (Filippenko 1989, Vanderriest 1990) — one may be able to put constraints on the geometry and kinematics of the BLR in those quasars; in particular, random or chaotic motion of the BLR could then be excluded. On the other hand, from a lack of such differences, one might be able to exclude, e.g., a model with large-scale rotation. Furthermore, detecting a significant difference in the broad emission line profiles of a multiply imaged quasar, which cannot be attributed to the time-delay between the images, may provide an alternative detection of microlensing, which does not depend on long-time monitoring of the source. Finally, our results yield a typical value for the variation in redshift (as measured from the broad emission lines) which can be tolerated in multiple quasars before they may be ruled out as lensing candidates.

References

Filippenko, A.V. (1989): "Evidence for Mg II λ2798 Profile Differences in the Gravitationally Lensed QSO 2237+0305A, B" Astrophys. J. 338, L49-L53

Irwin, M.J., Webster, R.L., Hewett, P.C., Corrigan, R.T., and Jedrzejewski, R.I. (1989): "Photometric Variations in the Q2237+0305 System: First Detection of a Microlensing Event" Astron. J. 98, 1989-1994

Vanderriest C. (1990): "Spectrophotometry of 0957+561 and the Microlensing Effect", in *Gravitational Lensing*, Proceedings, Toulouse, 1989, eds. Y. Mellier, B. Fort, and G. Soucail (Springer, Berlin), pp. 210-215

Schneider, P., and Wambsganss, J. (1990): "Are the Broad Emission Lines of Quasars Affected by Gravitational Microlensing?" Astron. Astrophys., in press

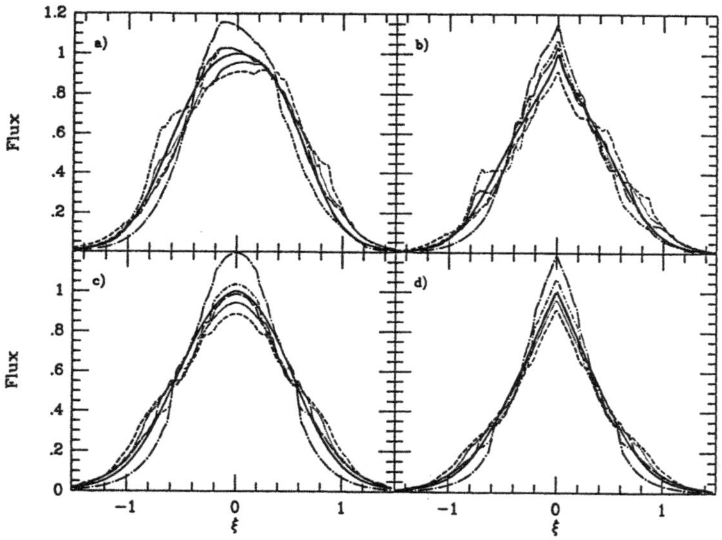

Fig. 1. Unlensed (solid lines) and microlensed profiles of broad emission lines as calculated according to the model of 2237+0305 described in the text. Each part of the figure contains five lensed profiles, according to five different mass functions for the microlensing stars. (a) and (b): BLR modeled by Keplerian rotation; (c) and (d): BLR modeled by gravitational infall. In all cases the lines are normalized for equal total flux.

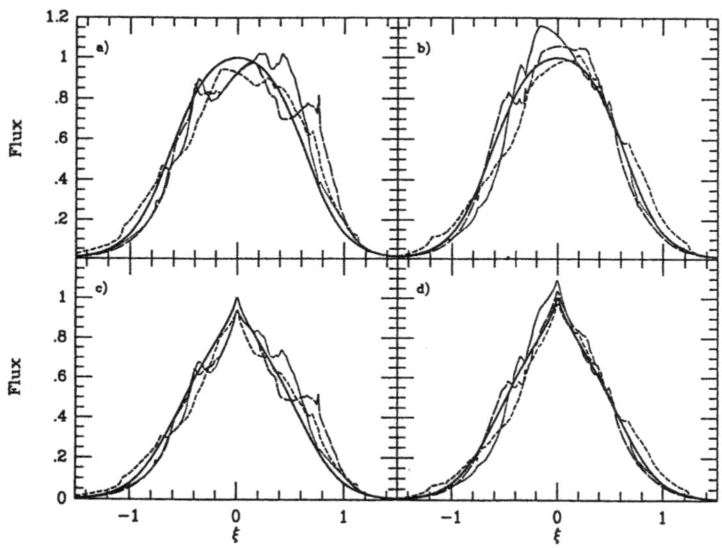

Fig. 2. For three microlensing models (low surface mass density without shear: long dashed; low surface mass density with shear: short dashed; high surface mass density with shear: dotted) lensed and unlensed profiles are shown for the case of Keplerian rotation. The physical parameters are the same for (a) and (b), and for (c) and (d).

Gravitational Microlensing and the Hamburg Quasar Monitoring Program

Ulf Borgeest*, Rainer Kayser, Sjur Refsdal, Jochen Schramm

and Thomas Schramm

Hamburger Sternwarte, Gojenbergsweg 112, 2050 Hamburg 80, Germany
* present address: Observatoire de Meudon, 5, Place Jules Janssen,
F-92195 Meudon Principal Cedex, France

Abstract: In May 1989 we started an extensive optical photometric monitoring program of quasars which are promising candidates for gravitational microlensing. We use the Calar Alto 1.2m telescope equipped with a CCD camera to obtain the quasar fluxes relative to stars included in the frames. The reduction is carried out immediately after the observation. Our main intention is to detect and analyse microlensing high amplification events. From the data obtained up to now, we find good evidence that quasars closely associated with foreground galaxies show a higher degree of variability than usual ones and interpret this result in terms of microlensing. There is, however, no evidence for a correlation between variability and the presence of strong heavy element absorption systems. In the first part of this contribution we give an overview on the basic microlensing theory and their applications to observed lightcurves. In the second part we discuss the monitoring program in some detail and present our first results.

1. Microlensing basics

In 1937 Zwicky proposed that multiple images of distant objects gravitationally lensed by foreground galaxies should exist (Zwicky 1937a,b). More than forty years later the first such object was indeed discovered: 0957+561A,B, the now classical *double quasar* (Walsh *et al.* 1979, Young *et al.* 1980).

This detection led to a substantial increase of theoretical and observational work on gravitational lensing, and some enthusiasm on using gravitational lenses as "the ultimate cosmological tool" appeared in the literature. Refsdal (1964a and b) was the first to point out that the time delay between lensed images can, in principle, be used to determine the Hubble parameter H_0 and the mass of the lens (see also Borgeest and Refsdal 1984, Borgeest 1986). Especially the determination of H_0 has later been criticized as being heavily dependent on the environment of the light path (Alcock and Anderson 1985, see also Falco *et al.* 1985).

It has been proposed that *microlensing* due to compact objects near the light path (Chang and Refsdal 1979) can be used to obtain information on the size and the structure of the energy source of quasars, as well as on the mass spectrum of compact objects in the lensing galaxy (Grieger *et al.* 1988, see also Canizares 1981).

Using the linearized theory of General Relativity (weak gravitational fields), as well as the usual thin lens, small deflection angle approximations, gravitational lensing is a mapping $\mathbf{D}\subseteq\mathbf{R}^2 \mapsto \mathbf{S}\subseteq\mathbf{R}^2$ of the deflector (or lens) plane onto the source plane, described by the *lens equation*

$$\zeta = \zeta(z) = \frac{D_s}{D_d}\, z + D_{ds}\gtrsim(z) \qquad , \tag{1}$$

where $\gtrsim$ is the deflection angle, given as the negative gradient of the twodimensional gravitational potential of the projected lens mass distribution, and D_d, D_s, and D_{ds} are the usual apparent size distances. The possibly multiple images of a point source located at ζ are observed at the locations given by the solutions z_I of Eq.(1), if they exist.

Since surface brightness is conserved under gravitational deflection of light, a source element dA_S is amplified by the ratio of dA_I and its image dA_S, which is given by the inverse Jacobian J^{-1} of Eq.(1). Obviously, a vanishing Jacobian leads to infinite amplification for point sources. Therefore $J(z) = 0$ defines a *critical curve* in the deflector plane . As in geometrical optics, the projection of the critical curve to the source plane by Eq.(1) is called *caustic*.

An appropriate way to describe the microlensing effect due to compact objects near the light path of an image I is the *normalized lens equation*, developed by Paczynski (1986), Kayser, Refsdal and Stabell (1986) and Kayser *et al.* (1989):

$$\zeta(z) = \begin{pmatrix} 1+\gamma & 0 \\ 0 & 1-\gamma \end{pmatrix} z + \mathrm{sign}(\sigma) \sum_i \frac{M_i}{M_0} \frac{z_i - z}{|z_i - z|^2} \qquad , \tag{2}$$

where M_0 is an (arbitrary) reference mass and the M_i are the masses of the microlenses. The positions of the compact microlenses are denoted by $z_i = (x_i, y_i)$. This equation has the advantage of describing microlensing in terms of only two parameters, the normalized surface density

$$\sigma = \frac{\sigma_s}{1-\sigma_c} \tag{3}$$

and the normalized shear

$$\gamma = \frac{\gamma'}{1-\sigma_c} \qquad , \tag{4}$$

where σ_s is the average surface density in compact objects and σ_c is the continuous surface density, both in units of the critical density

$$\sigma_0 = \frac{1}{\pi} \frac{c^2}{4G} \frac{D_s}{D_{ds} D_d} \qquad , \tag{5}$$

which alone would just lead to a focusing in the observer plane. Note that $(\sigma_s + \sigma_c)$ is the total smoothed out surface density of matter at the position of the macroimage I.

In this normalization the length units are

$$z_0 = \left(\frac{4G}{c^2}\frac{D_{ds}D_d}{D_s}\frac{M_0}{|1-\sigma_c|}\right)^{1/2} \tag{6}$$

in the deflector plane and

$$\zeta_0 = z_0(1-\sigma_c)\frac{D_s}{D_d} \tag{7}$$

in the source plane. For typical cosmological distances and $M_0 = M_\odot$ these length units are of the order of 10^{-2} parsec.

Microlensing may cause large variations of the amplification A, especially during high amplification events (HAEs) (Paczynski 1986, Kayser *et al.* 1986, Schneider and Weiss 1987), which occur whenever the source crosses a caustic of the microlenses (Chang and Refsdal 1979, 1984). The number of microimages changes by two during a HAE. The amplification near caustics is in general given by $A = K/\sqrt{d}$ (Chang 1984, see also Kayser and Witt 1989), where d is the distance from the caustic and the constant K is the so-called flux factor. Due to this simple law HAE can be used to analyze the brightness profile (perpendicular to the caustic) of sufficiently small sources (Grieger *et al.* 1988).

For small sources (radius $R_s \ll \zeta_0$) the maximum amplification during a HAE is given by (Schneider and Weiss 1987, Kayser and Witt 1989)

$$A_{max} = \frac{Kf}{\sqrt{R_s/\zeta_0}} \quad , \tag{8}$$

where K is the flux factor of the caustic at the crossing point, and f is a "form factor", depending on the brightness profile perpendicular to the caustic. The flux factor is of the order unity. The form factor equals one for a circular source with constant surface brightness, and can assumed to be of the order unity for reasonable sources (if R_s is appropriately defined). The maximum brightness change due to a HAE is thus related to the source radius by (Kayser and Witt 1989)

$$\Delta m_{max} \leq 2.5\log\left(1+\frac{Kf}{\sqrt{R_s/\zeta_0}}\right) \quad , \tag{9}$$

which is unbalanced since the amplification A_d on the "dark" side of the HAE is not known. This relation assumes $A_d \geq 1$, it does, e.g. not hold for over-focused images. We can use Eq.(9) to estimate the source radius:

$$R_s \leq \left(10^{0.4\Delta m_{max}} - 1\right)^2 \zeta_0 \quad . \tag{10}$$

Note that this equation can only be applied if the source is sufficiently small $(R_s \ll \zeta_0)$.

On the other hand, the "crossing time"

$$\Delta t_s = R_s/V_t \tag{11}$$

can be used to determine the effective transverse velocity between source and caustic, if the source radius is known, or to place constraints on the source size if V_t is known (Kayser *et al.* 1986, Grieger *et al.* 1986, 1988). In the latter case, Eq.(9) can be used to place constraints on the mass of the microlens.

This analysis can be done for each individual HAE, i.e. no long-term monitoring is necessary. The brightness profile of the source can be deconvolved from the light curve, if the data are of sufficient accuracy, thus it might even be possible to determine f for a single event. The flux factor K, however, remains unknown. If we observe several HAEs, we may use a mean value $\langle K \rangle$ obtained from numerical simulations for the analysis. This strategy is, however, not without problems: 1) the local parameters are usually not well known due to the non-uniqueness of the macromodel (Falco *et al.* 1985, Kayser 1990), 2) f and V_t may be quite different for different HAE, since the relative orientation source-caustic changes and since the motion of the caustics depends on the relative velocities of the stars, 3) the events may well be due to the caustics of different microlenses with different masses, i.e. either K or ζ_0 has to be changed, and 4) the distribution of the flux factor K depends on the (unknown) mass spectrum of the microlenses.

From theoretical considerations it has been clear that the quadruple *Einstein Cross* QSO 2237+0305 is the most promising candidate for the observation of microlensing events (Kayser and Refsdal 1989, see also Wambsganss *et al.* 1990). Indeed, a rapid change in the luminosity of image A has been reported (Irwin *et al.* 1989), which is most likely a HAE. In October 1988 image A has changed its brightness by $\Delta m = 0.4^m$ on a time scale of 12 days (Irwin, private communication). This time scale corresponds to a source radius of

$$R_s \approx 10^{-4}\,\mathrm{pc}\,\frac{V_t}{3000}\,\mathrm{km/s} \quad . \tag{12}$$

Since microlensing in image A of 2237+0305 can not be treated in the small optical depth approximation, it is not possible to draw any conclusion on the mass of the microlens involved (as, e.g., done by Irwin *et al.* (1989)). From numerical simulations it is, however, clear that it is not necessary to assume a lens mass below $0.1 M_\odot$ (Wambsganss *et al.* 1990, Witt *et al.* in preparation).

2. The Hamburg quasar monitoring program

In May 1989 we started an extensive optical photometric monitoring program in order to find and analyse microlensing HAEs in the lightcurves of a selected sample of quasars. We use the Max-Planck-Institute 1.2m telescope at the German-Spanish Astronomical Centre on Calar Alto equipped with a CCD camera. For our program, 15–30 per cent of the observation time at this telescope are available for a minimum of five years. For some objects observations exist from a 3 week test run in autumn 1988 so that we could search for the presence of variability on a relative long timescale. In most cases, however, we obtained lightcurves of good time resolution only for a short time span. Up to now we found no evidence for the presence of a HAE in any lightcurve and therefore restrict ourselves to a discussion of variability statistics at the end of this contribution.

In our program, the quasar fluxes are measured through different standard broad-band filters relative to reference stars in the CCD frames. A 0.01 mag accuracy (in relative photometry) can therefore be reached also for non-photometric conditions and a typical exposure time of 500 sec. Standard CCD reduction, identification of the objects and photometry is carried out automatically in about 5 min immediately after the observation on a MicroVax 3200 workstation. By this means, it is possible to follow interesting features in the lightcurves with higher time resolution and by using different filters. We want to obtain up to 30 data points for the annual observing period of those objects which showed only long-term variations or no variability up to now. Rapidly variable objects will be observed at least once each night during observing runs lasting a few weeks.

Our sample includes about 100 objects with different priority:

1) Quasars which are multiple due to macrolensing

These are the only objects for which we have a good chance to separate microlensing events from the intrinsic ones, if the time delay between the components is known. If the macrolens is a galaxy one may expect a relatively high optical depth for microlensing. Monitoring of these objects is also very important in order to obtain the time delays between the components from which limits on the Hubble parameter (Borgeest and Refsdal 1984) and a relative accurate estimate of the lens mass (Borgeest 1986) can be deduced. In Table 1, which presents our first results, we have denoted these multiple quasars by *GLE*.

2) Quasars with a closely associated foreground galaxy

On very deep exposures most high redshift quasars will show a foreground galaxy within 10" angular separation. In Table 1, we denote objects by *GAss* which have a foreground galaxy closer than 25 kpc ($h = 75$) to the line of sight and — if the galaxy redshift is below 0.05 — which shine through this galaxy within the Holmberg isophote. Only for these cases we expect a relatively high optical depth for microlensing.

3) Quasars with narrow heavy element absorption systems

From a uniform spectroscopic survey of 55 high redshift quasars, Sargent, Steidel and Boksenberg (1988) obtained 229 CIV and 40 MgII absorption systems (for a special discussion of the MgII sample see Sargent, Boksenberg and Steidel 1988). The large scale distribution of the redshifts – ignoring clustering on scales smaller 1000 km/s – is random and therefore consistent with the hypothesis that the heavy element absorption systems arise from intervening galaxies. This assumption is confirmed by the work of Bergeron (1988) who found in 10 out of 14 cases of MgII absorption within 10" a galaxy which agrees in redshift with the quasar absorption. The large average number of about 5 absorption systems per quasar found by Sargent et al. and the results of Bergeron indicate that in most cases the absorption occurs far away from the centre of the foreground galaxy, i.e. in its extended halo where one would expect a low density of stars. In our opinion, only metal rich systems showing a low degree of ionization and lines of large restframe equivalent widths W_r make it more probable that the quasar light passes through regions of higher star density. In Table 1 we denote such objects by *Abs* which show MgII absorption with $W_r > 1$Å for both lines of the doublet.

4) BL Lac objects and blazars
Ostriker and Vietri (1985, 1990) proposed that a significant fraction of the BL Lac objects are blazars whose continuum emission has been strongly amplified, relative to the line emission, by microlensing. This scenario is especially interesting for those BL Lacs which show a close foreground galaxy like 0235+164 and 0846+513 (Stickel, this workshop). Note, however, that blazars have – per definition – violent intrinsic variability; it will therefore be extremely difficult to find typical microlensing features in their lightcurves. The typical timescale for microlensing variability is months to several years if the relative velocity between source, lensing stars and observer is of the order of the peculiar velocities in galaxy clusters and therefore much larger than the short timescales known for blazar variability. Nevertheless, we have included some high redshift BL Lacs in our sample to monitor them with high time resolution in the near future.

Schneider and Weiss (1987) proposed (without going into details) that the blazar variability itself may in some cases be due to microlensing. A scenario which we have developed together with Camenzind (private communication) works as follows: Blazar lightcurves seem to be composed of at least two components, a long-term variability with timescales of months to years and a short-term one with timescales of hours to days. For some superluminal sources, which are a subclass of the blazars, it has been shown that the long-term *optical* variability is correlated with the appearance of superluminal *radio* knots (e.g. for 3C345 see Babadzhanyants and Belokon 1985). When these knots emerge they seem to have their maximum intensity at optical wavelengths and can even dominate the continuum. If the knots are small enough and a foreground galaxy exists the short-term variability can indeed be due to microlensing since then the relative velocities are of the order of c.

As instructive examples, we show in Figure 1 lightcurves of three different kinds. In two cases we also show lightcurves for control stars which were treated by the same photometric procedure as the quasar and which is of comparable brightness. All points correspond to 500 sec or 1000 sec exposures through a Johnson R filter. In all cases the magnitude scale is 0.1 mag. The first lightcurve is for 1700+642 (Reimers et al. 1989), one of the most luminous quasars known, having a closely (11") associated galaxy at redshift 0.086. This quasar seems not to be variable above our photometric error limit. The second one is for 0219+428, a well known medium redshift blazar, which appears as a BL Lac at high stages of the continuum. This curve shows the limit of the time resolution of our program. The lightcurve *seems* to be resolved, we have, however, detected a 0.22 mag variation between two nights. The last one is for 0235+164 (Stickel, this workshop), which is one of the most variable objects in our sample. We have detected a 0.7 mag variation within 18 hours. To resolve the structure of the lightcurve three or more exposures each night are necessary. Our fast data reduction facilities will help to follow the lightcurve with adequate accuracy.

All other results are listed in Table 1; it includes the PKS-like name of the objects, redshifts, and absolute visual magnitudes (for variable sources the lowest known value is given). In column 4 those objects are marked which were previously known to be variable, the next columns give the number n of our measurements, the total time span t in years of our observations, the corresponding time span t' in the restframe of the quasar and the maximum amplitude Δm of the variation found in our program. If no variability was found the standard deviation σ of the data points is given.

We have obtained variability data for 81 quasars, 47 of them have (to our knowledge) no published data in the literature. For most objects the number of data points is too small to recognize any features in the lightcurves. It is, however, interesting to discuss statistically the detection or nondetection of variability in the context of other quasar attributes. For this we restrict ourselves to those 47 quasars with previously unknown variability properties since some objects entered our object list because they were known to be variable in addition to the presence of a foreground galaxy or an absorption system. We classify an object as certainly or probably variable if our measured Δm is at least 0.1 mag; objects with lower Δm are classified as only possibly variable.

In Table 2 we concentrate on three classes of quasars corresponding to our a priori classification (see above): objects which are neither classified *Abs* nor *GAss* nor *GLE*, objects which are classified *Abs* (but not *GAss* and not *GLE*), and those classified *GAss* or *GLE*. In the first row we give the number of objects in each class, in the second one the percentage of objects being certainly or probably variable. We find good evidence that quasars closely associated with foreground galaxies show a higher degree of variability than usual ones. Besides a delusive statistical fluctuation this result has to be interpreted in terms of an effect non-intrinsic to the quasars but caused by the foreground galaxies. In our opinion, this is a strong hint that microlensing leads to a significant amount of variability for the class of *GAss* and *GLE* objects. There is, however, no evidence for a correlation between variability and the presence of strong heavy element absorption systems.

We have looked for systematic effects that could weaken our conclusions. There is no significant difference between the classes with respect to the average values of n, t or t'. But there is a selection effect with respect to the absolute visual magnitude. We have included in our sample a number of highly luminous quasars (HLQs, $M_v < -28.0$) since we are involved in an ESO key program to study the gravitational lens properties of these objects. One may argue that the HLQs show a lower degree of variability due to their in general larger redshift and probably larger continuum source. In the last row of Table 2 we give the percentage of HLQs in each class. Since there is only a small difference between the first and third class we think that these objects cannot account for the large difference in variability.

Our main result allows optimistic expectations for the future of our program. Only about 30 per cent of the quasars in our sample seem to have significant intrinsic variability. Therefore about 50 per cent of the objects showing multiple images or a foreground galaxy should show *only* microlensing variability so that HAEs can be recorded without contamination by intrinsic variations.

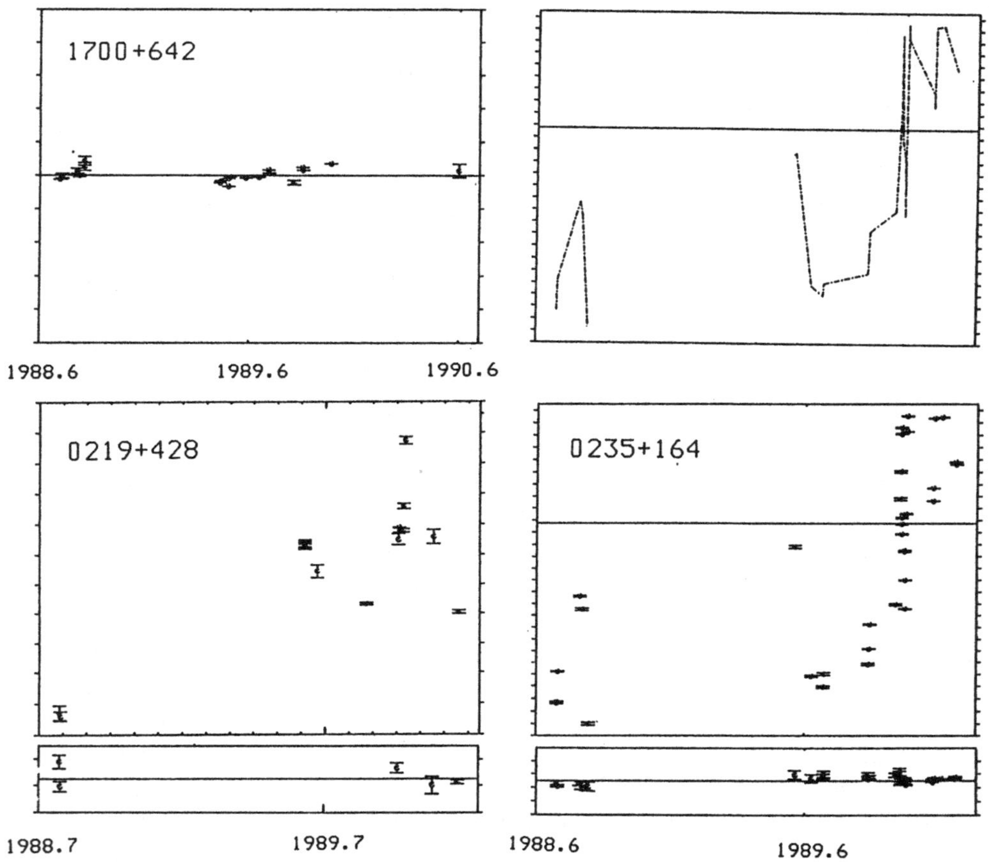

Fig. 1. Lightcurves for three objects of our sample. The magnitude scale is 0.1 mag. The pointed-dashed curve on the upper right is also for 0235+164; it illustrates the chronological order of the data points. The lower two curves are examples of control stars. For a further discussion see text.

Table 1. Photometric results

NAME	z_{em}	M_v	Var*	n	t	t'	Δm	Comments
0002+051	1.899	-29.2		3	0.9	0.3	< 0.02	Abs
0003+158	0.450	-25.7	X	7	1.3	0.9	0.11	
0007+106	0.089	-23.1	X	4	0.2	0.2	0.2	
0013−004	2.09	-28.2	X	2	0.9	0.3	<0.01	Abs
0014+813	3.380	-29.9		24	1.9	0.4	<0.02	Abs
0038−019	1.690	-28.1		14	1.9	0.7	0.15	GAss
0051+291	1.828	-27.5	X	4	0.3	0.11	<0.05	
0058+019	1.955	-28.9	X	9	1.9	0.7	0.15	Abs
0104+318	2.027	-26.9	X	3	0.4	0.13	0.08	GAss
0117+213	1.493	-28.6		14	1.9	0.8	<0.02	
0142−100	2.72	-29.5		6	1.2	0.3	0.1	GAss, GLE
0151+045	0.404	-24.8	X	11	1.3	0.9	0.3	GAss, Abs
0153+744	2.338	-29.8		13	1.9	0.6	<0.03	
0205+024	0.155	-23.5		3	1.1	1.0	0.09	
0215+015	1.715	-30.5	X	4	0.4	0.15	0.4	Abs
0219+428	0.444	-26.7	X	20	1.3	0.9	0.9	
0226−038	2.064	-28.7		3	0.3	0.10	<0.02	Abs
0235+164	0.934	-29.0	X	33	1.4	0.7	1.0	GAss, Abs
0248+430	1.316	-27.4		41	1.9	0.8	0.2	GAss, Abs
0420−014	0.915	-27.8	X	7	0.4	0.2	1.0	Abs
0446−208	1.896	-28.4		11	0.4	0.14	0.25	GAss
0454+039	1.345	-27.9		4	0.3	0.14	<0.02	Abs
0457+024	2.384	-27.2		5	0.4	0.12	0.2	
0731+653	3.035	-28.1		7	0.2	0.07	<0.02	
0735+178	0.42	-28.1	X	15	1.5	1.0	1.6	Abs
0809+483	0.871	-26.6	X	4	0.3	0.15	<0.02	
0836+710	2.17	-29.2		6	0.6	0.18	0.05	
0846+513	1.86	-29.6	X	12	0.6	0.2	0.9	GAss
0851+202	0.306	-28.6	X	13	0.6	0.5	0.7	
0903+175	2.756	-29.0		3	0.3	0.09	0.03	GAss
0953+549	2.584	-28.5		4	0.5	0.14	0.09	
0955+326	0.533	-26.8	X	3	0.3	0.22	<0.02	
0957+557	2.100	-28.2		5	0.3	0.09	<0.03	
0957+561	1.41	-27.9	X	20	1.3	0.6	0.1	GAss, GLE
1011+250	1.631	-29.6	X	4	0.3	0.13	<0.01	
1109+357	0.91	-25.1		3	0.2	0.13	<0.04	GAss
1120+019	1.465	-28.7		3	0.2	0.10	<0.03	GLE
1150+497	0.334	-24.7	X	5	0.6	0.3	0.4	GAss
1209+107	2.191	-28.0		3	0.05	0.02	0.07	GAss, Abs
1219+285	?	?	X	4	0.7	?	0.7	
1219+755	0.072	-22.0		11	1.9	1.8	0.2	GAss
1308+326	0.996	-29.7	X	9	0.9	0.4	0.6	
1317−122	0.33	-22.4		4	0.6	0.5	0.4	

Table 1. Photometric results, continued

NAME	z_{em}	M_v	Var*	n	t	t'	Δm	Comments
1327–206	1.169	-26.9		2	0.04	0.02	0.07	GAss, Abs
1332+552	1.249	-28.2	X	3	1.0	0.4	<0.02	GAss, Abs
1340+289	0.905	-29.4	X	3	1.0	0.5	0.05	
1345+584	2.039	-28.1		3	1.0	0.3	0.2	
1358+391	3.3	-29.3		5	0.9	0.21	0.16	GAss
1413+117	2.551	-29.0		7	0.7	0.19	0.3	GLE
1421+330	1.904	-28.7		4	1.1	0.4	0.07	
1441+522	1.57	-24.9		2	0.3	0.13	0.06	
1510–089	0.361	-29.3	X	8	1.0	0.7	0.7	
1520+413	3.1	-28.9		8	1.0	0.24	<0.03	
1522+101	1.321	-28.7		7	1.1	0.5	0.13	GAss
1543+489	0.400	-26.3		4	0.9	0.6	<0.02	
1604+290	1.95	-29.2		10	1.1	0.4	0.3	
1630+377	1.471	-28.7		6	0.9	0.4	<0.02	
1633+267	1.84	-28.3		7	1.0	0.3	<0.02	
1634+267	1.961	-27.5		7	1.0	0.3	0.5	Abs, GLE
1634+706	1.334	-29.6		10	0.9	0.4	0.07	Abs
1638+398	1.66	-28.7	X	23	1.8	0.7	1.7	
1641+399	0.595	-27.8	X	39	1.9	1.2	0.5	
1700+642	2.72	-30.2		22	1.8	0.5	<0.02	GAss
1701+610	0.164	-22.0		5	1.1	1.0	0.2	GAss
1704+608	0.371	-26.2	X	5	1.0	0.7	0.12	
1715+535	1.929	-29.1		17	1.6	0.6	0.15	
1718+481	1.084	-29.0		12	1.1	0.5	<0.01	
1749+701	0.76	-27.9	X	11	0.9	0.5	0.5	
1821+643	0.297	-26.3		7	0.5	0.4	0.08	
1857+566	1.595	-27.6		11	1.1	0.4	0.2	
2032+107	0.601	-23.8	X	8	1.1	0.7	<0.03	
2126–158	3.267	-29.1		8	1.9	0.45	<0.02	Abs
2134+004	1.936	-30.9	X	10	2.0	0.7	0.09	
2206–199	2.544	-28.3		5	1.0	0.3	0.07	Abs
2215–037	0.241	-24.1	X	21	0.9	0.4	0.26	GAss
2223–052	1.404	-29.5	X	30	1.9	0.8	2.2	
2237+030	1.695	-29.0	X	17	1.2	0.4	<0.04	GAss, GLE
2305+187	0.313	-24.7		7	0.9	0.7	0.46	GAss
2308+098	0.432	-26.3		6	1.8	1.3	0.10	GAss
2354+144	1.81	-28.7		3	1.0	0.4	0.25	GAss

* An 'X' in the row denoted 'Var' means that the variability of this object has been known prior to our project. These objects are not included in Table 2.

Table 2. Variability statistics

	nAbs nGAss, nGLE	Abs nGAss, nGLE	GAss or GLE
number of objects	21	7	19
fraction of variable objects	0.29	0.00	0.68
fraction of HLQs	0.67	0.86	0.58

Acknowledgements: This work has been supported by the Deutsche Forschungsgemeinschaft under Az. Re 439.

References

Alcock, C. and Anderson, N. 1985, *Ap.J.(Lett.)* **291**, L29

Babadzhanyants, M.K. and Belokon, E.T. 1985, *Astrofizika* **23**, 459

Bergeron, J. 1988, in *QSO Absorption Lines - Probing the Universe*, Ed. by Blades, J.C., Turnshek, D. and Norman C.A., Cambridge University Press, Cambridge, p. 127

Borgeest, U. 1986, *Ap.J.* **309**, 467

Borgeest, U. and Refsdal, S. 1984, *Astr.Ap.* **141**, 318

Canizares, C.R. 1981, *Nature* **291**, 620; corr. 1981, *Nature* **293**, 490

Chang, K. 1984, *Astr.Ap.* **130**, 157

Chang, K. and Refsdal, S. 1979, *Nature* **282**, 561

Chang, K. and Refsdal, S. 1984, *Astr.Ap.* **132**, 168

Falco, E.E., Gorenstein, M.V., Shapiro, I.I. 1985, *Ap.J.(Lett.)* **289**, L1

Grieger, B., Kayser, R., Refsdal, S. 1986, *Nature* **324**, 126

Grieger, B., Kayser, R., Refsdal, S. 1988, *Astr.Ap.* **194**, 54

Irwin, M.J., Webster, R.L., Hewett, P.C., Corrigan, R.T., Jedrzewski, R.I. 1989, *A.J.* **98**, 1989

Kayser, R. 1990, *Ap.J.* **357**, 309

Kayser, R. and Refsdal, S. 1989, *Nature* **338**, 745

Kayser, R., Refsdal, S., Stabell, R. 1986, *Astr.Ap.* **166**, 36

Kayser, R., Weiss, A., Refsdal, S., Schneider, P. 1989, *Astr.Ap.* **214**, 4

Kayser, R. and Witt, H.-J. 1989, *Astr.Ap.* **221**, 1

Ostriker, J.P. and Vietri, M. 1985, *Nature* **318**, 446

Ostriker, J.P. and Vietri, M. 1990, *Nature* **344**, 45

Paczynski, B. 1986 *Ap.J.* **301**, 503

Refsdal, S. 1964, *M.N.R.A.S.* **128**, 295

Refsdal, S. 1964, *M.N.R.A.S.* **128**, 307

Reimers, D., Clavel, J., Groote, D., Engels, D., Hagen, H.J., Naylor, T., Wamsteker, W. and Hopp, U. 1989, *Astr.Ap.*__218__, 71
Sargent, W.L., Steidel, C.C. and Boksenberg, A. 1988, *Ap.J.* **334**, 22
Sargent, W.L., Boksenberg, A. and Steidel, C.C. 1988, *Ap.J. Supl.*__68__, 539
Schneider, P. and Weiss, A. 1987, *Astr.Ap.* **171**, 49
Walsh, D., Carswell, R.F., Weymann, R.J. 1979, *Nature* **279**, 381
Wambsganss, J., Paczynski, B., Schneider, P. 1990, *Ap.J.* **358**, L33
Young, P.J., Gunn, J.E., Kristian, J., Oke, J.B., Westphal, J.A. 1980, *Ap.J.* **241**, 507
Zwicky, F. 1937, *Phys.Rev.* **51**, 290
Zwicky, F. 1937, *Phys.Rev.* **51**, 679

Lensing of BL Lac Objects

Manfred Stickel

Max-Planck-Institut für Astronomie, Königstuhl 17, 6900 Heidelberg 1, Germany

Abstract: Four highly variable, high redshift ($z \gtrsim 0.9$) BL Lac objects have been found to show a spatially resolved appearance on deep direct images. Because of the high redshift of the BL Lac objects, their non-stellar images are most likely due to an intervening galaxy with a much smaller redshift located nearly on the line of sight. These objects represent very good candidates for gravitationally (micro)lensed BL Lac objects. A number of properties of these objects and the BL Lac class in general are examined within the context of gravitational (micro)lensing.

1. Introduction

The most prominent characteristics of BL Lac objects are strong and rapid variability in all wavelength bands, high and highly variable optical and radio polarization and absent or weak emission lines in the optical spectra (Stein et al., 1976). These properties are generally considered to be due to relativistic beaming, where the observer is located at a small angle to the jet direction (Blandford and Rees, 1978; Antonucci and Ulvestad, 1985). In this model, the doppler-boosted, highly amplified optical continuum swamps the emission lines, resulting in the characteristic (near) featureless optical spectra of BL Lac objects

In contrast to this source intrinsic origin, gravitational microlensing has been suggested as an external mechanism for the strong and highly variable continuum emission of BL Lac objects. To explain the high volume density of local ($z \lesssim 0.2$) BL Lac objects, Ostriker and Vietri (1985,1990) proposed that these objects are actually distant highly variable quasars gravitationally microlensed by nearby elliptical galaxies, which then look like apparent BL Lac host galaxies.

On the contrary, Nottale (1986,1988) put forward the idea that the strong optical variability of the high-redshift ($z \gtrsim 0.5$) BL Lac objects may be due to gravitational microlensing of the optical continuum emission of non-variable quasars by intervening galaxies near the line of sight. Evidence for these galaxies, however, has generally not been found in direct images, but intervening absorption line systems observed in the spectra of some BL Lac objects can be considered as a direct hint for foreground matter.

More generally, a gravitational microlensing origin for at least part of the observed optical variability of Active Galactic Nuclei (AGN) and particularly of BL Lac objects has been proposed by Schneider and Weiss (1987).

On the basis of the observed lightcurves, a number of BL Lac objects have been considered to be potential candidates for gravitational microlensing (Schneider and Weiss, 1987; Kayser et al., 1989). Up to now, however, there were no observational attempts to support these theoretical considerations for individual objects. In the following, optical observations of four high redshift ($z \gtrsim 0.9$) BL Lac objects are summarized, which show a spatially extended appearance on deep direct images, most likely due to the presence of a foreground galaxy on the line of sight. In two of these sources, the continuum or line emission of the foreground galaxy is also present in the optical spectrum of the BL Lac object and allows a determination of its redshift. The properties of the four lensing candidate objects, and the BL Lac class in general, are examined within the context of gravitational (micro)lensing.

2. Observations of candidates for lensed BL Lac objects

Optical observations of three of the four (0235+164, 0537–441, 1308+326) gravitational microlensing candidates have been obtained during the study of the complete sample of radio selected 1 Jy BL Lac objects (Stickel et al., 1990). The fourth object (0846+513) has been observed because its properties are strikingly similar to that of the three 1 Jy candidates. A full account of the observations of 0235+164 and 0537–441 can be found in Stickel et al. (1988a,b), while the observational results for 0846+513 have been reported in Stickel et al. (1989).

2.1. 0235+164

This object is one of the best observed members of the BL Lac class and shows some outstanding properties. The historically recorded range of the variability exceds 5 mag (Pollock et al., 1979) and the optical lightcurve displays repeatedly strong optical outbursts of several magnitudes, which seem to be accompanied by simultaneous strong radio outbursts (Balonek, 1986) and the birth of new VLBI components (Bååth, 1984).

The optical spectrum reveals two intervening redshift systems at $z = 0.524$, seen in absorption and emission, and $z = 0.851$, detected only in absorption. A companion object only 2" to the south shows strong emission lines also at a redshift of $z = 0.524$. During a faint state, quasar-like emission lines of 0235+164 have been observed and a redshift of $z = 0.94$ has been determined (Cohen et al., 1987).

Despite the high redshift, the direct image of 0235+164 is not completely stellar, but shows an eastern elongation (fig. 1a), in addition to the southern companion object. Because no second brightness maximum is seen eastward of the BL Lac object, this eastern extension must be due to an intervening object very close to the line of sight to the unresolved background source. To separate the different components of 0235+1264, a two-dimensional image decomposition procedure was utilized, which models the the whole observed image as the sum of several two dimensional brightness distributions. It should be emphasized, that in this case a simple subtraction of a peak scaled point spread function from the observed image, as was done by Yanni et al. (1989), will certainly overestimate the distance between the background point source and the center of the intervening object.

The observed image of 0235+164 (fig. 1a) was found to be composed of the unresolved, high redshift BL Lac object (fig. 1b), the newly detected intervening galaxy (fig. 1c) and the well-known compact southern companion (fig. 1d). The distance between the intervening galaxy and the BL Lac point source was found to be $\sim 0.5''$, which corresponds to $\sim 3.5\,\mathrm{kpc}$ at $z = 0.524$. Thus the line of sight to the BL Lac object passes nearly through the center of the foreground galaxy, which moreover seems to form an interacting pair with the southern companion galaxy $\sim 15\,\mathrm{kpc}$ away. The extended [O II] $\lambda 3727$ emission detected by Cohen et al. (1987) and Yanni et al. (1989) can then be accounted for by recent star formation triggered by the tidal interaction.

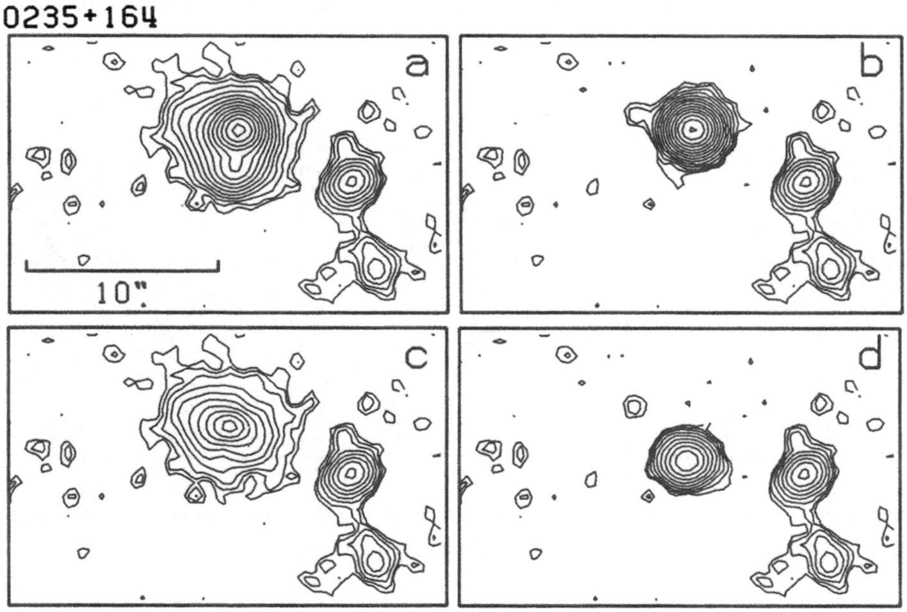

Fig. 1. Decomposition of the direct image of 0235+164 ($z = 0.94$) (a) into the unresolved high redshift BL Lac object (b), the intervening galaxy close to the line of sight (c) and the southern companion (d).

0235+164 is surrounded by a number of faint galaxies (Stickel et al., 1988a) and redshifts are now known for six of these galaxies. Including the southern companion and the foreground galaxy, four of them have a redshift of $z \simeq 0.524$, while two others have $z = 0.065$ and $z = 0.28$ (Cohen et al., 1987; Stickel et al., 1988a; Yanni et al., 1989; Burbidge et al., 1989). Thus there is considerable amount of foreground matter along the line of sight and the relatively large intervening galaxy may well be the central galaxy of a group or small cluster near z $\simeq$ 0.524.

2.2. 0537–441

0537–441 is highly variable in the optical with a total amplitude of more than 5 mag (Usher, 1975). Its redshift of z = 0.894 is derived from C III] λ1909 and Mg II λ2798 emission lines (Peterson et al., 1976), which are usually very weak or even totally absent. Despite the high redshift the appearance of 0537–441 on the deep direct image is not stellar (fig. 2a), but resembles a neighbouring galaxy 11" to the east the redshift of which is z = 0.186 (Stickel et al., 1988b).

Our image fitting procedure was used to analyze the direct image of 0537–441 (fig. 2a), which was found to be composed of an unresolved point source (the high redshift BL Lac object, fig. 2b), centered on a galaxy, which is very well fitted by an exponentional brightness distribution similar to disctype galaxies (fig. 2c). In addition, the neighbouring galaxy was also found to be very well modelled by a disc galaxy of similar scale length and apparent brightness (Stickel et al., 1988b).

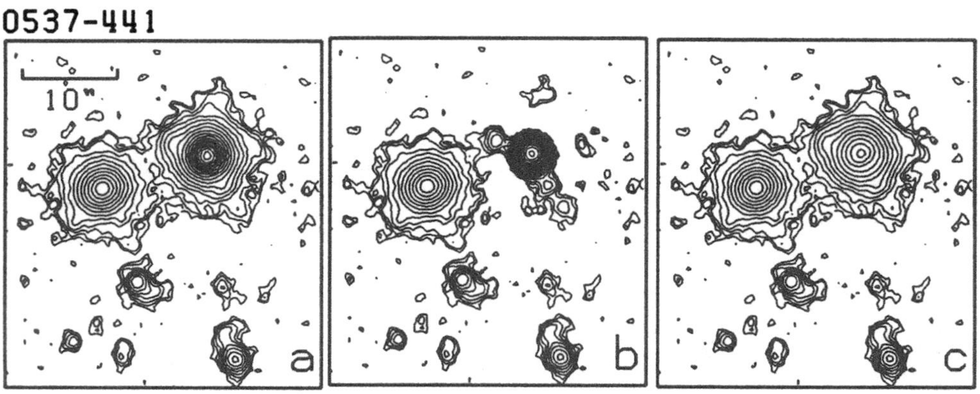

Fig. 2. The direct image of 0537–441 (z = 0.894), right of the center (a), decomposed into the BL Lac point source (b) and a disc component (c).

At a redshift of z = 0.894, 0537–441 is expected to be spatially unresolved. Therefore the extended disc component (fig. 2c) is not attributable to the host-galaxy of 0537–441, but is most likely due to an intervening galaxy with a much smaller redshift. The similarity of the apparent brightnesses and disc scale lengths of the neighbouring galaxy and the extended component of 0537–441 suggests that the intervening galaxy has also a redshift around z $\simeq$ 0.19. The absence of Mg II λ2798 as well as Ca II $\lambda\lambda$3933, 3968 in a high resolution spectrum of 0537–441 (Tytler et al., 1987) limits the possible redshift range of the suspected intervening galaxy to 0.18 $\lesssim$ z $\lesssim$ 0.32.

2.3. 0846+513

The optical counterpart of the radio source 0846+513 was found to be a highly variable stellar object lying only 12" south of a pair of interacting spiral galaxies (Arp et al., 1979). During faint states, weak broad emission lines of C IV λ1549 and C III] λ1909 at a redshift of $z = 1.86$ were detected (Arp et al., 1979; Stickel et al., 1989) whereas the spectrum was completely featureless, when the object was bright. During minimum light, an apparently steep optical spectrum is observed, which flattens considerably, when the objects gets brighter.

The direct image of 0846+513 (fig. 3a) shows that despite the high redshift inner isophotes around the BL Lac object are extended to the north. Using the image decomposition procedure described above, the image of 0846+513 was found to be composed of the unresolved BL Lac object (fig. 3b) and an intervening galaxy (fig. 3c) lying nearly exactly on the line of sight. The galaxy 2.5" southeast (fig. 3d) due its large diameter also intercepts the line of sight to the BL Lac object.

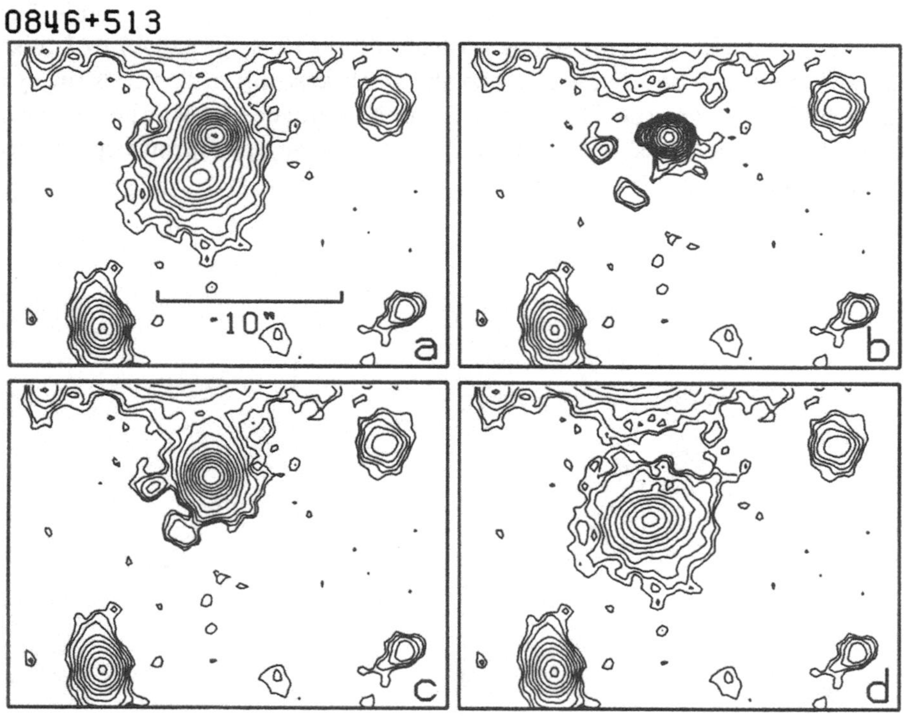

Fig. 3. Results of the image decomposition of 0846+513 ($z = 1.86$). The observed image (a) consists of the unresolved high redshift BL Lac object (b), the intervening galaxy (c) and the companion galaxy to the southwest (d).

The optical spectrum of 0846+513 was found to deviate considerably from a power law with a spectral index $\alpha \simeq -2.8$ observed by Arp et al. (1979). It is, however, very well represented by the energy distribution of an elliptical galaxy at a redshift of z = 0.235 added to a quasar power law with a spectral index of $\alpha \simeq -1$. At $\lambda \sim 6500$ Å, roughly 2/3 of the flux is due to the galaxy component, while only 1/3 comes from the quasar power law. This ratio is in excellent agreement with that derived from the decomposition of the direct image (Stickel et al., 1989).

2.4. 1308+326

This high redshift BL Lac object (z = 0.997; Miller et al., 1978; Stickel et al., 1990) shows strong optical variability with a total amplitude of more than 5 mag (Zekl et al., 1981) and is during its optical outbursts one of the most luminous extragalactic sources (Liller, 1976). In the optical spectrum, a very weak intervening absorption line system ($z_{abs} = 0.879$) has been detected.

The direct image of this object (fig. 4) shows, in addition to a close companion galaxy 5" southwest, that its outer isophotes are extended to the west. This non-stellar appearance can more easily be seen in the azimutally averaged surface brightness profile (fig. 4). By analogy to the other described BL Lac objects, a foreground galaxy close to the line of sight is most likely responsible for the non-stellar image. However, it seems rather unlikely that the redshift of this suspected foreground galaxy is identical to the weak absorption line redshift system at $z_{abs} = 0.879$ found in the BL Lac spectrum. No evidence for another intervening redshift system has as yet been found, mainly because appropriate spectroscopic observations are lacking.

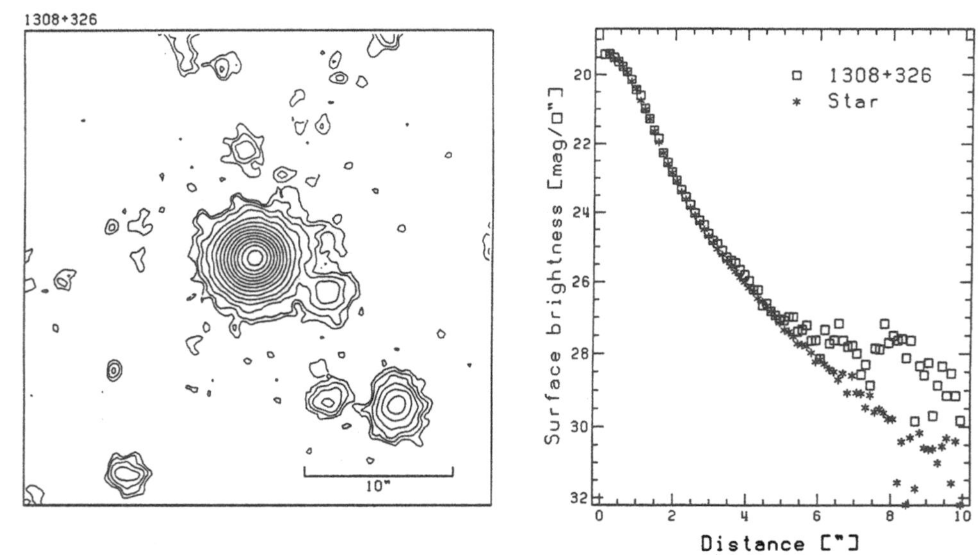

Fig. 4. Direct image (left) of 1308+326 (z = 0.997) showing the slightly extended isophotes to the west. The deviation from an unresolved stellar point source is more evident in the azimutally averaged surface brightnes profile (right).

3. Implications

The observational evidence for foreground galaxies near the line of sight to high redshift ($z \gtrsim 0.9$) BL Lac objects are in accord with the suggestion of Nottale (1986,1988) that some of these BL Lac objects may be affected by lensing. Moreover, this idea has the advantage of beeing consistent with the observations of quasar-like emission lines during faint states of these objects. At present no evidence has been reported in favour of the suggestion of Ostriker and Vietri (1985,1990) that nearby ($z \lesssim 0.2$) BL Lac objects are lensing candidates. Especially the expected redshifted emission lines of the background quasar have as yet not been seen in any nearby BL Lac object.

A foreground galaxy amplifies the background source as a whole, nearly regardless of the size of the emitting regions. In fact, the four lensing candidates not only have very high absolute magnitudes during optical outbursts, they also belong to the most luminous objects known with bolometric luminosities $\gtrsim 10^{13} L_{\odot}$ (Impey and Neugebauer, 1988).

All four BL Lac objects described are highly variable on short time scales and the total amplitude of the variability exceeds 5 mag in all cases. On the basis of the observational evidence for foreground galaxies on the line of sight, it is tempting to interpret the large outbursts of these BL Lac objects as a result of gravitational microlensing, but an unequivocal clue has still to be adduced. The only unquestionable proof is to observe the apparent brightness gradient on scales of some tenths of AU caused by microlensing (Grieger et al., 1988). Equally important is long term optical monitoring of candidate objects in order to study the lightcurves in more detail and to compare them with theoretical models.

Because the lensing candidate BL Lac objects are highly polarized, the background sources themselves have to be polarized. This seems to exclude low polarized quasars and favours highly polarized quasars (HPQs) as background sources. The class of HPQs is coincident with flat spectrum radio sources, which shows, with a few exceptions, variability with smaller amplitudes and longer time scales than BL Lac objects. Obviously very important will be the comparison of the optical polarization properties of candidates for lensed BL Lac objects with those of a group of HPQs, which are definitely not lensed. If the polarized optical continuum source is small enough to be affected by gravitational microlensing, it can be expected that the lensing candidates show higher peak polarization and a higher duty cycle of high polarization than the comparison group.

Under the assumption of a gravitational microlensing origin, the short time scales of the strong outbursts imply very high transverse velocities ($v \gtrsim c$). The only known sources moving with such high (superluminal) velocities are VLBI components observed emanating from the cores of flat spectum radio sources, which may be part of a relativistic jet pointing toward the observer. Because BL Lac objects in general have strong compact cores with flat radio spectra, the strong variability of some outstanding sources may well be the result of gravitational microlensing acting on superluminal VLBI components of the background sources. In the case of 0235+164, additional evidence comes from the birth of superluminal components coincident with strong optical oubursts. This weakens the arguments presented by Kayser (1988) against the gravitational microlensing scenario in this particular source. To make matters more complicated, the observed extremely high apparent superluminal motion ($v \simeq 45\,c$; Bååth, 1983) in this source may well be the result of a distorted source structure caused by gravitational lensing (Mutel, 1989).

As a conseqence of this connection of lensing and superluminal motion, extended steep spectrum sources may not show the strong and rapid variability as well as the characteristic (near) featureless spectra of lensing candidates during optical outbursts, even if a similar optical morphology with a foreground galaxy on the line of sight will be found. Clearly, such objects must exist but they may not be classified as BL Lac objects or as highly variable quasars.

On the other hand, some highly polarized quasars such as 3C345 or 3C 446 show brightness variations strikingly similar to those seen in highly variable BL Lac objects. It may be conjectured that a corresponding gravitational microlensing scenario may also be responsible for the variability of these sources, even if no direct evidence for foreground galaxies has as yet been reported. Although these two groups represent different classes of AGNs, their common property of optically violent variability (OVV; Webb et al., 1988) may well have a common origin. The tiny fraction of highly variable objects among HPQs is consistent with this picture.

It should be emphasized, that even in the candidates for lensed BL Lac objects the observed variability is certainly not exclusively attributable to gravitational microlensing alone, but is most likely the superposition of source intrinsic and lensing induced flux variations. Moreover, if the source diameter in the optical is below and in the radio regime above the critical size for microlensing, optical variability caused by microlensing will not necessarily be accompanied by similar radio variability.

In the four cases decribed above, direct evidence for foreground galaxies have been found on deep direct images, resulting in a non-stellar appearance. In fact, these sources were recognized as lensing candidates just because of their extended images. Therefore, a strong selection effect is at work in favour of finding the brighter foreground galaxies in front of the fainter background sources. Fainter ($M_R \gtrsim -21$ mag) foreground galaxies are easily hidden in the seeing disc of the point source and may not be detectable on direct images, even if the BL Lac object is in a faint state.

Optical and radio data can be used to rule out multiple images in the lensing candidates to a high degree of confidence, and this has been used by Narayan and Schneider (1989) to derive important parameters for the intervening galaxies of 0235+164 and 0537–441. If the intervening galaxy is massive enough to produce multiple images and has a sufficiently high ($z \gtrsim 0.1$) redshift, the light path no longer passes through the center of the galaxy, even if it is located nearly on the line of sight. The star density in the outskirts is much lower, which will lead to a more sparse amplification pattern. Thus the expected variability of a multiple imaged source is not as strong as in the case of a single image near the center of the foreground galaxy. A similar argument also applies to single imaged sources viewed through the outskirts of intervening galaxies.

Additional evidence for foreground matter along the line of sight is in any case the detection of an intervening absorption line system. By analogy to the early observations of 0235+164, where only absorption lines of the foreground galaxy but no emission line of the background source had been seen, BL Lac objects showing only intervening absorption line redshift systems can also be considered as potential gravitational lensing candidates. However, at present direct imaging searches for these suspected foreground galaxies are restricted to $z \lesssim 0.6$ for the brigther and to even smaller redshifts for the fainter galaxies. These limits will certainly be overcome with the advent of ground based new technology telescopes, allowing not only deep direct imaging but also spectroscopy with high spatial resolution.

As was shown for the case of 0846+513, the apparent steep optical spectrum during faint states of this BL Lac objects is most likely the result of the superposition of a quasar power law and the energy distribution of an elliptical galaxy. If the background quasar power law is amplified by microlensing, the relative fraction of the galaxy component decreases and the optical spectral index becomes flatter. A similar behaviour, bluer when brighter, has occasionally been seen in other BL Lac objects. In addition, foreground absorption taking place in the interstellar medium of the intervening galaxy may also be responsible for steep optical spectra observed in some BL Lac objects (e.g. 0235+164).

The presented evidence for foreground galaxies along the line of sight to a few high redshift BL Lac objects for the first time strongly supports the idea that the observed properties of some of these sources may be affected by gravitational (micro)lensing. Of great importance will be the detailed study of these candidates as well as an investiation of the frequency of similar candidates among other BL Lac objects and radio sources of high redshifts.

Acknowledgements: The observational results were obtained as part of a larger project carried out in close collaboration with Drs. J.W. Fried and H. Kühr. The author is grateful to Drs. P. Schneider, U. Borgeest and W. Fugmann for fruitful discussions.

References

Antonucci, R. R. J., Ulvestad, J. S. 1985, *Astrophys. J.* **294**, 158

Arp, H., Sargent, W.L.W., Willis, A.G., Oosterbaan, C.E.: 1979, *Astrophys. J.* **230**, 68

Bååth, L.B.: 1983, in *Highlights in Astronomy* **6**, 742

Bååth, L.B.: 1984, in *VLBI and Compact Radio Sources*, IAU Symposium **110**, eds. R. Fanti, K. Kellermann, G. Setti (Dordrecht: Reidel), p. 127

Balonek, T.J.: 1986, in *Continuum Emission in Active Galactic Nuclei*, ed. M.L. Sitko (Tucson: KPNO), p. 161

Blandford, R.D., Rees, M.J.: 1978, in *Pittsburgh Conference on BL Lac - Objects*, ed. A.M. Wolfe (Univ. of Pittsburgh Press), p. 328

Burbidge, E.M., Barlow, T.A., Cohen, R.D., Junkkarinen, V.T., Womble, D.S.: 1989, *Astrophysics & Space Science* **157**, 263
Cohen, R.D., Smith, H.E., Junkkarinen, V.T., Burbidge, E.M.: 1987, *Astrophys. J.* **318**, 577
Grieger, B., Kayser, R., Refsdal, S.: 1988, *Astron. Astrophys.* **194**, 54
Impey, C.D., Neugebauer, G.: 1988, *Astron. J.* **95**, 307
Kayser, R.: 1988, *Astron. Astrophys. Lett.* **206**, L8
Kayser, R., Weiss, A., Refsdal, S., Schneider, P.: 1989, Astron. Astrophys. 214, 4.
Miller, J.S., French, H.B., Hawley, S.A.: 1978, in *Pittsburgh Conference on BL Lac Objects*, ed. A.M. Wolfe (Univ. of Pittsburgh Press), p. 176
Mutel, R.L.: 1989, Proceedings of the Socorro Workshop on *Parsec Scale Radio Jets*, in press
Narayan, R., Schneider, P.: 1990, *Mon. Not. R. Astron. Soc.* **243**, 192
Nottale, L.: 1986, *Astron. Astrophys.* **157**, 383
Nottale, L.: 1988, in *Dark Matter*, eds. J. Audouze, J. Tran Thanh Van (Editions Frontières), p. 339
Liller, W.: 1976, IAU Circular No. 2939
Ostriker, J.P., Vietri, M.: 1985, *Nature* **318**, 446
Ostriker, J. P., Vietri, M.: 1990, *Nature* **344**, 45
Peterson, B.A., Jauncey, D.L., Wright, A.E., Condon, J.J.: 1976, *Astrophys. J. Lett.* **207**, L5
Pollock, J.T., Pica, A.J., Smith, A.G., Leacock, R.J., Edwards, P.L., Scott, R.L.: 1979, *Astron. J.* **84**, 1651
Schneider, P., Weiss, A.: 1987, *Astron. Astrophys.* **171**, 49
Stein, W.A., O'Dell, S.L., Strittmatter, P.A.: 1976, *Ann. Rev. Astron. Astrophys.* **14**, 173
Stickel, M., Fried, J.W., Kühr, H.: 1988a, *Astron. Astrophys. Lett.* **198**, L13
Stickel, M., Fried, J.W., Kühr, H.: 1988b, *Astron. Astrophys. Lett.* **206**, L30
Stickel, M., Fried, J.W., Kühr, H.: 1989, *Astron. Astrophys. Lett.* **224**, L27
Stickel, M., Padovani, P., Urry, C.M., Fried, J.W., Kühr, H.: 1990, *Astrophys. J.*, submitted
Tytler, D., Boksenberg, A., Sargent, W.L.W., Young, P., Kunth, D.: 1987, *Astrophys. J. Suppl.* **64**, 667
Usher, P.D.: 1975, *Astrophys. J. Lett.* **198**, L57
Webb, J.R., Smith, A.G., Leacock, R.J., Fitzgibbons, G.L., Gombola, P.P., Shepherd, D.W.: 1988, *Astron. J.* **95**, 374
Yanni, B., York, D.G., Gallagher, J.S: 1989, *Astrophys. J.* **338**, 735
Zekl, H., Klare, G., Appenzeller, I.: 1981, *Astron. Astrophys.* **103**, 342

GPSR Compliance

The European Union's (EU) General Product Safety Regulation (GPSR) is a set of rules that requires consumer products to be safe and our obligations to ensure this.

If you have any concerns about our products, you can contact us on ProductSafety@springernature.com

In case Publisher is established outside the EU, the EU authorized representative is:

Springer Nature Customer Service Center GmbH
Europaplatz 3
69115 Heidelberg, Germany

Batch number: 10640288

Printed by Printforce, the Netherlands